# Novel Analysis on Aroma Compounds of Wine, Vinegar and Derived Products

# Novel Analysis on Aroma Compounds of Wine, Vinegar and Derived Products

Editors

**Enrique Durán-Guerrero**
**Remedios Castro-Mejías**

MDPI • Basel • Beijing • Wuhan • Barcelona • Belgrade • Manchester • Tokyo • Cluj • Tianjin

*Editors*
Enrique Durán-Guerrero
University of Cádiz
Spain

Remedios Castro-Mejías
University of Cádiz
Spain

*Editorial Office*
MDPI
St. Alban-Anlage 66
4052 Basel, Switzerland

This is a reprint of articles from the Special Issue published online in the open access journal *Foods* (ISSN 2304-8158) (available at: https://www.mdpi.com/journal/foods/special_issues/Wine_Aroma).

For citation purposes, cite each article independently as indicated on the article page online and as indicated below:

LastName, A.A.; LastName, B.B.; LastName, C.C. Article Title. *Journal Name* **Year**, *Volume Number*, Page Range.

**ISBN 978-3-0365-1662-2 (Hbk)**
**ISBN 978-3-0365-1661-5 (PDF)**

# Contents

# About the Editors

**Enrique Durán-Guerrero**

Enrique Durán-Guerrero is an associate professor of the Analytical Chemistry Department of the University of Cádiz. His current research interests are focused on wine, vinegar, oenological products, polyphenols, volatile compounds, chromatography, food analysis, food quality, sensory analysis, extraction, among others.

**Remedios Castro-Mejías**

Remedios Castro-Mejías is an associate professor of the Analytical Chemistry Department of the University of Cádiz. Her current research interests are focused on enological products, chromatography techniques, polyphenols, volatile compounds, chromatography, food analysis, food quality, sensory analysis, extraction, food composition, food science and technology, among others.

# Preface to "Novel Analysis on Aroma Compounds of Wine, Vinegar and Derived Products"

Aroma compounds are some of the main compounds responsible for the acceptance of oenological products such as wine, vinegar and derived products. These kinds of compounds are produced during the winemaking process and they can be affected by natural, geographical and human factors: raw material, alcoholic and acetic fermentation, ageing, distillation, technological processes, etc. Therefore, it is very important to study and characterize the aromatic fraction of these oenological beverages in order to improve the quality of the final product.

This book is focused on some recent studies related to the study of the volatile composition of wine, vinegar and derived products, inmany different fields of science: oenology, chemistry, food science and technology, biochemistry, microbiology, biotechnology, engineering, sensory analysis, etc., and it is addressed to researchers from all these branches of science. It shows the great importance of sensory and analytical study of oenological products aroma and how they are influenced by the different stages and conditions under which they are elaborated.

The following authors have contributed to the scientific contents of the book:

Alexey Dobrydnev (Taras Shevchenko National University of Kyiv, Ukraine)
Álvaro Peña-Neira (University of Chile, Chile)
Ana Jiménez-Cantizano (University of Cádiz, Spain)
Ana Ruiz-Rodríguez (University of Cádiz, Spain)
Andrii Tarasov (Hochschule Geisenheim University, Germany)
Antonio Amores-Arrocha (University of Cádiz, Spain)
Antonio Castro Marín (University of Bologna, Italy)
Carmelo G. Barroso (University of Cádiz, Spain)
Carolina Muñoz-González (CSIC-UAM, Spain)
Celia Criado (CSIC-UAM, Spain)
Christoph Schuessler (Hochschule Geisenheim University, Germany)
Claudio Riponi (University of Bologna, Italy)
Cristina Úbeda (University of Seville, Spain)
Dominico A. Guillén-Sánchez (University of Cádiz, Spain)
Doris Rauhut (Hochschule Geisenheim University, Germany)
Elena Falqué (University of Vigo, Spain)
Enrique Durán-Guerrero (University of Cádiz, Spain)
Fabio Chinnici (University of Bologna, Italy)
Iván Vázquez-Pateiro (University of Vigo, Spain)
Jesús Simal-Gandara (University of Vigo, Spain)
José Manuel Mirás-Avalos (CITA, Spain)
M. Carmen Rodríguez-Dodero (University of Cádiz, Spain)
M. Jesús Ruiz-Bejarano (University of Cádiz, Spain)
M. Valme García-Moreno (University of Cádiz, Spain)
Manuel J. Valcárcel-Muñoz (Bodegas Fundador SLU, Spain)
María Ángeles Pozo-Bayón (CSIC-UAM, Spain)
María Carpena (University of Vigo, Spain)
Maria Fraga-Corral (Polytechnic Institute of Bragança, Portugal)

María Guerrero-Chanivet (Bodegas Fundador SLU, Spain)
María Ignacia Lambert-Royo (University of Chile, Chile)
María Pérez-Jiménez (CSIC-UAM, Spain)
Mariona Gil i Cortiella (Autonomous University of Chile, Chile)
Miguel A. Prieto (Polytechnic Institute of Bragança, Portugal)
Miguel Palma (University of Cádiz, Spain)
Mónica Schwarz (University of Cádiz, Spain)
Nicoló Giuliani (Hochschule Geisenheim University, Germany)
Pau Sancho-Galán (University of Cádiz, Spain)
Paula García-Oliveira (University of Vigo, Spain)
Paz Otero (University of Santiago of Compostela, Spain)
Rainer Jung (Hochschule Geisenheim University, Germany)
Ramón Natera (University of Cádiz, Spain)
Raquel A. Nogueira (University of Vigo, Spain)
Remedios Castro (University of Cádiz, Spain)
Rubén Del Barrio-Galán (University of Chile, Chile)
Uxía Arias-González (University of Vigo, Spain)
Víctor Palacios (University of Cádiz, Spain)
Yulian Volovenko (Taras Shevchenko National University of Kyiv, Ukraine)

We also would like to acknowledge the technical support performed by MDPI Editorial for the creation of this book.

**Enrique Durán-Guerrero, Remedios Castro-Mejías**
*Editors*

foods

MDPI

*Editorial*

# Novel Analysis on Aroma Compounds of Wine, Vinegar and Derived Products

**Enrique Durán-Guerrero * and Remedios Castro**

Analytical Chemistry Department, Faculty of Sciences-University Institute of Wine and Food Research (IVAGRO-CAIV), University of Cadiz, Agrifood Campus of International Excellence (CeiA3), Polígono Río San Pedro, s/n, 11510 Puerto Real, Spain; remedios.castro@uca.es
* Correspondence: enrique.duranguerrero@uca.es; Tel.: +34-956-016456

Aroma is one of the main responsible for the acceptance of oenological products such as wine, vinegar and derived products. Aroma compounds are produced during the winemaking process, and they can be affected by natural, geographical and human factors: raw material, alcoholic and acetic fermentation, aging, distillation, technological processes, etc. Therefore, it is very important the study and characterization of the aromatic fraction of these oenological beverages, in order to improve the quality of the final product.

Therefore, this special issue "*Novel Analysis on Aroma Compounds of Wine, Vinegar and Derived Products*" is focused on the recent research related to the study of the volatile composition of wine, vinegar and derived products from several different fields of science: oenology, chemistry, food science and technology, biochemistry, microbiology, biotechnology, chemical engineering, sensory analysis, etc. As a result, this special issue includes 12 valuable scientific contributions, 2 reviews and 10 original research works, which deal with the latest advances in both sensory and analytical tools to evaluate the effect of different techniques or winemaking stages on the oenological products' aroma.



**Citation:** Durán-Guerrero, E.; Castro, R. Novel Analysis on Aroma Compounds of Wine, Vinegar and Derived Products. *Foods* **2021**, *10*, 1245. https://doi.org/10.3390/foods10061245

Received: 24 May 2021
Accepted: 28 May 2021
Published: 30 May 2021



In this sense, Pérez-Jiménez et al. employed the in-mouth headspace sorptive extraction (HSSE) technique, which is based on the application of a polydimethylsiloxane (PDMS) coated bar in the mouth, after the sample intake, to perform the headspace intra-oral aroma extraction [1]. In this way, twenty-two wine aroma compounds were identified at three times: immediately after spitting out the sample, after 60 s, and finally, after 120 s. The different volatile compounds exhibited different release times, with low persistence for esters and linear alcohols. Muñoz-González et al. also used this methodology to evaluate the wine esters release and their perception under the influence of grape seed tannin extracts [2]. The authors concluded that the addition of this type of extract to wines could modify their aroma perception in a compound-dependent manner.

On the other hand, Tarasov et al. determined three sensory thresholds (detection, recognition, and rejection thresholds) for 1,1,6-Trimethyl-1,2-dihydronaphthalene (TDN) in Riesling wine using sensory analysis [3]. This volatile compound is related to kerosene aroma, and its recognition seems to be modulated by the wine serving temperature, with low temperatures facilitating it. Studying the winemaking of sweet wines by means of sensory analysis, Ruiz-Bejarano et al. established that the use of climate chambers in the elaboration of sweet wines seems to be an adequate alternative to the traditional method, allowing a total control of the process and producing very well-valued sweet wines [4].

From a sensory and instrumental point of view, Úbeda et al. studied the influence of bentonite added at different stages on traditional sparkling wines by solid-phase microextraction coupled to gas chromatography with mass spectrometric detection (SPME-GC-MS) [5]. Two times were considered, before and during tirage. The results showed that the addition of bentonite before this phase, in the base wine, had a lower effect on volatile compounds, with wines treated with 50% of the bentonite dosage in tirage with poor foam and aromatic characteristics. Also working on these type of wines (sparkling wines), and using solid-phase extraction-gas chromatography-mass spectrometry (SPE-GC-MS), Castro et al. established that the use of chitosan, a polysaccharide with fining, antimicrobial, antioxidant, and chelating properties, during the second fermentation, increased the protein (around 50%) and amino acid (9%) content of wines, without significant changes in polyphenols and organic acids. Esters increased as a consequence of this addition [6]. Amores-Arrocha, also thanks to the use of SPE-GC-MS, found that low bee pollen doses (0.1 and 0.25 g/L) before alcoholic fermentation could improve the aromatic profiles together with the odorant activity values levels in the production of red wines from Tintilla de Rota grape variety [7].

Ruiz et al., employing stir bar sorptive extraction coupled to gas chromatography and mass spectrometry (SBSE-GC-MS), studied the use of freezing techniques (ultrafast freezing, and liquid nitrogen freezing) in the wine-making of Muscat grapes [8]. The wines obtained using liquid nitrogen freezing exhibited higher levels of terpenoids, as well as higher levels of hydroxylic compounds and fatty acids than both the wines obtained through traditional methods and ultrafast freezing wines. In any case, both freezing techniques produced wines of a more intense aroma compared with those wines obtained by traditional methods.

The aging stage, both in wood and bottle, has also been considered in this special issue. Guerrero-Chanivet et al. carried out a study about the characterization of the aromatic profile of different wood chips (American oak, French oak, Spanish oak, Cherry and Chestnut) used for the aging of spirits and wines [9], whereas Vázquez-Pateiro et al. studied the evolution of volatile compounds, odor activity value-based aromatic notes, and sensory perception in Treixadura (*Vitis vinifera* L.) dry white wines during a 24-month bottle-aging period [10]. In this last study, most of the volatile compounds exhibited constant concentrations for 18 months in bottle, and after that significant and sharp decreases were observed.

In addition to these original research works, two reviews were also considered in this special issue. In one of them, Durán-Guerrero et al. have compilated all the different scientific works about the aroma of the Sherry oenological products (dry wines, sweet wines, vinegars, and brandies), with emphasis on the different analytical methodologies used [11]. In the other, the main secondary aroma compounds present in wine and the microorganisms involved in their presence were contemplated by Carpena et al. [12].

In summary, the Special Issue "*Novel Analysis on Aroma Compounds of Wine, Vinegar and Derived Products*" shows the great importance of sensory and analytical study of oenological products aroma and how they are influenced by the different stages and conditions under which they are elaborated.

**Funding:** This research received no external funding.

**Institutional Review Board Statement:** Not applicable.

**Informed Consent Statement:** Not applicable.

**Conflicts of Interest:** The authors declare no conflict of interest.

## References

1. Pérez-Jiménez, M.; Muñoz-González, C.; Pozo-Bayón, M.A. Oral release behavior of wine aroma compounds by using in-mouth headspace sorptive extraction (HSSE) method. *Foods* **2021**, *10*, 415. [CrossRef] [PubMed]
2. Muñoz-González, C.; Criado, C.; Pérez-Jiménez, M.; Pozo-Bayón, M.Á. Evaluation of the effect of a grape seed tannin extract on wine ester release and perception using in vitro and in vivo instrumental and sensory approaches. *Foods* **2021**, *10*, 93. [CrossRef] [PubMed]
3. Tarasov, A.; Giuliani, N.; Dobrydnev, A.; Schuessler, C.; Volovenko, Y.; Rauhut, D.; Jung, R. 1,1,6-Trimethyl-1,2-dihydronaphthalene (TDN) Sensory Thresholds in Riesling Wine. *Foods* **2020**, *9*, 606. [CrossRef] [PubMed]
4. Ruiz-Bejarano, M.J.; Durán-Guerrero, E.; Castro, R.; Barroso, C.G.; Rodríguez-Dodero, M.C. Use of Sensory Analysis to Investigate the Influence of Climate Chambers and Other Process Variables in the Production of Sweet Wines. *Foods* **2020**, *9*, 424. [CrossRef] [PubMed]

5. Ubeda, C.; Lambert-Royo, M.I.; Gil i Cortiella, M.; Del Barrio-Galán, R.; Peña-Neira, Á. Chemical, Physical, and Sensory Effects of the Use of Bentonite at Different Stages of the Production of Traditional Sparkling Wines. *Foods* **2021**, *10*, 390. [CrossRef] [PubMed]
6. Castro, A.; Riponi, C.; Chinnici, F. Chitosan in sparkling wines produced by the traditional method: Influence of its presence during the secondary fermentation. *Foods* **2020**, *9*, 1174. [CrossRef] [PubMed]
7. Amores-Arrocha, A.; Sancho-Galán, P.; Jiménez-Cantizano, A.; Palacios, V. Bee Pollen Role in RedWinemaking: Volatile Compounds and Sensory Characteristics of Tintilla de Rota Warm Climate Red Wines. *Foods* **2020**, *9*, 981. [CrossRef] [PubMed]
8. Ruiz-Rodríguez, A.; Durán-Guerrero, E.; Natera, R.; Palma, M.; Barroso, C.G. Influence of two different cryoextraction procedures on the quality of wine produced from muscat grapes. *Foods* **2020**, *9*, 1529. [CrossRef] [PubMed]
9. Guerrero-Chanivet, M.; Valcárcel-Muñoz, M.J.; García-Moreno, M.V.; Guillén-Sánchez, D.A. Characterization of the Aromatic and Phenolic Profile of Five Different Wood Chips Used for Ageing Spirits and Wines. *Foods* **2020**, *9*, 1613. [CrossRef] [PubMed]
10. Vázquez-Pateiro, I.; Arias-González, U.; Mirás-Avalos, J.M.; Falqué, E. Evolution of the aroma of treixadura wines during bottle aging. *Foods* **2020**, *9*, 1419. [CrossRef] [PubMed]
11. Durán-Guerrero, E.; Castro, R.; García-Moreno, M.D.V.; Rodríguez-Dodero, M.D.C.; Schwarz, M.; Guillén-Sánchez, D. Aroma of Sherry Products: A Review. *Foods* **2021**, *10*, 753. [CrossRef] [PubMed]
12. Carpena, M.; Fraga-Corral, M.; Otero, P.; Nogueira, R.A.; Garcia-Oliveira, P.; Prieto, M.A.; Simal-Gandara, J. Secondary aroma: Influence of wine microorganisms in their aroma profile. *Foods* **2021**, *10*, 51. [CrossRef] [PubMed]

*foods* MDPI

*Review*
# Aroma of Sherry Products: A Review

**Enrique Durán-Guerrero [1,*], Remedios Castro [1], María de Valme García-Moreno [1], María del Carmen Rodríguez-Dodero [1], Mónica Schwarz [2,3] and Dominico Guillén-Sánchez [1]**

1 Analytical Chemistry Department, Faculty of Sciences-IVAGRO, Agrifood Campus of International Excellence (ceiA3), Campus Universitario de Puerto Real, University of Cadiz, s/n, Puerto Real, 11510 Cadiz, Spain; remedios.castro@uca.es (R.C.); valme.garcia@uca.es (M.d.V.G.-M.); maricarmen.dodero@uca.es (M.d.C.R.-D.); dominico.guillen@uca.es (D.G.-S.)
2 "Salus Infirmorum" Faculty of Nursing, University of Cadiz, 11001 Cadiz, Spain; monica.schwarz@uca.es
3 Nutrition and Bromatology Area, Faculty of Medicine, University of Cadiz, Plaza Falla, 9, 11003 Cadiz, Spain
* Correspondence: enrique.duranguerrero@uca.es; Tel.: +34-956-016-456

**Abstract:** Jerez (Sherry) is a well-known wine-producing region located in southern Spain, where world-renowned oenological products such as wines, vinegars, and brandies are produced. There are several factors that provide characteristic physical, chemical, and sensory properties to the oenological products obtained in this Sherry region: the climate in the area with hot summers, mild winters, and with limited rainfall; the raw material used consisting on Palomino Fino, Moscatel, and Pedro Ximénez white grape varieties; the special vinification with fortified wines; and aging techniques such as a dynamic system of biological or oxidative aging. These special organoleptic characteristics are responsible for, among others, the aromatic profile of the wines, vinegars and brandies from the area, which explains why this is a subject that has been extensively researched over the years. This bibliographic review aims to compile the different scientific contributions that have been found to date, in relation with the aroma of the oenological products from the Sherry area (dry wines, sweet wines, vinegars, and brandies). We have mainly focused on the different analytical methodologies used and on the main analytes of interest.

**Keywords:** Sherry; wine; vinegar; brandy; aroma



**Citation:** Durán-Guerrero, E.; Castro, R.; García-Moreno, M.d.V.; Rodríguez-Dodero, M.d.C.; Schwarz, M.; Guillén-Sánchez, D. Aroma of Sherry Products: A Review. *Foods* **2021**, *10*, 753. https://doi.org/10.3390/foods10040753

Academic Editor: Christopher John Smith

Received: 5 March 2021
Accepted: 31 March 2021
Published: 1 April 2021



## 1. Introduction

The winemaking tradition in the agricultural areas within the Jerez (Sherry) region dates far back in time. This is an eminent wine-producing region located in the south of Spain, surrounded by mountains and coastal lands that condition the climate in the area, which together with its particular aging methods, are determinant to attain the highly desirable organoleptic characteristics of its oenological products [1]. Worldwide renowned oenological products such as wines, vinegars, and brandies are the result of this unique combination of factors.

Sherry wines are considered among the most highly appreciated products in the world of oenology [2]. Diversity is undoubtedly one of the distinctive features of Sherry's identity, where just three grape varieties (Palomino, Moscatel, and Pedro Ximénez) give rise to different wines that clearly differ in terms of color, aroma, flavor, and texture depending on their elaboration process. [3].

Those wines that are subjected exclusively to biological aging—i.e., those which are protected from any direct contact with the air by the natural flor velum—retain their initial color, and display a series of distinctive aromatic and gustatory notes derived from the yeasts that form that essential flor velum [4]. On the other hand, other Sherry wines are aged by oxidative or physicochemical means, in direct contact with the oxygen in the air. These gradually acquire a darker hue, and exhibit more complex aromas and flavors [5].

Furthermore, the type of fermentation, which can be either complete or partial allows the production of highly dry wines (fortified wines) or extraordinarily sweet wines (natural sweet wines). By mixing these two types in different proportions, new wines with varying levels of sweetness (liqueur fortified wines) are also obtained [6,7].

With regard to Sherry vinegars, these are obtained from the grapes grown in the local vineyards. The authorized grape varieties for the production of Sherry vinegar are the same that those employed for Sherry wine. The Sherry vinegar production process basically consists in the acetic fermentation of local wines, as a result of the transformation of alcohol in acetic acid by acetic bacteria (*Mycoderma aceti*) and its subsequent aging in wooden casks. The final product presents a color between old gold and mahogany, with an intense aroma, lightly alcoholic, with notes of wine and wood predominating, and a pleasant taste, despite the acidity, with a long aftertaste [8,9].

On the other hand, Sherry Brandy is the product resulting from the distillation of wines (mainly Airén and Palomino ones) and its subsequent aging to confer the final product its distinctive organoleptic qualities [10].

All these products share in common a singular and dynamic aging process that is characteristic of the Sherry area: 'Criaderas y Solera'. This aging process uses oak casks, generally American oak (*Quercus alba*), that may vary between 250 and 600 L volume depending on the product to be obtained. The porosity of the American oak is ideal to allow the contact of the aging product with the oxygen in the air, thus facilitating its oxidation and favoring the aging process. The evolution of all the product physicochemical parameters is largely due to the impact of wood on the aging process. In fact, wood is a definite determinant of the organoleptic properties achieved by all the Sherry oenological products [5,11]. Moreover, the high level of aromatic content of these Sherry products is also influenced by the high level of aromatic composition of the American oak, compared to other types of oaks, such as French oak (*Quercus petraea, Quercus robur*).

During the aging phase in the winemaking process, the capacity of the wood to release certain compounds is essential and will vary according to the size and age (previous uses) of the cask. Thus, the smaller the cask size, the greater the wood surface in contact with the liquid. In this sense, the use of small barrels is not always convenient, since the effect of the wood on the final product could be greater than desirable [12]. Based on experience, 500–600 L barrels seem to be the most appropriate size for the aging of Sherry products, since they provide the ideal balance between wood surface and content volume.

Another characteristic of these wines is that they are aged in preconditioned casks, i.e., casks that have previously contained sherry wine. They are known as *"barricas envinadas"* (casks in which Sherry wine has been aged). This significantly contribute to providing these products with different nuances depending on the type of preconditioning undergone by the casks [13].

The aforementioned 'Criaderas y Solera' aging method could be defined as a dynamic aging process, as opposed to the static aging by vintages. In the latter system, the oenological product to be aged remains in the same barrel during the entire aging period, while in the Criaderas y Solera method, however, the oenological product is stored in casks classified into groups, known as 'scales', according to the age of the product that they contain. The scale that contains the oldest oenological product is called 'solera' and it is located at ground level. This is topped, according to its younger age, by the first criadera, the second, the third and so on (Figure 1). A small amount of the product, which must be the same from each of the casks that make up the solera, is extracted for bottling and distribution. The resulting empty space is replenished with the equivalent volume of the oenological product from the first criadera. The same procedure is applied to the first and second criadera, which are refilled with the product from the corresponding topping criadera. In this way, a uniform product is obtained in terms of flavor, aroma and color. The same organoleptic characteristics are obtained, since the amount of refilling product is rather reduced in comparison with the larger amount of product in the receiving cask. Thus, the small amount of product added to the cask acquires the characteristics of the predominant older product it is mixed with [14,15].

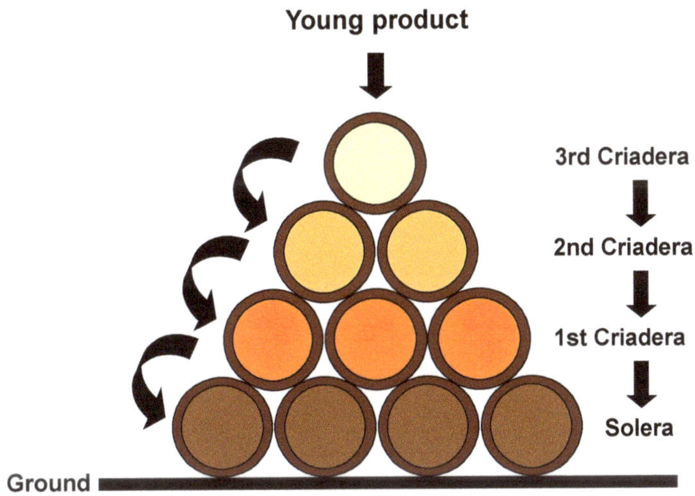


**Figure 1.** Criaderas y Solera aging method.

Because of this peculiar aging process, it is quite difficult to estimate the exact age of the oenological product, so it is usually referred to as average age. This parameter is defined as the ratio between the total volume of product in the system and the annual volume that is taken out for its commercialization. Depending on their average age, they will be classified into different categories, which will exhibit different characteristics, depending on the original oenological matrix that was used (wine, vinegar, or Brandy).

All the aforementioned features in the elaboration of Sherry products provide them with their own qualities that will constitute their seal of quality. Thus, such characteristics like polyphenolic compounds content [10,16–18], chromatic attributes [17,19,20], organic acids [13,17], or sugars contents [14] have been suggested to be determinant parameters regarding the ultimate quality of Sherry wine, vinegar, or brandy.

The aroma of oenological products, in general, represents an important determinant of their quality, and there are numerous studies that support this point [21–24]. Although not all volatile compounds contribute to aroma perception [25], the study of aromatic profile is still of major importance, since the acceptance of the final product by the consumer depends on them to a great extent [26]. Consequently, in recent years, significant technological advances have been made in terms of extraction methods and the subsequent analysis of these compounds [27,28]. In parallel, sensory analysis has been consolidated as an essential tool to perform a complete investigation that covers all the aspects related to aroma. An increasing number of studies propose sensory analysis as a crucial tool to determine the quality of the final product [29]. Moreover, a recent study by Cruces-Montes et al. [30] presented the perception of the attributes of Sherry wine and its consumption in young people in the south of Spain. Their results showed that the consumption of Sherry wine was recognized to different dimensions, and flavor was especially important for some types of Sherry wine.

Figure 2 shows the growing progression in the number of studies that address the subject of aroma in the typical products from Jerez (Sherry) area: wine, vinegar, and brandy. This rising number of studies and publications is explained by the importance of the content of volatile compounds regarding the aroma of wine products, as well as by the socioeconomic relevance of these products in the region. Also, the evolution of analytical technologies and their innovations contribute for the increment of this kind of studies. On these bases, we have considered the importance of a literature review that would cover the most prominent aspects associated to this tandem: aroma and Sherry oenological products.

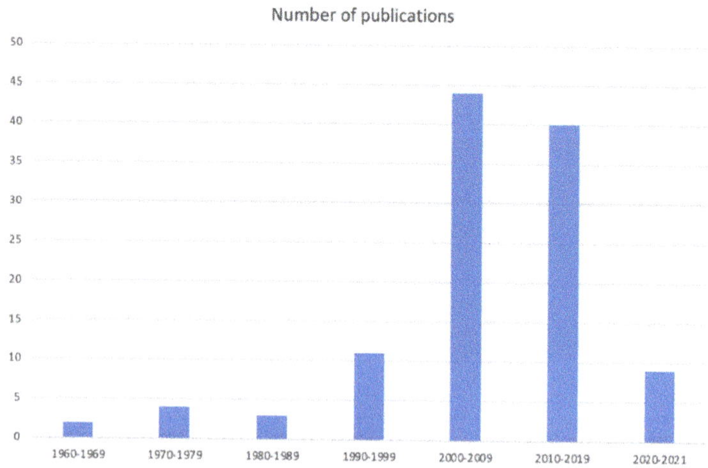


**Figure 2.** Number of publications addressing the aroma of Sherry oenological products. Source: Scopus.

## 2. Study of the Aroma of Dry Sherry Wines

The aromatic content of Sherry wines and, in particular, that of Fino wines, has been extensively studied by employing the analytical methods previously described. Table 1 shows the main aromatic compounds detected in Fino, Amontillado, and Oloroso Sherry wines, together with the bibliographic references where these compounds are mentioned. Sensory descriptors and concentration ranges also appear in Table 1.

**Table 1.** Volatile compounds identified in dry Sherry wines, sensory descriptors, concentration ranges, and bibliographic references.

| Volatile Compounds | Sensory Descriptors | Concentration (mg/L) | References Fino | References Amontillado | References Oloroso |
|---|---|---|---|---|---|
| *Carbonyls* | | | | | |
| Acetaldehyde | Overripe apple | 85–545 | [31–36] | [32,37] | [32,38] |
| Acetoin | Butter | 0.011–74 | [31–36] | [32,37] | [32] |
| Benzaldehyde | Bitter almond/cherry | 0.013–0.076 | [33,36] | | [39] |
| 2,3-Butanedione | Butter-cookie | 0.170–2.1 | [33,34,36] | [37] | [38] |
| Furfural | Sweet/woody/almond/baked/bread | 0.179–7.14 | [32] | [32] | [32] |
| β-Ionone | Balsamic/rose/violet/berry/phenolic | 0.062 | [32,35] | [32] | |
| Neral | Sweet/citrus/lemon peel | | [33] | | |
| Octanal | Herbaceous | 0.090–0.390 | [32–36] | [37] | |
| *Acids* | | | | | |
| Butanoic acid | Cheese/butter | 0.607–14.6 | [31–36] | [32,37] | [32,38] |
| Decanoic acid | Rancid | 0.004–0.370 | [31,33,36] | | [39] |
| Dodecanoic acid | Mild fatty/coconut/bay oil | | [33,36] | | |
| Hexanoic acid | Fatty/sweat/cheese | 0.635–2.39 | [31–36] | [32] | [32] |
| Isobutanoic acid | Acidic/cheese/dairy/buttery/rancid | 2.2–22.1 | [31,33,36] | | |
| Isobutyric acid | Acidic/cheese/dairy/buttery/rancid | 0.002–4.58 | | [32] | [32,39] |
| 3-Methylbutanoic acid | Cheese | 1.5–679 | [31–33,35,36] | [32,37] | [38] |
| Nonanoic acid | Waxy/cheesy/dairy | 0.003–0.011 | | | [39] |
| Octanoic acid | Fatty/waxy/rancid/oily/cheesy | 0.001–1.6 | [31,33,34,36] | | [39] |
| *Alcohols* | | | | | |
| Benzyl alcohol | Floral/rose/phenolic/balsamic | 0.045–3.3 | [31–33,36] | [32] | [32] |
| 1-Butanol | Fusel oil/sweet/balsam/whiskey | 0.001–19.9 | [31–36] | [32] | [32,39] |
| 2-Butanol | Sweet/apricot | 1.1–4.4 | [31–33,35,36] | [32] | [32] |
| 2,3-Butanediol | Fruity/creamy/buttery | | [33,36] | | |
| 1-Decanol | Fatty/waxy/floral | 0.124–1.26 | [32,33,35,36] | [32] | [32] |
| 3-Ethoxy-1-propanol | | 0.250–0.490 | [31,33,35] | | |
| 1-Heptanol | Musty/pungent/leafy green/apple/banana | 0.300–0.870 | [33] | [32] | [32] |
| Hexanol | Fusel oil/fruity/alcoholic/sweet/green | 0.001–2.5 | [31–33,35,36] | [32] | [32,39] |
| E-3-Hexenol | Green/cortex/floral/oily/earthy | 0.055–0.085 | [31,32,35] | | |
| Z-3-Hexenol | Green/grassy/melon rind | 0.055–0.085 | [31–33,35] | | |
| Isoamyl alcohols | Vinous/solvent | 0.020–444 | [31–36] | [32,37] | [32,38] |
| Isobutanol | Vinous/solvent | 25.7–102 | [31–36] | [32,37] | [32] |
| Isopropyl alcohol | Alcohol/musty/woody | 1.4–2.7 | [31] | | |
| Methanol | Slight alcoholic | | [33,36] | | |

**Table 1.** *Cont.*

| Volatile Compounds | Sensory Descriptors | Concentration (mg/L) | References Fino | References Amontillado | References Oloroso |
|---|---|---|---|---|---|
| 2-Methyl-1-butanol | Roasted/fruity/fusel oil/alcoholic/wine/whiskey | | | | [39] |
| 2-Methyl-1-pentanol | | 0.020–0.090 | [32] | [32] | [32] |
| 3-Methyl-1-pentanol | Pungent/fusel oil/brandy/wine/cocoa | 0.110–18 | [31–33,35,36] | [32] | [32] |
| 4-Methyl-1-pentanol | Nutty | 0.029–0.135 | [31–33,36] | | [32] |
| 1-Octanol | Waxy/green/citrus/aldehydic/floral/coconut | | [33,36] | | |
| 1-Pentanol | Pungent/fermented/bready/yeasty/fusel oil/winey/solvent | 0.060–102 | [32,33] | [32] | [32] |
| Phenethyl alcohol | Rose | 0.003–99 | [31–36] | [32,37] | [32,38,39] |
| Propanol | Alcoholic/fermented/musty/yeasty/apple/pear | 12.3–16.3 | [31,33,35,36] | | |
| *Volatile phenols* | | | | | |
| 4-Ethylguaiacol | Toasted/clove | 0.002–0.740 | [32–36] | [32,37] | [32,39] |
| 4-Ethylphenol | Smoke/phenolic/creosote | 0.004–0.094 | [33,36] | | [32,39] |
| Eugenol | Cinnamon/clove | 0.002–0.477 | [31–36] | [32,37] | [32,39] |
| Guaiacol | Phenolic/smoke/spice/vanilla/woody | 0.280–0.434 | | | [39] |
| Methyleugenol | Spicy/cinnamon/clove/musty/waxy/phenolic | 0.157 | | | [32] |
| *Esters* | | | | | |
| Butyl acetate | Fruity/solvent/banana | 0.091–0.161 | [33] | | [32,39] |
| Diethyl malate | Brown sugar/sweet/wine/fruity/herbal | 0.800–23.6 | [31–33,35,36] | [32] | [32] |
| Diethyl succinate | Mild fruity/cooked apple/ylang | 0.001–55.4 | [31–33,35,36] | [32] | [32,39] |
| Ethyl acetate | Pineapple/varnish | 13.9–260 | [31–36] | [32,37] | [32] |
| Ethyl benzoate | Fruity/dry/musty/sweet/wintergreen | 0.180–0.215 | [32] | | [32] |
| Ethyl butanoate | Banana/apple | 0.172–3.5 | [31–36] | [37] | [32,38,39] |
| Ethyl decanoate | Sweet/waxy/fruity/apple/grape/oily/brandy | 0.22 | | | [32] |
| Ethyl furoate | Balsamic | | [33,36] | | |
| Ethyl heptanoate | Fruity/pineapple/brandy/rum/wine | 0.021–0.109 | [32,35] | [32] | [32] |
| Ethyl hexanoate | Almond/apple | 0.078–0.280 | [31,33–36] | [37] | [38,39] |
| Ethyl 3-hydroxybutanoate | Fruity/green grape/tropical | 0.030–0.747 | [31,33,35,36] | | |
| Ethyl 3-hydroxyhexanoate | Rubber | | [33] | [37] | |
| Ethyl isobutanoate | Apple/pineapple | 0.028–1.660 | [31–36] | [32,37] | [32,38] |
| Ethyl isovalerate | Fruity/sweet/apple/pineapple/tutti frutti | 0.001–0.009 | | | [39] |
| Ethyl lactate | Raspberry/milky | 12–854 | [31–36] | [32,37] | [32,38] |
| Ethyl laurate | Sweet/waxy/floral/soapy/clean | 0.024–0.140 | [32,35] | [32] | [32] |
| Ethyl myristate | Mild waxy/soapy | 0.099–0.119 | [32,33,35,36] | [32] | |
| Ethyl octanoate | Pear | 0.008–1.3 | [31–36] | [32,37] | [39] |
| Ethyl propanoate | Sweet/fruity/rum/juicy fruit/grape/pineapple | 0.109–1.92 | [31–33,35,36] | [32] | [32] |
| Ethyl palmitate | Mild waxy | 0.042–0.070 | [32,35] | [32] | |
| Ethyl pyruvate | Fruity/sweet/rum | 0.081–0.201 | [31–33,35,36] | [32] | [32] |
| Ethyl valerate | Sweet/fruity/apple/pineapple/green/tropical | 0.001–0.010 | | | [39] |
| Hexyl acetate | Green/fruity/sweet/fatty/fresh/apple/pear | 0.001–0.008 | | | [39] |
| Hexyl hexanoate | Green/sweet/waxy/fruity/berry | 0.247 | | | [32] |
| Hexyl lactate | Sweet/floral/green/fruity | 1.1 | | | [32] |
| Isoamyl acetate | Banana | 0.050–0.855 | [31,33,34,36] | [37] | [38,39] |
| Isoamyl laurate | Winey/alcoholic/fatty/creamy/yeasty/fusel oil | 0.357 | | [32] | |
| Isobutyl acetate | Sweet/fruity/banana | 0.025–0.137 | [31,33] | | [32,39] |
| Isobutyl isobutanoate | Fruity tropical/fruit pineapple/grape skin/banana | 0.066 | | [32] | |
| Isobutyl lactate | Faint buttery/fruity/caramel | 0.034–0.242 | [32,33,35,36] | | [32] |
| Methyl acetate | Solvent/fruity/winey/brandy/rum | 6.6 | | | [32] |
| Methyl butanoate | Strawberry/butter | 0.486–4.86 | [33,34,36] | [37] | [32,38] |
| Monoethyl succinate | Odorless | | [33,36] | | |
| Phenethyl acetate | Flowers | 0.100–1.1 | [31,33,36] | [37] | [38] |
| Phenethyl octanoate | Sweet/waxy/slightly cocoa/caramel/winey/brandy | 0.190–0.275 | [32,33,36] | [32] | |
| Propyl acetate | Solvent/fusel oil/sweet/fruity | 0.042–0.162 | [31–33,36] | [32] | [32] |
| Propyl butanoate | Pungent/rancid | 0.112–0.150 | [32,35] | [32] | |
| *Terpenes* | | | | | |
| β-Citronellol | Rose | 0.280–1.33 | [31–33,35,36] | [32] | |
| Farnesol | Sweet/floral | 0.282–5.79 | [32,35] | [32] | [32] |
| Linalool | Citrus/orange/floral/terpy/waxy/rose | 0.009–0.032 | [31] | | [38] |
| Nerol | Floral/green | 0.151–0.176 | [32,35] | [32] | |
| E-Nerolidol | Floral/green/citrus/woody/waxy | 0.076–0.213 | [32,35] | [32] | [32] |
| Z-Nerolidol | Waxy/floral | 0.696 | [32,33,36] | | |
| 4-Terpineol | Pine | 0.777 | [32] | | |
| α-Terpineol | Pine/woody/resinous/lemon/citrus | 0.006–0.015 | [32,33,35] | | [39] |

**Table 1.** *Cont.*

| Volatile Compounds | Sensory Descriptors | Concentration (mg/L) | References Fino | References Amontillado | References Oloroso |
|---|---|---|---|---|---|
| *Lactones* | | | | | |
| γ-Butyrolactone | Creamy/oily | 0.004–40.8 | [31–36] | [32] | [32,39] |
| γ-Decalactone | Peach | 0.043 | [32–36] | | |
| Pantolactone | | 0.470–5.22 | [31–33,35,36] | [32] | [32] |
| Sotolon | Walnut/cotton candy/curry | 0.100–0.670 | [33,34,36] | [37] | [38] |
| cis-Whiskeylactone | Burnt/wood/vanilla/coconut | 0.009–0.410 | [31,34,36] | [37] | [38,39] |
| trans-Whiskeylactone | Sweet/spicy/coconut/vanilla | | [33] | | [39] |
| *Miscellaneous* | | | | | |
| Methionol | Cooked potato/cut hay | 0.063–3.4 | [31–36] | [32,37] | [32,38] |
| p-Cymene | Citrus/terpene/woody/spice | | [33,36] | | |
| 1,1-Diethoxyethane | Green fruit/liquorice | 8.4–58.8 | [31–36] | [32,37] | [32,38] |

All types of dry Sherry wines are produced from the same grape, 'Palomino fino', and it is the subsequent elaboration of the product (biological or oxidative aging), the main responsible of obtaining wines with different organoleptic characteristics. Also, the extracted wood components induce these changes. Therefore, it is mainly the aging process that will determine the differences between the three types of dry wines: Fino, Oloroso, and Amontillado. Thus, the aroma of Fino wines will be conditioned by the flor velum yeast, which, in addition to shielding the wine from oxygen, will contribute with a series of compounds derived from its metabolism. At the other end, we have the Oloroso wine that undergoes oxidative aging and contains higher levels of alcohol, so that during this aging stage the compounds that were initially present in the wine aroma will evolve due to oxidation, esterification, and other reactions. Finally, Amontillado wines undergo a first stage of biological aging and then an oxidative one [5].

A large number of volatile compounds are common to all of them, including acetaldehyde, acetoin, eugenol, and 1,1-diethoxyethane, among others. Acetaldehyde may come from different sources, although it appears particularly as a secondary product resulting from the aerobic metabolism of the flor velum yeasts responsible for the biological aging process [40,41]. This compound is also the precursor of a large number of other compounds that are involved in the aroma of Sherry wines, either as a result of biological or oxidative aging. In particular, it is the precursor of 1,1-diethoxyethane, one of the main acetals in Sherry wines, which is formed through chemical and biochemical reaction with ethanol [42]. This compound contributes to the fruity aromas and balsamic notes of these wines.

Acetoin is one of the other acetaldehyde-derived compounds with aromatic significance in Sherry wines. This compound is preferentially formed by a condensation reaction of two acetaldehyde molecules [42]. Acetoin is one of the compounds responsible for the bitter notes of Fino wines. The reduction of the acetoin gives rise to 2,3-butanediol, another aromatic compound involved in the aroma of Sherry wines.

The reaction between acetaldehyde and α-ketobutyric acid during the anaerobic metabolism of the yeasts in the flor velum gives rise to sotolon. This compound has a high impact on the aroma of these wines, particularly in the nutty, curry, and cotton candy notes that are present in all the Sherry wines [42].

It should be noted that Sherry wines from exclusively biological aging—i.e., Fino wines—have a particularly high acetaldehyde content, which is actually attributable to their biological aging. This compound is not only responsible for the sharp character of Fino wines' aroma, but also contributes enriches it with the notes of overripe or ripe apples [33,35,43] that are inherent to this wine.

According to the bibliography, other major volatile compounds to be found in the wine are isoamyl alcohols, ethyl lactate, and 1,1-diethoxyethane (Table 1). A certain number of volatile compounds clearly differentiate Fino wines from other types of Sherry wines, among them E-3-hexenol, Z-3-hexenol, γ-decalactone, terpinen-4-ol, Z-nerolidol, farnesol, and octanal. This suggests that their origin may be linked to the biological aging process that characterizes this wine, and that they do not remain as part of the composition of other wine types, like Amontillado, which undergoes a subsequent oxidative aging procedure.

Other aromatic compounds that play a significant role in the aroma of biologically aged wines are β-citronellol and β-ionone. These compounds are responsible for the citrus and balsamic notes in the aroma of these wines, although they are present at concentration levels of μg/L (Table 1). Other compounds that also stand out are phenethyl octanoate, ethyl palmitate, nerol, propyl butanoate, and ethyl myristate. All of these compounds have been detected in both Fino and Amontillado wines (Table 1).

Amontillado wines, which are obtained through an initial biological aging stage and a subsequent oxidative process as above mentioned, exhibit certain characteristics of their own. For example, they do not contain ethyl benzoate in their composition; they are the only types of Sherry wines that present detectable concentrations of isobutyl isobutanoate (0.066 mg/L) and isoamyl laurate (0.357 mg/L), and present lower concentration levels of 1,1-diethoxyethane, isobutanol, and phenethyl alcohol, while their levels of E-nerolidol are higher with respect to that in Fino or Oloroso wines [32]. The main volatile compounds that can be found in Amontillado wines are ethyl lactate, acetaldehyde, isoamyl alcohols, diethyl succinate, and ethyl acetate, all of them at levels of concentration of dozens or even hundreds of mg/L (Table 1). It has long been known that oxidative aging results in a higher concentration of esterified compounds in Amontillado wines, since their greater concentration of ethanol results in evident increment in ethyl lactate and ethyl acetate concentrations during the aging phase [32]. However, the compound that contributes the most to the aroma of Amontillado wines is ethyl octanoate, that is usually present at concentrations below 1 mg/L [37], followed by ethyl butanoate, eugenol, ethyl isobutanoate, and sotolon, which maintain their relative contributions to the wine aroma throughout the period of oxidative aging, even though their concentrations increase with time. It is precisely this second aging stage, the oxidative one, which confers Amontillado wines their main odorant characteristics.

Considerable levels of acetaldehyde are also found in Oloroso and Amontillado wines, although in lower concentrations than in Fino wines (around five times lower) [42]. The most abundant compounds in Oloroso wines are isoamyl alcohols, ethyl lactate, ethyl acetate, acetaldehyde, and diethyl succinate. Other compounds such as ethyl butyrate, ethyl caproate, ethyl decanoate, ethyl isovalerate, ethyl valerate, guaiacol, hexyl acetate, hexyl hexanoate, hexyl lactate, methyl acetate, 2-methylbutan-1-ol, methyleugenol, β-methyl-γ-octalactone, nonanoic acid, 2-phenylethanol, and 2-phenylethanol acetate tend to be more characteristic of Oloroso wines, and are not found either in Amontillado or Fino wines.

The narrow correlation between the aromatic composition of Sherry wines and the type of cask wood as well as the degree of toasting of the wood has already been studied [39]. The wines aged in French oak and chestnut casks undergo greater changes in their volatile compound composition during the oxidative aging process. American and Spanish oak, on the other hand, modify to a lesser degree the volatile compound profile of these wines during their aging. In relation to the wood toasting degree, it is the medium-toasted casks that produces the wines with the greatest volatile composition. These results are similar to those reported by other authors with regard to fortified and sweet wines aged in wood [44,45]. Eugenol and guaiacol are compounds derived from the degradation of lignin and their content increases during the aging in contact with wood. β-methyl-γ-octalactone was only identified in Oloroso wines aged in contact with oak wood, but not in those aged with chestnut. High concentrations of γ-butyrolactone were also determined in all the samples studied, similarly to those already reported by Hevia et al. [44]. Ethyl valerate, hexyl acetate, or ethyl octanoate (compounds that contribute with floral and fruity notes to the aroma of the wines) decreased with aging, except for the wines aged in French oak casks, which saw their concentration increased along with other compounds such as isobutyl acetate, ethyl valerate or isoamyl acetate.

## 3. Study of the Aroma of Natural Sweet Wines

According to the specifications in the Protected Denomination of Origin "Jerez-Xérès-Sherry" [6], Natural Sweet Wines are those produced using musts from very ripe or sun-dried grapes, generally of the Pedro Ximénez (PX) or Moscatel varieties. These musts, which are rich in sugars as a result of the raisining process, are only partially fermented in order to preserve most of their original sweetness. During this sweet vinification, the musts are fortified with wine alcohol as soon as the fermentation process starts, to reach a minimum alcohol content of 15% vol. The wines produced through this method are subsequently aged in direct contact with atmospheric oxygen, which favors a progressive aromatic concentration and increases their complexity while an intense color and a dense appearance is acquired, although with no negative impact on the typical freshness of these varieties. The alcohol content should range between 15° and 22° vol.

Table 2 presents the different volatile compounds determined in natural sweet wines, their sensory descriptors as well as the concentration ranges reported in the bibliographic references.

**Table 2.** Volatile compounds identified in natural sweet wines, sensory descriptors, and concentration ranges reported in the bibliographic references

| Volatile Compounds | Sensory Descriptors | Concentration (mg/L) | References |
|---|---|---|---|
| *Alcohols* | | | |
| (E)-2-Hexenol | Herbaceous/green/green tomato | 0.001–0.36 | [46–51] |
| 2,3-Butanediol (levo/meso) | Ripe fruit/butter | 0.001–4015.0 | [47,49,52,53] |
| 2-Butanol | Vinous/medicinal | 0.003–0.12 | [46–51] |
| 2-Methylbutanol | Roasted/fruity/ alcoholic/fusel oil/ wine/whiskey | 1.40–1.66 | [45] |
| 2-Phenylethanol | Rose/talc/honey | 0.12–78.88 | [45,47–49,51–55] |
| 2-phenylethyl alcohol | Rose/honey | 0.002–25.91 | [46,50,56,57] |
| 3-Ethoxypropanol | Overripe pear | 0.30–17.37 | [47,49] |
| 3-Hexenol (E/Z) | Herbaceous/green/grass | 0.001–0.079 | [46–49,52,56] |
| 3-Methyl-2-butanol | | | [54] |
| 3-Methylpentanol | Pungent/fusel oil/brandy/wine/cocoa | 0.022–0.030 | [47] |
| Benzyl alcohol | Roasted/toasted/disifectant/fruity/walnut/floral/rose/phenolic/balsamic | 0.001–0.772 | [45–52] |
| Butanol | Vinous/medicinal | 0.001–1.76 | [45–51] |
| Heptanol | Oily | 0.006–0.037 | [46,57] |
| Hexanol | Cut grass/resinous/herbaceous/wood | 0.001–1.02 | [45–51,54,55,57] |
| Isoamyl alcohols | Solvent/cake/fusel alcohols/nail polish/ripe fruit | 0.003–146.72 | [46–53,55–57] |
| Isobutanol | Alcohol/solvent/vinous/nail polish | 0.003–40.90 | [45–51,53,56,57] |
| Methanol | Solvent/pungent fruity | 57.5–163.0 | [49,53,57] |
| Pentanol | Bitter almond/synthetic | 0.001–0.014 | [49,51] |
| Propanol | Fusel alcohol/ripe fruit | 8.4–88.0 | [49,53,57] |
| *Aldehydes* | | | |
| (E)-2-Hexenal | Herbaceous | 0.012–0.308 | [48,51,55] |
| 2-Hexanal | | | [54] |
| 3-Methylbutanal | Ethereal/aldehydic/chocolate/peach/fatty | 0.094 | [56] |
| Acetaldehyde | Stewed apple/pungent | 13.29–347.0 | [49,53,56,57] |
| Benzaldehyde | Roasted/bitter almond/nutty/smoky | 0.003–0.151 | [45–50,54,55] |
| Decanal | Soapy/green lemon | | [57] |
| Hexanal | Fatty/herbaceous/green apple | 0.004–0.444 | [46–50,52,54,55] |
| Nonanal | Waxy/aldehydic/rose/orange peel fatty | | [54,55] |
| Octanal | Herbaceous | 0.046–0.127 | [47,54,55] |
| Phenylacetaldehyde | | 0.068 | [56] |
| *Ketones* | | | |
| 2,3-Butanedione | Buttery/ripe fruit/yogurt/cake | 0.004–5.07 | [46–51,56,57] |
| 2,3-Pentanedione | Buttery/cream/cake | 0.004–0.435 | [46–48,50,57] |
| 2-Octanone | Floral/over ripe fruit | 0.002–0.022 | [51] |
| 6-Methyl-5-hepten-2-one | | | [54,55] |
| Acetoin | Buttery/cream/sour yogurt/sour milk | 0.070–1228.52 | [46–51,53,57] |
| *Furans* | | | |
| 2-Furaldehyde | Fusel alcohol/cake/burnt/almond/ripe fruit/toasted bread/incense/floral | 0.001–5.002 | [45–52,54–57] |
| 5-Hydroxymethyl-2-furaldehyde | Rancid/toasted | 0.003–102.40 | [45,51] |
| 5-Methyl-2-furaldehyde | Toasted/bitter almond/cake/burnt/caramel | 2.4 | [45–52,54,55,57] |
| Ethyl 2-furoate | Balsamic | | [57] |
| Furfuryl alcohol | Varnish | 0.005–0.023 | [47,49,57] |

**Table 2.** *Cont.*

| Volatile Compounds | Sensory Descriptors | Concentration (mg/L) | References |
|---|---|---|---|
| *Acids* | | | |
| 2-Ethyl-hexanoic acid | | | [54,55] |
| 2-Methylbutanoic acid | Rancid | 0.003–0.009 | [48] |
| 3-Methylbutanoic acid | Lactic/rancid/cheese | 0.001–2.495 | [45–48,50,56,57] |
| Acetic acid | Fatty | 3.31–4.08 | [52] |
| Butanoic acid | Aged cheese/rancid | 0.003–0.627 | [46–50,56] |
| Decanoic acid | Rancid/cheese/wax/plasticine | 0.005–0.185 | [45,47,49,51,52,54,55,57] |
| Dodecanoic acid | Fatty/coconut/bay | | [54,55] |
| Hexadecanoic acid | Waxy/fatty | | [54,55] |
| Hexanoic acid | Cheese/rancid | 0.030–0.069 | [49] |
| Isobutanoic acid | Cheese/rancid/fat | 0.003–5.623 | [45,49] |
| Nonanoic acid | Waxy/dirty/cheese/dairy | 0.011–0.033 | [45,54,55] |
| Octanoic acid | Rancid/cheese/fatty | 0.002–0.506 | [45,47–49,51,52,54,55,57] |
| Propanoic acid | Fat | 0.080–1.371 | [49] |
| Tetradecanoic acid | Waxy | | [54,55] |
| *Esters* | | | |
| 2-Phenylethyl acetate | Fruity/honeyed/floral/rose | 0.001–0.094 | [46–50,52,54–58] |
| 2-Phenylethyl hexanoate | | 0.007–0.015 | [47] |
| 2-Phenylethyl octanoate | Cocoa/caramel/winey/brandy | | [57] |
| 3-Methylpropyl acetate | | 0.037 | [56] |
| 4-Methyl-2-pentyl acetate | | 0.181 | [52,55] |
| Benzyl acetate | Floral/fruity/jasmine/fresh | 0.0416 | [45] |
| Butyl acetate | Solvent/fruity/banana | 0.016–0.154 | [45,47,54,55] |
| Butyl lactate | | | [57] |
| cis-3-Hexenyl acetate | | 0.001–0.002 | [45] |
| Diethyl malate | Green | 0.003–0.531 | [47,49,51] |
| Diethyl pentanedioate | | | [55] |
| Diethyl succinate | Overripe fruit/lavender | 0.101–1.76 | [45,47,49,52,54,55,57,58] |
| Ethyl 2-methylbutanoate | | 0.0041 | [56] |
| Ethyl 2-methylpentanoate | | | [56] |
| Ethyl 2-methylpropanoate | | 0.054 | [56] |
| Ethyl 3-hydroxybutanoate | Grape/green apple/marshmallows | 0.005–0.062 | [47,49,57] |
| Ethyl 3-methylbutanoate | | 0.0075 | [45,54–56] |
| Ethyl 3-methylpentanoate | | 0.001 | [56] |
| Ethyl 4-methylpentanoate | | 0.001 | [56] |
| Ethyl acetate | Pineapple/varnish/balsamic/fruity/solvent/pungent/glue | 0.031–113.33 | [46–50,52,53,57] |
| Ethyl benzoate | Fruity/medicinal/wintergreen/anise | 0.002–0.005 | [49,57] |
| Ethyl butanoate | Banana/pineapple/strawberry | 0.012–0.386 | [45,52,54–58] |
| Ethyl cyclohexanoate | | | [56] |
| Ethyl decanoate | Synthetic/rancid | 0.015–0.162 | [52,54,55,57,58] |
| Ethyl dihydrocinnamate | | 0.001 | [56] |
| Ethyl dodecanoate | Waxy/floral/soapy/clean | 0.077–0.106 | [54,55,58] |
| Ethyl furoate | Plum/floral | 0.0001 | [49] |
| Ethyl heptanoate | Strawberry/banana | 0.005–0.046 | [46,47,58] |
| Ethyl hexadecanoate | | 0.008 | [54,55,58] |
| Ethyl hexanoate | Banana/green apple | 0.005–0.147 | [45,47,49,54–58] |
| Ethyl isobutanoate | Apple/pineapple | 0.002–3.869 | [45,58] |
| Ethyl lactate | Lactic/yogurt/strawberry/raspberry/buttery | 0.001–93.8 | [46–48,50–52,57] |
| Ethyl octadecanoate | | | [54,55] |
| Ethyl octanoate | Pineapple/pear/soapy/banana | 0.002–0.174 | [45,49,52,54–58] |
| Ethyl pentanoate | Fruity/apple/pineapple/green/tropical | 0.005–0.071 | [45,52] |
| Ethyl propanoate | Banana/apple | 0.005–0.152 | [46,47,58] |
| Ethyl succinate | Toffee/coffee | 0.029–70.0 | [47,49] |
| Ethyl tetradecanoate | Mild waxy/soapy | 0.002 | [54,55,58] |
| Hexyl acetate | Apple/pear/banana/floral | 0.001–2.14 | [45–48,50,52,57] |
| Isoamyl acetate | Banana | 0.008–0.019 | [49,54,55] |
| Isoamyl butanoate | Banana/fruity | 0.012–0.089 | [47,48] |
| Isobutyl lactate | Faint buttery/fruity/caramel | | [57] |
| Methyl acetate | Solvent/fruity/winey/brandy/rum | 0.064–0.085 | [58] |
| Methyl butanoate | Strawberry/butter | | [57] |
| Methyl octanoate | | 0.001 | [52] |

**Table 2.** *Cont.*

| Volatile Compounds | Sensory Descriptors | Concentration (mg/L) | References |
|---|---|---|---|
| *Terpenes* | | | |
| 4-Terpineol | Moldy | 0.002 | [48,54,55] |
| Carvacrol | Thyme | | [54,55] |
| Farnesol | Floral/fruity/balsamic/clove | 0.002–0.080 | [46–48,50,54,57] |
| Geranial | Citrus | 0.002–0.078 | [46,47,50] |
| Geraniol | Floral/fruity/rose/waxy/citrus | | [54] |
| γ-Terpineol | | 0.034–2.99 | [52] |
| Linalool | Muscat/rose/lavender | 0.006–1.62 | [52,54–57] |
| Linalool oxide | | | [54,55] |
| Nerol | Citrus/magnolia | 0.013 | [47,54,55] |
| Nerol oxide | | | [54,55] |
| Nerolidol | Floral/green/citrus/woody/waxy | | [54,55] |
| p-Cymene | Fresh/citrus/lemon/woody/spicy | 0.23–0.58 | [52] |
| Thymol | Herbal/thyme/phenolic/medicinal/camphor | | [54,55] |
| α-Terpineol | Lily/cake | 0.004–0.016 | [45,47,48,54,55] |
| β-Citronellol | Rose | | [54,55] |
| β-Myrcene | | | [54,55] |
| *Lactones* | | | |
| 4-Caprolactone | Herbaceous/coconut | 0.001–0.005 | [49] |
| γ-Butyrolactone | Cake/caramel/fruity/empyreumatic/coconut/toasted | 0.003–37.90 | [45–51,57] |
| γ-Decalactone | Peach/coconut | 0.001–0.129 | [46–50,58] |
| γ-Heptalactone | Fruity/coconut/herbaceous/caramel | 0.001–0.120 | [46–48,50] |
| γ-Hexalactone | Cake/fruity/peach | 0.003–0.023 | [47,48] |
| γ-Nonalactone | Over-ripe fruit | 0.015–0.372 | [51,58] |
| γ-Pentalactone | Cut hay | 0.002–0.006 | [49] |
| Pantolactone | Toasted bread/smoked | 0.065–0.190 | [47,49,57] |
| Sotolon | walnut/cotton candy/curry | 0.176 | [56] |
| cis-Whiskeylactone | Burnt/wood/vanilla/coconut | 0.011–0.028 | [47,56,57] |
| trans-Whiskeylactone | Spicy/coconut/vanilla | 0.004–0.049 | [45,47,57] |
| *Mercaptans* | | | |
| 2-Methyl-3-furanthiol | Fried | 0.035 | [56] |
| 3-Mercaptohexanol | Green/lemon | | [56] |
| 4-Mercapto-4-methyl-2-pentanone | Broom/cat urine/black currant sprout | | [56] |
| Dimethyl disulphide (DMDS) | | 0.0098 | [56] |
| Methional | Boiled vegetables/oxidized | 0.02 | [56] |
| Methionol | Cooked potato/cut hay | 0.001–0.070 | [46,47] |
| *Methoxypyrazines* | | | |
| 3-Isobutyl-2-methoxypyrazine | Green pepper/asparagus/potato | | [56] |
| 3-Isopropyl-2-methoxypyrazine | | | [56] |
| 3-sec-Butyl-2-methoxypyrazine | | | [56] |
| *Miscellaneous* | | | |
| 1,1,6-Trimethyl-1,2-dihydronaphthalene (TDN) | Gasoline | | [54,55] |
| 1,1-Diethoxyethane | Green fruit/licorice/cake/fruity/over-ripe fruit | 0.023–4.795 | [46–48,50,51,57] |
| β-Damascenone | Fruity/rose/plum/grape/raspberry | 0.01 | [56] |

The volatile composition of a selection of sweet Andalusian PX and Moscatel wines was studied by Márquez et al. [52]. The major compounds identified included ethyl acetate, isoamyl alcohols, ethyl lactate, acetic acid, 2-furaldehyde, linalool, diethyl succinate, α-terpineol, and 2-phenylethanol. Both varietals presented elevated contents of isoamyl alcohols, ethyl acetate and ethyl lactate, fatty acids such as hexanoic, octanoic, and decanoic acids. Norisoprenoid 1,1,6-trimethyl-1 and 2 dihydro naphthalene (TDN) at low levels were also confirmed. Muscat presented very high concentrations of linalool, α-terpineol and limonene, and higher ones than PX in TDN. On the other hand, 2-furaldehyde and 5-methyl-2-furaldehyde were detected at significant levels in PX. With respect to PX, and according to the data provided by Campo et al. [56], who analyzed different types of dessert wines, PX also contains significant concentrations of 3-methylbutanal, phenylacetaldehyde, methional, sotolon, and the ethyl esters 2-, 3-, as well as 4-methylpentanoic acids, all of them with high aromatic activity. Nevertheless, the compounds that best differentiated the PX from the other wines were 3-methylbutanal, furfural, β-damascenone, ethyl cyclohexanoate, and sotolon.

The aromatic profile of the natural sweet wines from the Jerez-Xérès-Sherry Protected Designation of Origin (PDO) are the result of different contributions in the course of their production; from the grapes' cultivation to the aging of the wine. It is necessary to clarify that, given that the musts obtained from raisined PX grapes fortified with wine alcohol from the neighboring production area of Montilla-Moriles PDO can be used, we have included in this bibliographic research the works that have also studied those musts.

While the sugar enrichment of the grapes can be achieved through the overripening of the grapes on the vines by twisting their stems without cutting them off, the traditional system in the Jerez (Sherry) region is the so called 'asoleo', which consists on drying the bunches of grapes in the sun for several days in order to partially dry or raisin the grapes (Figure 3).

**Figure 3.** Traditional sun drying process for the raisining of the grapes.

A certain concentration of the compounds is to be expected, but Franco et al. [48], compared the aromatic profiles of sun-dried raisins and fresh grapes' musts and were able to confirm the decrease in concentration of farnesol and of some 6-carbon alcohols and aldehydes responsible for herbaceous aromas (hexan-1-ol, (E)-hex-3-en-1-ol, (Z)-hex-3-en-1-ol, (E)-hex-2-en-1-ol, hexanal, and (E)-hex-2-enal). The authors attributed this reduction in specific compounds to the inactivation by exposure to light of the lipoxygenase enzymes responsible for the production of C6. They also detected very marked increments in the content of some other volatiles: isobutanol; benzyl alcohol; 2-phenylethanol; 5-methylfurfural; $\gamma$-butyrolactone, and $\gamma$-hexalactone, all of them related to the anaerobic metabolism of sugar, which encouraged the authors to suggest the promotion of this mechanism during the 'asoleo' traditional overripening system of the grapes, as it is known to occur in freshly harvested grapes [59]. In addition, high temperatures favor the formation of products derived from Maillard reactions which are responsible for roasted coffee or cocoa aroma notes. The complexity of these phenomena that affect the aromaticity of raisined grape must was analyzed by López de Lerma et al. [51]. They hypothesized that the criterion for determining the optimum raisining length of time perhaps should not be determined by aiming at a sugar concentration of around 400 g/L. In fact, they observed that some of the aromatic families of interest related to fruity and toasted notes started to decrease in concentration at an earlier stage, so they recommended reducing dehydration, and opted for rapid response tools such as the electronic nose to control the process. For Ruiz et al. [50] however, raisining consists of two stages: during the first 4 days, slight changes occur in the chemical and sensory aromatic profiles, and thereafter the raisins are substantially enriched in aromas.

During the 'asoleo' traditional overripening system a number of risks are faced, such as the possibility of rain or nighttime moisture, which may result in a loss of quality due to fungal attacks [60]. Several researchers have studied an alternative of great interest such as the use of climatic chambers to keep the control on temperature and humidity conditions (Figure 4). This method would also allow the raisining stage to be shortened.

**Figure 4.** Climatic chamber with temperature and moisture control.

Ruiz et al. [50] compared the volatile compositions of raisined grape musts obtained by "asoleo" traditional overripening system or in a climatic chamber. The data obtained were processed as aroma values and grouped into aromatic families, according to their contribution to characteristic olfactory notes. The caramel note was the highest value in both cases (associated with increases in 3-hydroxy-2-butanone, known as acetoin, $\gamma$-butyrolactone, and 2,3-butanedione) and together with the floral note related to concentration increases in geranial, phenethyl acetate, phenethyl alcohol and farnesol, were perceived more clearly in the musts obtained from climatic chamber raisins. The same authors [46] analyzed the effect of chamber temperature and drying time on the PX grapes and a combination of 40 °C for 96 h was established as ideal to obtain more intense caramel and floral notes (mainly due to important increases in phenethyl alcohol) together with a characteristic and highly appreciated milky note associated to an increment in methylbutanoic acid. However, these results do not agree with those obtained by Serratosa et al. [61], who considered 50 °C as a better option that allowed them to obtain a must that was sensorially very similar to that produced by traditional raisining methods. Ruiz-Bejarano et al. [62] evaluated the sensory profile of PX and Muscat grapes, from three different harvests, which had been raisined either through 'asoleo' traditional overripening system or by means of a climatic chamber under temperature and moisture control. The results were very enlightening with regard to the considerable possibilities exhibited by the alternative raisining method. Particularly, the grapes from one of the harvests, which had been affected by rain falls during the days before their cropping, produced musts marked by more intense fungal or humidity notes as well as weaker fruity and aroma intensity when the grapes had undergone the 'asoleo' traditional overripening system than when the must was produced by means of a climatic chamber. The analysis of ochratoxin A (OTA) in the musts confirmed a 4-fold fungal contamination in the raisins obtained by 'asoleo' traditional overripening system (up to 28.8 g/kg) [63].

As already discussed, the sweet wines from the Jerez-Xérès-Sherry PDO require some degree of fermentation. Fermentation brings complexity and acidity, while balancing the intense sweet notes (fruit, raisin) that are predominant in wines that are simply the result of adding wine alcohol to the raisined grape must [58]. This was confirmed by Ruiz et al. [47], who carried out a study on the aromatic characterization of wines obtained from raisined PX grape musts as a result of the different degrees of fermentation. In another paper, Ruiz-Bejarano et al. [55] studied the volatile composition of sweet wines obtained from raisined Muscat musts under different vinification conditions, including as experimental variables the type of yeast (*S. cerevisiae* vs. *S. bayanus*), the fermentation temperature (room vs. chilled), the addition of ammonium phosphate nutrient, and the prefermentative pellicular maceration with pectolytic enzymes. According to their results, the concentrations of esters are favored by the addition of nutrients, by the practice of pellicular maceration with enzymes, and especially by the combination of these practices with the use of *S. bayanus* yeast. On the other hand, the concentration of acetates was encouraged by fermentation with *S. cerevisiae* at room temperature. Moreover, certain alcohols and aldehydes (1-hexanol, hexanal, benzaldehyde, 2-phenylethanol) increased their presence in those assays that included skin maceration with enzymes. From a sensory point of view [62,63], the sweet Muscat wines fermented at low temperature (< 10 °C) with *S. bayanus* yeast without nutrients and pectolytic enzymes, were characterized by intense citrus and floral notes and were the best rated, while the ones obtained using nutrients were granted the lowest scores. In a follow-up study [64], the same authors observed that the use of *S. bayanus* significantly decreased ethyl carbamate content in the wines—a compound declared to be carcinogenic—while the use of nutrients and pectolytic enzymes increased its content levels. PX wines, with their characteristic amino acid profile, presented lower concentrations of this compound than Muscat wines.

The high sugar concentration in raisin musts, as much as 400 g/L, causes some difficulties to the production of sweet wine. Espejo et al. [65] tested the use of pectolytic enzymes combined with prefermentative maceration to facilitate the pressing and improve must extraction yields. They succeeded to obtain wines with improved aromatic and taste characteristics. The use of osmo-resistant yeasts has been the subject of study of several researchers [49,53,66,67]. As an example of this, a study with *Torulaspora delbrueckii* [67], a yeast of low volatile acidity production capacity with concentrated musts, high aroma revealing capacity, but low alcohol resistance, produced wines with higher citrus notes, lower raisin notes, and better overall ratings than those fermented using *S. cerevisiae*. The concentrations of isoamyl alcohol, 2-phenylethanol, isobutyl alcohol, benzaldehyde, 2,2-diethoxyethyl benzene, and 2-phenylethyl isobutyrate increased, while those of ethyl butyrate, some acetates, and certain fatty acids decreased.

No work has been found in the literature on the aromatic evolution of natural sweet wines from the "Jerez-Xérès-Sherry" PDO during their aging by means of the Criaderas y Solera method. Only a limited number of related works have been found [57,68,69], but the production of the wine was carried out in a different way from those established for the "Jerez-Xérès-Sherry" PDO.

On the other hand, Ruiz-Bejarano et al. [54], analyzed the evolution of 51 volatile substances during the static aging of sweet wines made from PX and Moscatel grape musts from two different vintages in 30 L American oak barrels. With respect to aging time, several ethyl esters (ethyl 3-methylbutanoate, diethyl pentanedioate, and diethyl succinate) increased significantly, while ethyl decanoate and ethyl dodecanoate decreased, which is explained by hydrolysis and esterification phenomena. The acetates, n-butyl acetate, isoamyl acetate and phenylethyl acetate; the terpenes, nerol oxide, linalool, thymol, carvacrol and β-myrcene; the alcohols, 3-methyl-2-butanol and 1-hexanol; aldehydes such as benzaldehyde, nonanal, octanal, hexanal and 2-hexenal, 2-furaldehyde (originating from raisining) and 1,1,6-trimethyl-1,2-dihydronaphthale, increased significantly with aging time, probably as a result of their concentration. Some of the compounds detected and that mainly derive from contact with oak were eugenol, 4-ethylphenol, and 5-methylfuraldehyde. In a previous study [55], the same authors had investigated the effect of the type and time aging length on sweet Moscatel wines, by comparing aging in medium-toasted 30 L American oak barrels with the aging carried out through contact with chips of the same oak variety at doses of 4 g/L, as well as in the absence of wood. The levels of most compounds were affected by the presence or absence of wood and, to a large extent, also by the type of contact, i.e., barrel or chips. The sensory analyses [63], according to expectations, detected greater oak notes as aging time grew longer, although their intensity levels were higher in the cask-aged wines. It also established a clear preference for cask-aged wines over those aged in contact with chips, where an aromatic defect could be perceived. Cask-aging was confirmed as an improving agent and one that was particularly effective with grapes coming from less optimal harvests from a sensory point of view [62].

Ruiz et al. [47] studied the accelerated aging of sweet wines from raisined PX grapes in contact with American oak chips at doses of 1 and 2 g/L at 20 °C, together with other alternatives to the traditional method. They confirmed significant increases in 2,3-butanedione, isoamyl acetate, eugenol, vanillin, furfural, and 5-methylfurfural, and volatile phenols such as guaiacol, 4-ethylguaiacol, 4-ethylphenol, syringol and isoeugenol, as well as (E) and (Z) isomers of β-methyl-γ-octalactone.

Herrera et al. [45] monitored the static aging of a natural sweet PX wine in 16 L casks made of American, French and Spanish oak, as well as of chestnut wood. Some wood-derived compounds—such as eugenol, *trans*-whiskeylactone, benzaldehyde, or 5-methyl-2-furaldehyde among others—increased their concentrations with time, regardless of the botanical origin of the wood. The same happened with certain other compounds such as isobutyl acetate and isobutanol, which, as expected, also increased their concentration as a result of the evaporation of water through the wood pores.

## 4. Study of the Aroma of Sherry Vinegar

Sherry vinegar is a product resulting from the acetic fermentation of the wines produced in the Sherry region. It is produced and aged using traditional practices and must display certain organoleptic and analytical characteristics. Depending on the aging times to which the vinegars are subjected, the following are distinguished: Sherry Vinegar (six months minimum aging time), Reserva Sherry Vinegar (two years minimum aging time), and Gran Reserva Sherry Vinegar (10 years minimum aging time). In addition, there are also semi-sweet or sweet Sherry vinegars (depending on the amount of sugar), namely Pedro Ximénez Sherry Vinegar and Moscatel Sherry Vinegar, which have one of these types of sweet wines added during the aging process [8].

Vinegar aroma has been a subject of study for several decades, and Table 3 shows the different volatile compounds studied in Sherry vinegar, their sensory descriptors and the concentration ranges found in the bibliographic references.

**Table 3.** Volatile compounds identified in Sherry vinegars, sensory descriptors, and concentration ranges reported in the bibliographic references

| Volatile Compounds | Sensory Descriptors | Concentration (mg/L) | References |
|---|---|---|---|
| *Acetates* | | | |
| Benzyl acetate | Sweet/floral/fruity/jasmine/fresh | 0.013–0.224 | [70–78] |
| Bornyl acetate | Woody/pine/herbal cedar/spice | | [79] |
| 2,3-Butanediol diacetate | | | [79,80] |
| n-Butyl acetate | Solvent/fruity/banana | 0.1–2.8 | [71,73,75,77,80–82] |
| Ethyl acetate | Fruity/sweet/weedy/green | 0.1–3.9 | [16,72,73,75,79,80,82–88] |
| Ethyl 2-phenyl acetate | Sweet/floral/honey/rose/balsamic/cocoa | 25–132 | [70,71,73,74,79–82,87,89] |
| Geranyl acetate | Floral/rose/lavender/ green/waxy | | [79,89] |
| (E)-2-Hexen-1-ol acetate | Green/fruity | | [79,90] |
| (Z)-3-Hexen-1-ol acetate | Green/fruity/banana/apple/grassy | 0.01–0.03 | [73,78–80,91] |
| Hexyl acetate | Fruity/green apple/banana/sweet | 0.007–0.09 | [71–73,75,78,79,83,87,92,93] |
| Isoamyl acetate | Sweet/fruity/banana | 2.7–16.3 | [71–75,78–80,82,83,86,93] |
| Isobutyl acetate | Sweet/fruity/banana | 1.0–4.3 | [71–73,75,78–80,82,83,87] |
| Methyl acetate | Sweet/fruity | 0.011–0.05 | [71,72,75,82,84–86,88,90] |
| 4-Methyl-2-pentyl acetate | Sweet/fruity/banana | | [79,87,89,93] |
| 2-Methyl-1-propyl acetate | Sweet/fruity/apple banana | 9.97 | [84,89] |
| Neryl acetate | Floral/rose/citrus/pear | | [79] |
| 3-Oxobutan-2-yl acetate | Pungent/sweet/creamy/buttery | | [90] |
| Phenylethyl acetate | Floral/rose/sweet/honey/fruity/tropical | 0.5–4.8 | [70–72,74,75,79–83,87,93] |
| Phenyl methyl acetate | Sweet/floral/honey/spicy/waxy/almond | | [79,94] |
| 1,2-Propanediol diacetate | Fruity/acetic | | [79] |
| Propyl acetate | Solvent/fruity/fusel/raspberry/pear | 0.06–0.2 | [71–73,75,78,82,85] |

**Table 3.** *Cont.*

| Volatile Compounds | Sensory Descriptors | Concentration (mg/L) | References |
|---|---|---|---|
| *Esters* | | | |
| Diethyl succinate | Mild fruity/cooked apple | 0.007–2.44 | [70–73,75,78–84,86–89,92,93] |
| Dihydroxymethyl jasmonate | | | [79] |
| Ethyl benzoate | Fruity/dry musty/sweet | 0.006–0.013 | [71,72,75,77,80] |
| Ethyl butanoate | Sweet/fruity/tutti frutti | 0.05–0.3 | [71–75,78–80,83,84,87,93] |
| Ethyl ciclohexanoate | | | [80] |
| Ethyl decanoate | Sweet/waxy/fruity/apple/grape/oily | 0.008–0.054 | [72,73,79,81,83,89,93] |
| Ethyl dodecanoate | Sweet/waxy/floral/soapy/clean | | [72] |
| Ethyl-3-ethoxypropanoate | | | [72,90] |
| Ethyl formate | Green/alcohol/rose/cognac | 24.3–194 | [85,86,88] |
| Ethyl heptanoate | Fruity/pineapple/cognac/rum/wine | | [80] |
| Ethyl hexanoate | Sweet/fruity/pineapple/waxy/green banana | 0.05–75 | [70–73,75,78,79,81,83,87,92,93] |
| Ethyl hydrogensuccinate | | | [70,72] |
| Ethyl isobutyrate | Sweet/ethereal/fruity/alcoholic/fusel | 0.006–1 | [71–75,77,78,82,92] |
| Ethyl isovalerate | Fruity/sweet/apple/pineapple/tutti frutti | 0.03–1.1 | [71–75,78–80,82,87,89,92] |
| Ethyl lactate | Sharp/tart/fruity/buttery/butterscotch | 0.007–63 | [70,71,73,82,85,86,88] |
| Ethyl levulinate | Sweet/fruity/floral/ berry/green pineapple/rhubarb | | [79] |
| Ethyl 2-methyl butanoate | Sharp/sweet/green/apple/fruity | 0.07–0.15 | [71–75,79,80,82,83,87,89,93] |
| Ethyl 3-methylpentanoate | Pineapple/fruity/tropical | | [80] |
| Ethyl nonanoate | Fruity/rose/waxy/rum/wine/tropical | | [80] |
| Ethyl octanoate | Fruity/wine/waxy/sweet/apricot/banana/brandy/pear | 0.02–0.05 | [71–73,75,79,80,83,89,92] |
| Ethyl propanoate | Fruity/banana/pineapple | 0.6–1.5 | [71,72,75,79,80,82,83] |
| Ethyl vanillate | Phenolic/burnt/smoky/powdery/metallic | | [70,79] |
| Ethyl valerate | Sweet/fruity/apple/pineapple/green | 0.002–0.67 | [71–73,75,78,79,81,82,92,93] |
| Isobutyl isothiocyanate | Green | | [79] |
| Methyl butyrate | Fruity/apple/sweet/ banana/pineapple | | [80] |
| Methyl hexadecanoate | Oil/waxy/fatty | | [89,90] |
| Methyl hexanoate | Fruity/pineapple | | [79] |
| Methyl nonanoate | Sweet/fruity/pear/waxy/tropical/wine | | [90] |
| Methyl 9-octadecanoate | | | [90] |
| Methyl salicylate | Mint | | [70,72,77,79,87] |
| *Acids* | | | |
| Acetic acid | Sharp/pungent/sour/vinegar | | [16,72,79,80,83,87,89,93] |
| Benzoic acid | faint balsam/urine | | [70,79,80,89] |
| Butanoic acid | Sharp/cheesy/rancid/butter | | [70,72,77,79,80,83] |
| Decanoic acid | Unpleasant/ rancid/sour/fatty | 0.03–0.5 | [70–75,78,79,81–83,87,89,92,93] |
| Dodecanoic acid | Fatty/coconut/bay | | [90,93] |
| 2-Ethylhexanoic acid | | | [72] |
| Formic acid | Pungent/vinegar | | [79] |
| Heptanoic acid | Rancid/sour/cheesy/sweat | 0.10–0.15 | [71,77] |
| Hexadecanoic acid | Waxy/fatty | | [70,72,79,89] |
| 9-Hexadecenoic acid | | | [79,89] |
| Hexanoic acid | Sour/fatty/sweat/cheese | 1.3–2.2 | [70–73,75,77,79–83] |
| (4-Hexyl-2,5-dioxo-2,5-dihydro-3-furanyl) acetic acid | | | [90] |
| Isobutyric acid | Acidic/sour/cheese/dairy/buttery/rancid | 0.06–0.15 | [72–74,78,80,90,92] |
| Isopentanoic acid | Stinky feet/sweaty/cheese | 49–60 | [70–75,78–83,87,89,90,92] |
| Nonanoic acid | Waxy/dirty/cheese/dairy | 0.01–0.04 | [71,72,77–79,87,92,93] |
| Octadecanoic acid | Fatty/waxy | | [79] |
| 9-Octadecenoic acid | | | [79] |
| Octanoic acid | Fatty/waxy/rancid/oily/vegetable/cheesy | 0.7–2.6 | [70–73,75,78–83,87,89,92] |
| Oleic acid | Faint fatty/waxy | | [79] |
| Pentadecanoic acid | Waxy | | [72,79,89] |
| Pentanoic acid | Acidic/sweaty/rancid | | [70,72] |
| Phenylacetic acid | Sweet/honey/floral/honeysuckle/sour/waxy/ | | [70] |
| Propanoic acid | Pungent/acidic/cheesy/vinegar | | [70,72,80] |
| Sorbic acid | | | [72] |
| Tetradecanoic acid | Waxy | | [70,72,79,89] |

**Table 3.** *Cont.*

| Volatile Compounds | Sensory Descriptors | Concentration (mg/L) | References |
|---|---|---|---|
| *Alcohols* | | | |
| 2-Acetoxy-1-propanol | | | [70] |
| Benzyl alcohol | Floral/rose/phenolic/balsamic | 81–1980 | [70,71,75,78–83,92,93] |
| Borneol | Balsamic/camphoreous/herbal/woody | | [79] |
| 2,3-Butanediol | Fruity/creamy/buttery | 353–95 | [80,81,85,88,91] |
| 1-Butanol | Fusel/oily/sweet/ balsamic/whiskey | | [70,93] |
| Butoxyethoxyethanol | | | [70] |
| Ethanol | Alcoholic/medical/strong | 1.03–9000 | [72,75,80,82,85,86,88] |
| 3-Ethoxy-1-propanol | Fruit | | [70,79] |
| 4-Ethyl resorcinol | | | [76] |
| γ-Eudesmol (2-naphthalene methanol) | Waxy/sweet | | [77,93] |
| Eugenol | Sweet/spicy/clove/woody | 0.01–0.1 | [70,71,73,75,77,78,92] |
| Fenchyl alcohol | Camphoreous/pine/woody/dry/rooty/ sweet/lemon | | [79,89] |
| 1-Heptanol | Musty/leafy/violet/herbal/green/sweet/ woody/peony | | [90] |
| 2-Heptanol | Fresh/lemon/grass/herbal/sweet/floral/ fruity/green | | [90] |
| 3-Heptanol | Herbal | | [90] |
| 1-Hexanol | Ethereal/fusel/oily/fruity/alcoholic/ sweet/green | 0.002–0.4 | [71–73,78,87,92] |
| 2-Hexanol | Chemical/winey/fruity/fatty/cauliflower | | [89,93] |
| trans 2-Hexen-1-ol | Fresh/green/leafy/fruity/unripe banana | | [73,90] |
| cis 3-Hexen-1-ol | Fresh/green/grassy/foliage/vegetable/ herbal/oily | 0.04–0.05 | [71,75,80,82,87] |
| Methanol | Alcoholic | 11–67 | [16,75,82,85,86,88] |
| 2-Methyl-1-butanol | Roasted winey onion/fruity/fusel/alcoholic/whiskey | 560–13,000 | [71–75,78,79,81–86,88,89,92,93] |
| 3-Methyl-1-butanol | Fusel/alcoholic/pungent/cognac/ fruity/banana | 5000–60,000 | [16,70–75,79–86,88–90,93] |
| 2-Methyl-1-hexadecanol | | | [90] |
| 2-Methyl-1-propanol | Winey/whiskey | 3.5–14.3 | [71,73–75,80–82,85,86,88,93] |
| 1-Nonanol | Fresh/clean/fatty/floral/rose/orange/ dusty/wet/oily | | [89,90] |
| Phenylethyl alcohol | Sweet/floral/fresh/rose | 0.013–27.1 | [70–72,74,75,78–90,92,93] |
| 1-Propanol | Alcoholic/fermented/fusel/tequila/musty/ sweet/fruity/apple/pear | 0.66–13.1 | [75,80,82,84,86,88] |
| Propano-1,2,3-triol | | 3200–21,600 | [16] |
| Vanillyl alcohol | Sweet/creamy/vanilla/caramellic/ cracker/milky/ | | [80] |
| *Phenols* | | | |
| 4-Acetyl-2-methylphenol | | | [79] |
| 2/4-Ditertbutyl phenol | | | [79] |
| 4-Ethylguaiacol | Spicy/smoky/bacon/phenolic/clove | 0.6–2.9 | [70–74,78–81,89,92] |
| 4-Ethylphenol | Phenolic/smoky | 0.02–1.6 | [70,71,73–75,78–82,87,89,90,92,93] |
| Guaiacol | Phenolic/smoky/spicy/vanilla/woody | 0.009–0.016 | [75,78,80,82,90] |
| 4-Methylguaiacol | Spicy/clove/vanilla/phenolic/ medicinal/leathery | | [80] |
| Phenol | Phenolic/plastic/rubber | | [79,83] |
| *Terpenes* | | | |
| Camphene | Woody/herbal/fir needle/camphor | | [89,90] |
| Citronellene | Floral/rose/herbal/citrus/ | | [79] |
| β-Citronellol | Floral/leathery/waxy/ rose/citrus | | [87,89] |
| Cymene | Fresh/citrus/lemon/woody/spicy | | [79] |
| Eucalyptol | Eucalyptus/herbal/camphoreous/medicinal | | [79,90] |
| Geraniol | Sweet/floral/fruity/rose/waxy/citrus | | [79,89] |
| Limonene | Pine/herbal/peppery | | [79,87,89] |
| Linalool | Citrus/floral/sweet/bois de rose/woody/green/blueberry | | [79,87,89] |
| trans p-Mentha-2,8-dienol | | | [90] |
| Nerol | Sweet/citrus/magnolia | | [79,89,93] |
| Perillaldehyde | Fresh/green/oily/grassy/fatty/minty/cherry | | [79] |
| Safranal | Fresh/herbal/phenolic/metallic/rosemary/ tobacco/spicy | | [90] |
| δ-Selinene | | | [79] |
| γ-Terpinene | Oily/woody/lemon/ lime/tropical/herbal | | [79] |
| 4-Terpineol | Peppery/woody/earthy/musty/sweet | | [79,87,89] |
| α-Terpineol | Pine/lilac/citrus/woody/floral | 0.007–0.1 | [71,73,77–79,87,89,90,92,93] |
| β-Terpineol | Pungent/earthy/woody | | [89] |
| Thymol | Herbal/thyme/phenolic/medicinal/camphor | | [89] |

**Table 3.** *Cont.*

| Volatile Compounds | Sensory Descriptors | Concentration (mg/L) | References |
|---|---|---|---|
| *Aldehydes* | | | |
| Acetaldehyde | Pungent/ethereal/aldehydic/fruity | 5–61 | [16,75,84,85,88] |
| Benzaldehyde | Strong/sharp/sweet/bitter/almond/cherry | 0.05–1070 | [70–73,75,77,78,80,81,83,84,87,89–94] |
| Butanal | Pungent/cocoa/musty/green/malty/bready | | [94] |
| 2-Butenal | Floral | | [94] |
| Cuminaldehyde (4-(1-methylethyl)-benzaldehyde) | Spicy/cumin/green/herbal | | [77] |
| Decanal | Sweet/aldehydic/waxy/orange peel/citrus/floral | | [94] |
| trans 2-Decenal | Fatty/orange/rose/aldehydic/floral/green | | [90] |
| Dodecanal | Soapy/waxy/aldehydic/citrus/green/floral | | [94] |
| Heptanal | Fresh/aldehydic/fatty/green/herbal/cognac/ | | [94] |
| Hexanal | Fresh/green/fatty/aldehydic/grassy | 0.009–0.05 | [71,73,94] |
| Isobutyraldehyde | Fresh/aldehydic/floral/green | | [94] |
| Isovaleraldehyde | Aldehydic/chocolate/ peach/fatty | | [79] |
| 2-Methylbutanal | Musty/cocoa/coffee/nutty | | [74,94] |
| 3-Methylbutanal | Ethereal/aldehydic/chocolate/peach/fatty | | [74,94] |
| 3-Methylpropanal | | | [94] |
| Nonanal | Waxy/aldehydic/rose/orange peel/fatty | | [74,87,89,93,94] |
| (E)-2-Nonenal | Fatty/green/cucumber/aldehydic/citrus | | [74,80,94] |
| Octanal | Aldehydic/waxy/citrus/orange peel/green/fatty | 0.011–0.014 | [74,79,87,90,93,94] |
| 3-Octanal | | | [89] |
| Pentanal | Fermented bready/fruity/nutty/berry | | [94] |
| Propanal | Earthy/alcoholic/winey/whiskey/cocoa/nutty | | [94] |
| Undecanal | Waxy/soapy/floral/aldehydic/citrus/green/fatty/fresh laundry | | [94] |
| Vanillin | Sweet/vanilla/cream/chocolate | 2.5–4.4 | [70,71,75,80] |
| *Furanic componds* | | | |
| 5-Acetoxymethyl-2-furaldehyde | Baked bread | | [72,73,79] |
| 2-Acetyl-2,5-dimethylfuran | | | [79] |
| 2-Acetylfuran | Sweet/balsamic/almond/cocoa/caramellic/coffee | $0.6 \times 10^{-5}$–$1.7 \times 10^{-5}$ | [70,79,82,90] |
| 2-Acetyl-5-methylfuran | Musty/nutty/hay/coconut/milky | | [70,73,79] |
| 5-Ethoxymethylfurfural | | | [79] |
| Ethyl furoate | | 0.03–0.2 | [71,72] |
| Furfural | Sweet/woody/almond/fragrant/baked bread | 0.1–2.2 | [70–73,75,77–79,81,83,87,91,92,94] |
| Furfuryl alcohol | Alcoholic/chemical/musty/sweet/caramel/bread/coffee | 0.3–1.04 | [70,71,75,80,82] |
| 5-Hydroxymethylfurfural | Fatty/buttery/musty/waxy/caramellic | | [70,72,79] |
| 5-Methylfurfural | Sweet/caramellic/bready/coffee | 0.005–0.02 | [70–72,75,78–80,92] |
| 1-(5-Methyl-2-furyl)-1-propanone | | | [79] |
| *Ketones* | | | |
| Acetoin | Sweet/buttery/creamy/dairy/milky/fatty | 0.28–708 | [16,70–73,75,77,79–81,83–86,88,90,93] |
| Benzophenone | Balsamic/rose/metallic/geranium | | [79,90] |
| 2,3-Butanodione | Buttery/sweet/creamy/pungent/caramellic | 17–42 | [71,72,74,75,80,86,90,91] |
| β-Damascenone | Sweet/fruity/rose/plum/grape/raspberry/sugar | | [80,90] |
| 3-Heptanone | Green/fatty/fruity | | [90] |
| 2-Heptanone | Fruity/spicy/sweet/herbal/coconut/woody | | [90] |
| Hydroxyacetone | Pungent/sweet/caramellic/ethereal | 5.34–70 | [86] |
| 3-Hydroxy-3-methyl-2-butanone | | | [70] |
| α-Ionone | Sweet/woody/floral/violet/tropical/fruity | 0.018–0.038 | [81] |
| β-Ionone | Floral/woody/sweet/fruity/berry/tropical | | [80] |
| Isovalerone | Green/fruity/metallic/pineapple/banana | | [79] |
| 5-Methyl-3-hexanone | | | [79] |
| 3-Nonanone | Fresh/sweet/jasmin/spicy/herbal/fruity | | [72] |
| 1-Octen-3-one | Herbal/mushroom/earthy/musty/dirty | | [74,80] |
| 1-(2,3,6-Trimethylphenyl)-3-buten-2-one | | | [79] |

**Table 3.** *Cont.*

| Volatile Compounds | Sensory Descriptors | Concentration (mg/L) | References |
|---|---|---|---|
| *Lactones* | | | |
| γ-Butyrolactone | Creamy/oily/fatty/caramellic | 0.005–0.38 | [70,71,75,78,82,84–86,88,92] |
| δ-Decalactone | Fresh/oily/waxy/peach/coconut/buttery/sweet | | [79] |
| δ-2-Decenolactone | | | [79] |
| γ-Dodecalactone | Fatty/peach/sweet/metallic/fruity | | [80] |
| γ-Heptalactone | Sweet/coconut/nutty/caramellic/ | | [79] |
| δ-Laurolactone | | | [79] |
| α-Methyl-γ-crotonolactone | | | [79] |
| Pantolactone | Cotton/candy | | [80] |
| Solerone | | | [70] |
| Sotolone | Sweet/caramellic/maple/sugar burnt/sugar/coffee | 0.748 | [75,80] |
| cis-Whiskeylactone | Coconut/toasted/nutty/burnt | 0.1–1.5 | [70,71,75,79,80,82] |
| trans-Whiskeylactone | Coconut/toasted/nutty/celery/burnt | 0..07–0.3 | [70,71,75,78,79,82,92] |
| *Enolic derivatives* | | | |
| Cyclotene | Sweet/caramel/maple/sugar/coffee/woody | | [70,79,90] |
| 2,3-Dihydro-3,5-dihydroxy-6-methyl-4H-pyran-4-one | | | [79] |
| 3-Ethyl-2-hydroxy-2-cyclopenten-1-one | Sweet/caramellic/maple | | [70,79] |
| Cyclotene | Sweet/caramel/maple/sugar/coffee/woody | | [70,79,90] |
| *Miscellaneous* | | | |
| 2-Butyl-4-methyl-1,3-dioxolane | Nutty/fatty | | [90] |
| Cadalene (1,6-dimethyl-4-(1-methylethyl)-naphthalene) | | | [77] |
| Cyclotetradecane | | | [79] |
| Dibutyl formamide | | | [79] |
| N,N-Dimethylformamide | Slight amine | | [79] |
| Methyl styrene | | | [79] |
| N-(3-Methylbutyl) acetamide | | | [79] |
| Pentadecane | Waxy | | [79] |
| Styrene | Sweet/balsamic/ floral/plastic | | [79] |
| Tetradecane | Mild waxy | | [79] |
| Tridecane | | | [79] |
| 1,1,6-Trimethyl-1,2-dihydronaphthalene (TDN) | Gasoline | $4.4 \times 10^{-5}$–$10.5 \times 10^{-5}$ | [72,82,90] |
| Vitispirane | Floral/fruity/earthy/woody | | [89] |

In the 1990s, Blanch et al. [83] found no major differences between the volatile composition of the Sherry vinegars studied and other non-aged wine vinegars that were also considered in the study. However, it was observed in this work that the Sherry vinegars generally exhibited higher concentrations of most compounds and particularly of acetaldehyde, a compound that had already been found in previous studies also in aged vinegars [95]. Guerrero et al. [96] reached similar conclusions after analyzing Sherry vinegars and other unaged vinegars, which in this latter case had been produced by means of submerged culture acetification methods (quick acetification). This study was conducted according to the standardized analysis methods of the time. Morales et al. [84] showed that the use of NaOH or MgO to neutralize the high acetic acid content of vinegars prior to their analysis by gas chromatography significantly reduced the content of many of the volatile compounds that were originally present. Natera et al. [81] analyzed Sherry vinegars by means of solid phase microextraction (SPME) and the volatile compounds found in higher proportions were 2-methyl-1-propanol, 2- and 3-methyl-1-butanol, 3-hydroxy-2-butanone, 2-phenylethanol, isoamyl acetate, 2,3-butanediol, and isopentanoic acid.

More recently, Guerrero et al. [97,98] were able to identify and successfully quantify 47 volatile compounds by means of stir bar sorptive extraction (SBSE). This extraction methodology prevented sample interferences and increased the analytical sensitivity (Figure 5). Callejón et al. [71] analyzed volatiles in Sherry and Rioja vinegars employing headspace sorptive extraction (HSSE) and observed that the latter allowed to determine up to 53 volatile compounds, with 5 of them detected for the first time in this matrix: ethyl 2-methylbutyrate, ethyl heptanoate, ethylfuroate, ethyl benzoate, and acetophenone. Even though the volatile profiles of both types of vinegars were qualitatively similar, the Sherry vinegars contained greater amounts of some of them, including ethyl butyrate, ethyl isovalerate, ethyl lactate, isovaleric acid, and 4-ethylphenol.

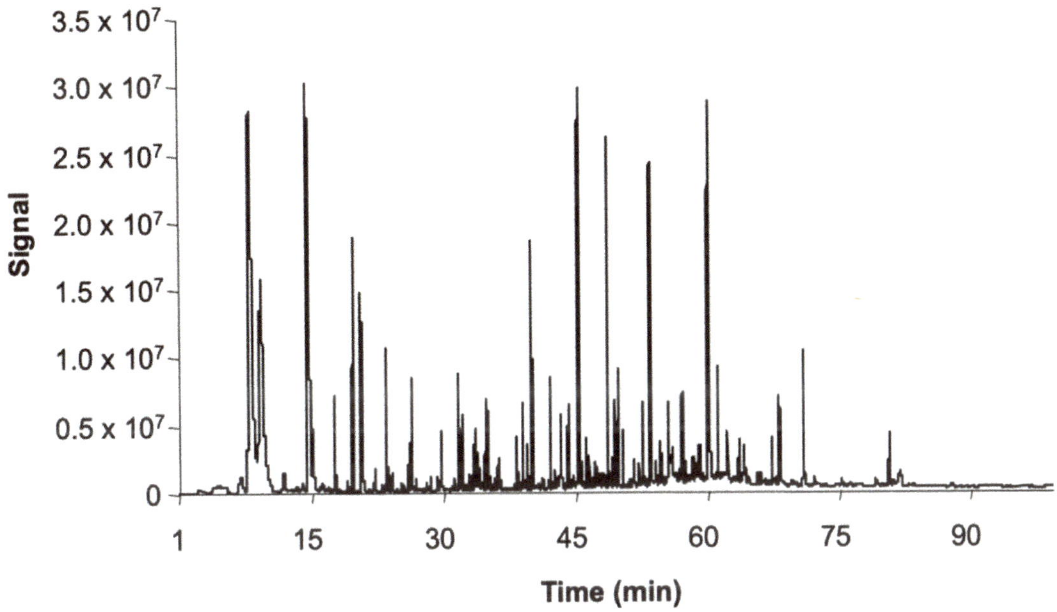


**Figure 5.** Example of chromatogram of a Sherry vinegar obtained by SBSE-GC-MS and retention times of some relevant compounds: ethyl isobutyrate (13.62); propyl acetate (13.99); isobutyl acetate (15.76); ethyl butyrate (16.84); ethyl isopentanoate (18.46); hexanal (18.70); isopentyl acetate (20.57); ethyl pentanoate (20.77); 1-butanol (21.84); 3-methyl-1-butanol (23.84); 2-methyl-1-butanol (24.12); ethyl hexanoate (24.65); hexyl acetate (25.80); 3-hydroxy-2-butanone (26.62); *cis* 3-hexenyl acetate (27.59); ethyl lactate (28.51); 1-hexanol (28.87); ethyl octanoate (31.87); 2-furaldehyde (32.87); benzaldehyde (35.15); isobutyric acid (36.84); 5-methyl-2-furaldehyde (36.95); butyric acid (38.89); isovaleric acid (40.28); diethyl succinate (40.58); α-terpineol (41.51); benzyl acetate (42.64); ethyl-2-phenyl acetate (44.59); phenylethyl acetate (45.95); hexanoic acid (46.57); benzyl alcohol (47.03); 2-phenylethanol (49.21), 2-ethyl hexanoic acid (50.17); octanoic acid (53.75); eugenol (57.21); decanoic acid (60.39); 5-hydroxymethyl-2-furaldehyde (68.90).

When comparing Sherry vinegars to vinegars from other Protected Designation of Origins (PDO), Ríos-Reina et al. [72] carried out a study for the discrimination of vinegars from the three vinegar Protected Designation of Origin (PDO)s in Spain ('Sherry Vinegar', 'Vinegar of Condado de Huelva', and 'Vinegar of Montilla-Moriles'). Other authors evidenced that the volatile content in vinegar is influenced not only by the production process, which is similar for Sherry vinegars and vinegars from Huelva, but also by the raw material, in this case, the grape variety used, Palomino for Sherry PDO and Zalema for Huelva PDO, as well as by geographical factors associated to each PDO [73]. Other authors [90], compared Sherry vinegars to vinegars from Huelva PDO and from Montilla-Moriles PDO, and some of their volatile compounds, namely 1-heptanol, methyl nonanoate, 2-methylbutanoic acid, 2,2,6-trimethyl-cyclohexanone, trans-2-decenal, eucalyptol, and α-terpineol allowed the differentiation of Huelva PDO vinegars from those produced under the Sherry PDO and Montilla-Moriles PDO, while diacetyl, acetoin, ethyl 3-ethoxypropanoate, 2- and 3-heptanone, 2-methyl-1-hexadecanol, 1-octen-3-ol, p-cresol, and camphene allowed to differentiate the vinegars from the Montilla-Moriles PDO. Moreover, Sherry PDO vinegars could be differentiated by their β-damascenone, 5-hydroxymethylfurfural, 3-heptanol, trans-2-hexen-1-ol, and trans-2-hexen-1-yl acetate contents. All of this not only corroborates the conclusions reported by previous studies, but also demonstrates that PDO vinegars can be classified based on their volatile profiles.

These differences were also observed in the studies carried out by means of Fourier transform mid-infrared spectroscopy (FTIR) with attenuated total reflectance (ATR) on Sherry and Huelva vinegars, both PDOs from Andalusia. These vinegars are produced following similar oenological practices that include different periods of aging in oak wood using the well-known Criaderas y Solera system. The authors concluded that aging in oak wood by means of Criaderas y Solera presented a series of bands in the region of 1500–900 $cm^{-1}$ of the spectrum that enabled their differentiation according to the aging time of the vinegars from both PDOs. Aging in wood led to significant changes in the ATR-FTIR spectra due to a greater presence in the vinegars of compounds such as acetic acids, alcohols, esters, and ethers [99]. This spectroscopic technique has also been successfully applied to the differentiation of vinegars derived from different raw materials and production processes, including Sherry vinegars [100]. The percentage of successful classification achieved was similar to that obtained based on their volatile content.

Casale et al. [91] who observed that the determination of the spectral fingerprint of 17 Sherry vinegars together with other vinegars of different nature or origin (white wine, red wine, balsamic vinegar, apple vinegar, etc.) by Heaspace mass spectrometry without a previous chromatographic separation, allowed to differentiate them from the rest of the vinegars. Other study allowed the differentiation of the Sherry vinegars studied from other white and red wine vinegars, as well as from apple and balsamic vinegars, based only on 14 compounds among which eugenol (2-methoxy-4-prop-2-enyl-phenol), furfural (2-furancarboxaldehyde), several organic acids (isobutyric acid, nonanoic acid, etc.), some aldehydes, and esters (benzyl acetate, ethyl benzeneacetate, and ethyl benzoate) were the most relevant [77].

Benito et al. [101] carried out the characterization and differentiation of 66 vinegars from wines from the PDO "Rioja" and 18 from the Sherry PDO on the basis of different analytical parameters including glycerol and acetoin content along with other parameters such as organic acids, pH, acidity, Cu, Fe, etc. For this purpose, they used both classical statistical techniques (cluster analysis, principal component analysis) and others of later development, such as neuronal networks. These authors observed that, although a significant variability was observed in both groups of vinegars in terms of the parameters considered, given the wide range of aging times applied to the vinegars, they could be clearly differentiated by means of either set of chemometric techniques.

However, not only the raw material used which could determine the volatile composition, but also aging process, environmental conditions, microbiological activity could also induce different volatile profiles. The differences found between the different types of vinegars, including Sherry vinegars, and according to the studies that have been considered, seem to be due to both the starting raw material and the special and specific circumstances under which the production processes are carried out. In order to differentiate between relevant and irrelevant factors in the production of Sherry vinegar, Morales et al. [88] carried out a study in which they addressed the acetification stage by means of a submerged culture, as a factor that could determine the composition of the vinegar obtained, as opposed to the raw material used. The results revealed very significant changes in the volatile profile of the product as a consequence of the acetification process, even though the polyphenolic compounds content was not altered by this process. Therefore, the raw material used was considered to be the predominant factor. Durán et al. [102] also studied the changes that take place in the volatile composition over the acetification process of Sherry vinegars and succeeded to correlate it with the FTIR signal obtained.

Chinnici et al. [70] studied the possibility of differentiating between Sherry and Modena vinegars from different categories (traditional Modena balsamic vinegar "extravecchio", traditional Modena balsamic vinegar "affinato", and Modena balsamic vinegar). In their study they reported 93 volatile compounds detected and identified by Solid Phase Extraction (SPE). The study revealed the differentiation between the different vinegars on the basis of several parameters such as the extent of Maillard reactions, alcoholic, or non-alcoholic fermentation, or the length of wood aging. In the same line of work, Marrufo et al. [79] using in this case SBSE-GC-MS, obtained a 100% separation between traditional Modena balsamic vinegars, Modena balsamic vinegars, and Sherry vinegars on the basis of furanes, terpenes, acetates, and esters (Figure 6). Durán et al. [94] observed a significant differentiation between Sherry vinegars and Modena balsamic vinegars according to their aldehydic compounds content.

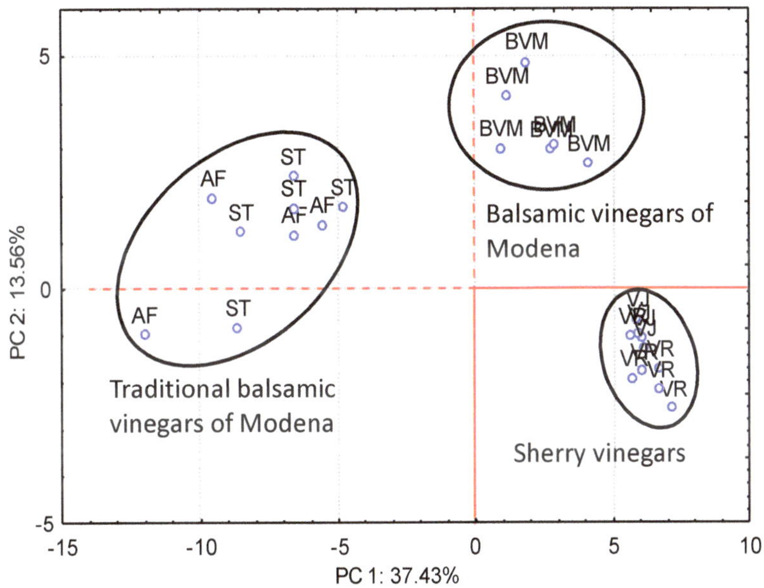


**Figure 6.** Differentiation of Sherry vinegars from Italian vinegars, based on their volatile content. According to Marrufo et al. [79] with modifications. AF: Affinato Traditional Balsamic Vinegar of Modena; ST: Stravecchio Traditional Balsamic Vinegar of Modena; BVM: Balsamic Vinegar of Modena; VJ: Sherry Vinegar; VR: Reserva Sherry Vinegar.

For an in-depth characterization of the volatile profile of Sherry vinegars and its contribution to the perceived aroma, Aceña et al. [74] conducted a study by Gas Chromatography-Olfactometry (GC-O) on extracts from commercial Sherry vinegars obtained by HS-SPME. Among the 37 odorants found, some of them presented OAVs (odor activity values) greater than 1 (ethyl isovalerate, ethyl isobutyrate, isoamyl acetate, isovaleric acid, 2-phenylethanol, 4-ethylguaiacol, isobutyric acid, 2-phenylethyl acetate, and 4-ethylphenol), which suggests their significant contribution to the vinegar aroma.

Callejón et al. [75] were able to detect 108 aromatic notes by GC-O in Sherry vinegars, and identified 64 of them. In addition, they found that the mixture of compounds whose aroma most resembled the aroma of Sherry vinegar was a combination of diacetyl, ethyl acetate and sotolon. A more recent study has investigated the olfactometric profile of Sherry vinegars (dry and sweet Pedro Ximénez), together with vinegars from other Spanish denominations of origin (Montilla-Moriles and Condado de Huelva) and concluded that the most abundant aromas in the Sherry vinegars identified by GC-O belonged to the "grassy vegetal" family, while the "spicy" family of compounds was more characteristic of the sweet PX vinegars [80]. These authors were able to satisfactorily correlate the values obtained by GC-MS-O with those obtained by sensory analysis. Therefore, it seems clear that the volatile compound composition of vinegar is closely related to the aroma perceived by sensory analysis, which is why the latter discipline has become in recent years a clear complement to the analysis of the aromatic profile of oenological products in general, and of vinegars in particular.

The first methodological approach to sensory analysis applied to Sherry vinegars was carried out by González-Viñas et al. [103]. These authors conducted a study in which they determined the taste group thresholds (geometric mean of the individual best-estimate thresholds (BETs)) in organic acid solutions and in vinegars. This study demonstrated that the aromatic profile of the sample has an influence on the perception of the different descriptors, as was the case with the acid descriptors. On the other hand, Tesfaye et al. [104] developed a methodology for the sensory analysis of vinegars and applied it to the characterization of Sherry vinegar aroma after aging in wood. In that study, they observed that, a significant improvement in the quality of the vinegar aroma could be perceived after the first six months of aging. Later, these authors perfected the sensory analysis methodology applied to Sherry vinegars and succeeded in considerably reducing the deviations between judgments and the increment in the number of descriptors [105]. To date, the aroma of sherry vinegar has been characterized in detail from the sensory point of view [82] and the descriptors "glue", "wood", and "pungent" are typical of this type of vinegar, regardless of the aging method applied. On the other hand, the descriptors "raisin" and "alcohol/liquor" tend to be more characteristic of longer-aged vinegars (Gran Reserva), while the descriptor "wine character" at higher values is generally associated to younger Sherry vinegars [82].

As we have established, the aging process has a strong influence on the aromatic profile of oenological products. In the case of vinegar, it has been proven that there are numerous chemical and biochemical transformations that take place during the aging process, and that are similar to those that occur during the aging of Sherry wines either during their biological or oxidative aging. This fact has made of this stage a target for many studies on Sherry vinegar, as we have already seen. Thus, Palacios et al. [16] reported significant increases in acetic acid and other compounds such as acetoin, due to water loss by evaporation. However, other compounds, such as higher alcohols, decreased as a consequence of the synthesis of acetates. Similarly, the high concentration level of the residual alcohol that can be found in Sherry vinegars, together with their high acidity, favors higher concentrations of ethyl acetate to be developed during the aging process in comparison with other types of vinegars. This fact has been corroborated by other authors [86], who described significant rises in ethyl acetate concentrations during the aging of Sherry vinegars with a residual alcohol content of around 2%. These authors also described increments in other compounds—such as methyl acetate, methanol, diacetyl or $\gamma$-butyrolactone—that took place during the aging of Sherry vinegars. However, it cannot be ignored that, as already mentioned, other factors—such as the acetification system used—may modify, even more than the actual aging process, the volatile content of Sherry vinegars [85].

The type of wood used for aging also seems to have an impact on the volatile composition of Sherry vinegars. American oak (*Quercus alba*) is the most commonly wood used, but other types of wood such as French oak (*Quercus petraea*), Spanish oak (*Quercus pyrenaica*), or chestnut (*Castanea sativa*) have also been employed [92]. It has been demonstrated that chestnut wood provides a significantly different volatile profile with respect to that obtained from oak woods, and that Spanish oak and French oak woods provide a similar content of volatile compounds in aged vinegars. Moreover, from the sensory point of view, it has been observed that French oak wood provides highly favorable organoleptic characteristics to aged vinegars, while Spanish oak wood generates vinegars that are quite similar to those traditionally aged in American oak casks [92].

On the other hand, the aging of Sherry vinegar in wood containers is a lengthy and costly process that is susceptible of shortening. However, in order to preserve the typicity of this product, it is essential to verify that the volatile profile of the product obtained by accelerating methods does not differ from that obtained by traditional aging procedures. Hence, some studies have dealt with the sensory profile of vinegars aged in an accelerated manner using American and French oak chips [106]. The authors concluded that the differences between the samples were mostly due to the pungency of the samples rather than to the character provided by oak wood. Generally speaking, Sherry vinegars elaborated in a traditional way showed higher scores for the attributes studied: aromatic intensity, richness in aroma, ethyl acetate, woody odor, wine character, pungent sensation, coconut, vanillin, clove odor, and general impression. In addition, woody odor was very similar for both samples, traditionally aged and infused with oak chips. On the other hand, Durán Guerrero et al. [107] presented a method to accelerate the aging of Sherry vinegars by the joint application of micro-oxygenation and wood shavings while trying to resemble the natural aging process that takes place in wooden casks. Using an oxygen dose of 70 mL/L/month and 5 g/L of American oak chips they were able to obtain, in just 14 days, vinegars with a volatile profile similar to those aged by traditional methods in 105 days (86% time reduction). More recently, Jiménez-Sánchez et al. [78] used a combination of micro-oxygenation, wood shavings and ultrasound energy to further accelerate the aging process of Sherry vinegars. In this case, different types of wood were used (American, French, and Spanish oak), and it was observed that Spanish oak provided a greater amount of volatile compounds. In addition, with the combined use of ultrasound, wood shavings and micro-oxygenation, the vinegars obtained in just 4 days, had similar volatile profiles to those of vinegars aged by the traditional method for 6 months.

Finally, it is worth mentioning that, although Sherry vinegar is a product of ancient tradition, it is also open to innovation and has recently been used in the development of new products. Aroma is a key factor in the elaboration of such new products derived from Sherry vinegar, and has therefore been studied in different occasions. For example, the effect that the maceration with peels from different fruits (orange, lemon, lime, grapefruit, strawberry) exerts on the aroma of Sherry vinegar has been studied (Figure 7) and a product with a marked fruity character has been obtained by using peel concentrations at 200 g/L and 3-day maceration time [89]. From a sensory point of view, descriptors 'fruity', 'sweet', and 'aroma intensity' were directly correlated with olfactory impression, which means that the preference of the vinegars was mainly based on these three descriptors. Moreover, the descriptors that allowed the best discrimination among vinegars macerated with different fruits were fruity, citric, and sweet.

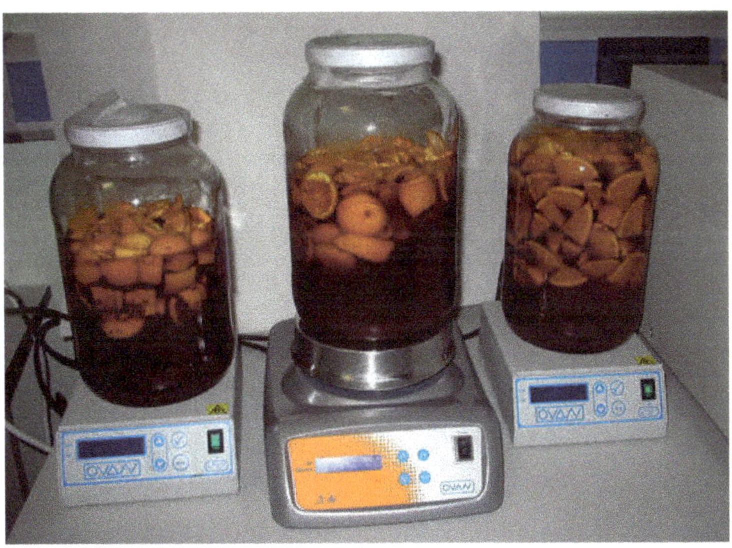

**Figure 7.** New products derived from Sherry vinegar: maceration with citrus fruits.

In a subsequent study, the maceration time was reduced to a few minutes by applying accelerating energies, such as microwaves or ultrasound [93]. The aroma of this type of product obtained by maceration was studied by GC-MS-O, and it was observed that there was a significant increase mainly in compounds with 'floral' aromas. Vinegars macerated presented high content in alcohols, aldehydes, and terpenoids, and from a sensory point of view, the lowest values of floral, greasy and citric categories were obtained for vinegars without maceration [87]. Another example of innovation concerning Sherry vinegar is that proposed by Marrufo-Curtido et al. [76] where dietary fiber from citrus fruits was added to the vinegars with an increase in the sensory descriptor 'citrus' observed in the final product. In addition, these fiber-enriched vinegars were very highly valued from a sensory point of view. Finally, Sherry vinegars have also been used in the development of other novel products by adding small quantities to fruit juices in order to produce soft drinks [108]. The character provided by the addition of vinegar improved the sensory properties of the fruit juices, which were favorably rated in a subsequent consumers' survey. Based on the olfactory and gustatory impression, and purchase intent, the acetic beverages made from peach and pineapple juices were the most appreciated, followed by apple juice, while those obtained from orange juice were the least preferred by consumers.

## 5. Study of the Aroma of Sherry Brandy

Sherry brandy displays certain characteristics that differentiate it from other aged spirits. Such characteristics derive from their aging according to the dynamic system known as Criaderas y Solera, and from the requirement to age in preconditioned 500–600 L capacity oak casks, mostly American oak [109]. According to their minimum average aging time, Sherry brandies are classified into three categories: Solera Brandy (6 month minimum aging time), Solera Reserva Brandy (1 year minimum aging time), and Solera Gran Reserva Brandy (3 year minimum aging time).

The composition of a Sherry brandy is determined by:

(1) The grape variety from which the initial wine distillate is obtained (mainly Airén, Palomino, and Pedro Ximénez grapes) [110,111];
(2) The fermentation and production conditions of the base wine [112];
(3) The processing and nature of the initial distillate, a mixture containing varying quantities of *holanda* (low-grade spirit), medium-grade spirits and distillates (high-grade spirit), with at least 50% of the total ethanol content coming from medium and low grade spirits [19,113,114];
(4) The origin and conditioning of the wood cask, i.e., the type of oak and its toasting intensity [115,116];
(5) The preconditioning of the cask with wine, i.e., the type of wine that it has previously contained and for how long [13];
(6) The previous length or frequency of use of the barrel, i.e., whether it is used to produce brandy for the first time after its preconditioning with wine or it has been used several times to hold and produce brandy [117].

All of these factors have an impact on the physicochemical and organoleptic characteristics of Sherry brandies and provide them with a rich and varied aroma. However, with regard to their aromatic profile scarce bibliography is available. Table 4 presents the volatile compounds determined in Sherry brandy, their sensory descriptors and the concentration ranges found in the bibliographic references.

**Table 4.** Volatile compounds identified in Sherry brandy, sensory descriptors and concentration ranges reported in the bibliographic references.

| Volatile Compounds | Sensory Descriptors | Concentration (mg/L) | References |
|---|---|---|---|
| *Alcohols* | | | |
| 2-Butanol | Vinous/medicinal | 1.8 | [117] |
| 2-Methylbutanol | Roasted/fruity/fusel oil/alcoholic/wine/whiskey | 80.9–181.8 | [117–119] |
| 2-Phenylethanol | Rose/talc/honey | 4.99–22.4 | [118,119] |
| 2-Phenylethyl alcohol | Rose/honey | 2.16–2.52 | [117] |
| 3-Hexenol (E/Z) | Herbaceous/green/grass | 0.238–2.245 | [118,119] |
| Butanol | Vinous/medicinal | 7.92–9.36 | [117] |
| Hexanol | Cut grass/resinous/herbaceous/wood | 3.99–10.44 | [117–119] |
| Isoamyl alcohols | Solvent/cake/fusel alcohols/nail polish/ripe fruit | 193–678 | [117–119] |
| Isobutanol | Alcohol/solvent/vinous/nail polish | 119.88–133.92 | [117] |
| Methanol | Solvent/pungent fruity | 238.32–245.16 | [117] |
| *Aldehydes* | | | |
| Acetaldehyde | Stewed apple/pungent/ | 78.84–86.76 | [117] |
| Benzaldehyde | Roasted/bitter almond/nutty/smoky | 2.91–35.3 | [118,119] |
| *Furans* | | | |
| 2-Furaldehyde | Fusel alcohol/cake/almond/toasted bread/incense/floral | 0.19–14.54 | [10,116,117,120] |
| 5-Hydroxymethyl-2-furaldehyde | Rancid/toasted | 0.072–87.09 | [10,116,117,120] |
| 5-Methyl-2-furaldehyde | Toasted/bitter almond/cake/burnt/caramel | 0.062–1.94 | [10,116,117,120] |
| *Acids* | | | |
| Acetic acid | Fatty | 210.1–307.6 | [116] |
| Decanoic acid | Rancid/cheese/wax/plasticine | 5.12–15.1 | [118,119] |
| Dodecanoic acid | Fatty/coconut/bay | 1.51–7.18 | [118,119] |
| Octanoic acid | Rancid/cheese/fatty | 0.007–13.4 | [118,119] |
| *Esters* | | | |
| 2-Phenylethyl acetate | Fruity/honeyed/floral/rose | 0.013–0.119 | [118,119] |
| Diethyl succinate | Overripe fruit/lavender | 0.071–5.40 | [118,119] |
| Ethyl 2-methylbutanoate | | 0.103–0.241 | [119] |
| Ethyl 2-methylpropanoate | | 0.064–0.454 | [118] |
| Ethyl acetate | Pineapple/varnish/balsamic/fruity/solvent/pungent/glue | 134.28–236.52 | [117] |
| Ethyl butanoate | Banana/pineapple/strawberry | 0.327–14.9 | [118,119] |
| Ethyl decanoate | Synthetic/rancid | 0.64–4.93 | [117–119] |
| Ethyl dodecanoate | Sweet/waxy/floral/soapy/clean | 0.160–1.08 | [117–119] |
| Ethyl heptanoate | Strawberry/banana | 0.057–0.104 | [118,119] |
| Ethyl hexadecanoate | Mild waxy | 1.44 | [117] |
| Ethyl hexanoate | Banana/green apple | 0.46–1.79 | [117–119] |
| Ethyl isopentanoate | Fruity/sweet/apple/pineapple/tutti frutti | 0.090–0.443 | [118,119] |
| Ethyl lactate | Lactic/yogurt/strawberry/raspberry/buttery | 48.24–50.76 | [117] |
| Ethyl nonanoate | Fruity/rose/waxy /rum/wine/tropical | | [118,119] |
| Ethyl octanoate | Pineapple/pear/soapy/banana | 0.63–5.4 | [117–119] |
| Ethyl pentanoate | Sweet/fruity/apple/pineapple/green | 0.041–0.398 | [118,119] |
| Ethyl succinate | Toffee/coffee | 3.96–7.2 | [117] |
| Ethyl tetradecanoate | Mild waxy/soapy | 0.36 | [117] |
| Hexyl acetate | Apple/pear/banana/floral | 0.0004–0.003 | [118,119] |
| Isoamyl octanoate | | 0.002–0.018 | [118,119] |
| Isoamyl acetate | Sweet/fruity/banana | 0.101–1.098 | [118,119] |
| (E)-Methyl-2-octenoate | | 0.0007–0.0027 | [118,119] |
| Methyl decanoate | | 0.001–0.007 | [118,119] |
| *Terpenes* | | | |
| Linalool | Muscat/rose/lavender | 0.053–0.590 | [118,119] |
| Nerolidol | Floral/green/citrus/woody/waxy | 0.002–0.004 | [118,119] |
| α-Terpinene | | 0.0017 | [118,119] |
| α-Terpineol | Lily/cake | 0.007–0.097 | [118,119] |
| *Volatile phenols* | | | |
| 4-Ethylguaiacol | Spicy/smoky/bacon/phenolic/clove | 0.046–0.210 | [118,119] |
| Eugenol | Cinnamon/clove | 0.007–0.071 | [118,119] |
| Vanillin | Vanilla | 0.13–5.94 | [10,116,117,120] |
| *Miscellaneous* | | | |
| 1,1-Diethoxyethane | Green fruit/licorice/cake/fruity/overripe fruit | 105.84–115.56 | [117] |
| β-Damascenone | Fruity/rose/plum/ grape/ raspberry | 0.001–0.084 | [118,119] |

Durán et al. [118,119], after the analysis of 48 Sherry brandies, emphasized the quantitative importance of isoamyl alcohol, 2-methyl-1-butanol, benzaldehyde, diethyl succinate, 2-phenylethanol, octanoic acid, decanoic acid, lauric acid, ethyl decanoate, and ethyl octanoate, with concentration levels above mg/L. Several of the compounds identified seemed to increase with aging time, although only ethyl esters, 2-phenylethyl acetate, linalool and eugenol did so significantly. A number of the compounds identified, such as ethyl laureate, ethyl myristate, ethyl palmitate, and lauric acid, were derived from the initial distillate, and their starting acids (lauric acid, myristic acid, caprylic acid, ... ) may also be present, since they are the precursors of the esterification reactions with ethanol that give place to the appearance of the above mentioned esters. Some of the compounds may also have their origin in the wood itself (caprylic acid, myristic acid, or palmitic acid, among others) or in the wine preconditioning process [13] such as ethyl lactate or ethyl succinate. The furfuryl compounds may also come from two sources, since they are generated in the thermal processes during the distillation, but also during the toasting of the cask wood and then transferred to the spirits [10,120]. Other compounds such as vanillin and certain coumarins have also been identified [121].

Multivariate statistical techniques have been used to determine the discrimination accuracy between the three types of Sherry brandies based on their aromas. The results pointed towards a clear differentiation of Solera from both Solera Reserva and Solera Gran Reserva [119], where the last ones showed a widely dispersed pattern. These results are in agreement with those from other works on the polyphenolic composition of brandies in which the discrimination of the intermediate Solera Reserva only reached 57% [10]. The reason for this characteristic pattern could be the lack of a minimum aging time. Nevertheless, when an electronic nose that allows the analysis of global aromatic profiles was used, higher discrimination percentages were achieved for the different categories of brandies [122].

Although the concentration of some of these polyphenolic compounds over time tends to increase mainly due to either wood extraction or water losses during the aging process, it has been demonstrated that the brandies that are aged in old casks—i.e., casks that were not used for the first time for this purpose—continue to evolve and gradually improve the complexity of their aroma [117]. This takes place at an evidently slower rate mostly due to the Criaderas y Solera system which involves a periodic supply of air that favors oxidative phenomena. We should point out the long aging times for Solera Gran Reserva Sherry brandies which is generally in the order of 20 years or more in currently commercialized brandies [123].

As previously mentioned, the aging of Sherry brandy is considered its most characteristic production stage, i.e., the one that provides it with its distinctive character, and since the associated costs are rather high, as it was seen for other Sherry products above, considerable interest has been shown to investigate alternative methods to accelerate the process while preserving the product's chemical and sensory profile. Among such methods, those that use wood chips and ultrasound as the accelerating energy, with or without the addition of air, are the ones that have gained most of the attention [124–126], since they can shorten aging times by 6 to 18 times (Figure 8).

These are the tools that have been used at laboratory or pilot plant scale to evaluate the suitability of different varietal spirits to be aged as Sherry brandy [110]. It has been concluded that the effect of aging is different depending on the grape variety, thus the aroma profile of the worst rated young brandies improved (as occurred with Ugni Blanc and Corredera), while the aged Muscat of Alexandria and Garrido brandies were awarded lower sensory ratings compared to their unaged samples. In the same study, the spirits that had been made from Jaén Blanco and Zalema grapes were the most appreciated, both young and aged, which were equally characterized by clear fruity notes and high aromatic intensities. This accelerated aging system has also been used to determine the potential of woods from different botanical origin (American, French, and Spanish oak, chestnut and cherry) for the aging of brandies [127], and to evaluate the use of Colombard, Moscatel, Palomino fino, Pedro Ximénez, and Zalema varietal holandas distilled by means of a rotary evaporator [114] to produce brandy. Some of the products obtained were rated high by a tasting panel.

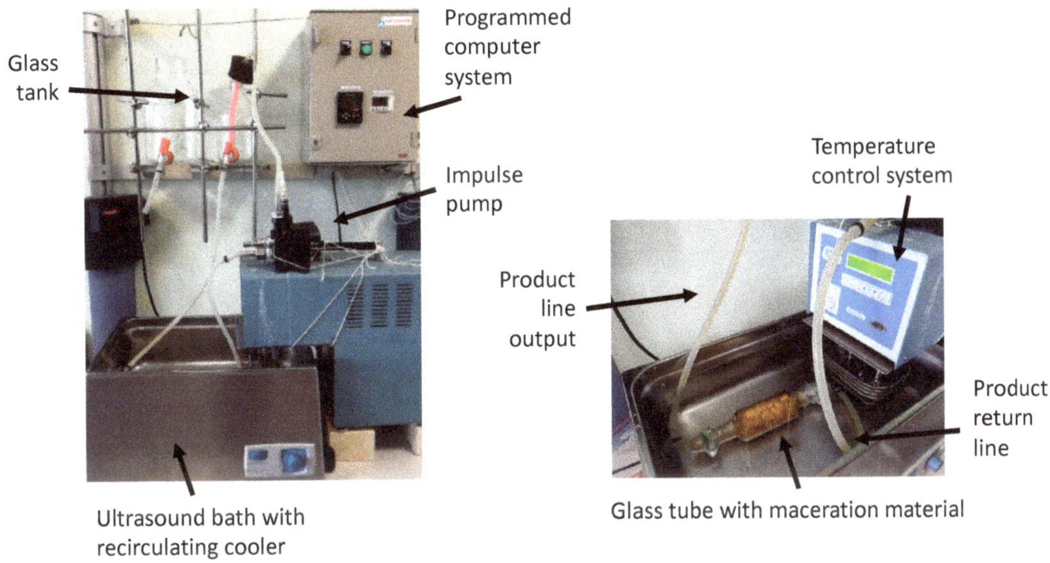


**Figure 8.** Ultrasound system for the accelerated aging of oenological products [124].

## 6. Final Remarks

As can be seen from all this research, the uniqueness of Sherry oenological products, in terms of their aromatic composition, is determined both by the raw materials used and by each and every one of the significantly conditioning factors in their production processes. This also includes a number of environmental factors and, in particular, the aging stage. All of these factors contribute to the highly distinctive aroma displayed by Sherry wines, vinegars, and brandies and make of them the superior oenological products that are internationally acclaimed. Moreover, although aroma of Sherry products has been widely studied to date, due to the complexity of these special products, further innovation in analytical methodologies and advanced instrumentation is still needed. The reliable analysis of volatile compounds may contribute to a better knowledge and quality control of Sherry products, and therefore to meet the high levels of consumer demand, in an increasingly competitive sector.

**Author Contributions:** Conceptualization, E.D.-G., R.C., M.d.V.G.-M., M.d.C.R.-D., M.S., and D.G.-S.; Methodology, E.D.-G.; Investigation, E.D.-G., R.C., M.d.V.G.-M., M.d.C.R.-D., M.S., and D.G.-S.; Writing—original draft preparation, E.D.-G., R.C., M.d.V.G.-M., M.d.C.R.-D., M.S., and D.G.-S.; Writing—review and editing, E.D.-G., R.C., M.d.V.G.-M., M.d.C.R.-D., M.S., and D.G.-S.; Visualization, R.C., M.d.V.G.-M., M.d.C.R.-D., and M.S.; Supervision, E.D.-G.; Project administration, D.G.-S.; Funding acquisition, D.G.-S. All authors have read and agreed to the published version of the manuscript.

**Funding:** The work was supported by the project FEDER INNTERCONECTA "INNTER-VINANDAL ITC-20131018", from the Spanish Ministry of Economy and Competitiveness (MINECO) and the European Regional Development Fund (ERDF); and by the University of Cadiz and Bodegas Fundador, S.L.U. through contract OT2019/108.

**Institutional Review Board Statement:** Not applicable.

**Informed Consent Statement:** Not applicable.

**Acknowledgments:** The authors wish to dedicate this work to the memory of Carmelo García Barroso and Ramón Natera Marín, former members of the research group, without whose invaluable scientific and human contributions, many of the works presented in this bibliographic review would not have been carried out.

**Conflicts of Interest:** The authors declare no conflict of interest.

## References


1. Pardo-Calle, C.; Segovia-Gonzalez, M.M.; Paneque-Macias, P.; Espino-Gonzalo, C. An approach to zoning in the wine growing regions of "Jerez-Xérès-Sherry" and "Manzanilla-Sanlúcar de Barrameda" (Cádiz, Spain). *INIA Span. J. Agric. Res.* **2011**, *9*, 831–843. [CrossRef]
2. Johnson, H.; Robinson, J. *The World Atlas of Wine*, 8th ed.; Mitchell Beazley: London, UK, 2019.
3. Pozo-Bayón, M.A.; Moreno-Arribas, M.V. Sherry Wines: Manufacture, Composition and Analysis. In *Encyclopedia of Food and Health*; Elsevier Inc.: Amsterdam, The Netherlands, 2015; pp. 779–784.
4. Marin-Menguiano, M.; Romero-Sanchez, S.; Barrales, R.R.; Ibeas, J.I. Population analysis of biofilm yeasts during fino sherry wine aging in the Montilla-Moriles, D.O. region. *Int. J. Food Microbiol.* **2017**, *244*, 67–73. [CrossRef]
5. Pozo-Bayón, M.A.; Moreno-Arribas, M.V. Sherry Wines. In *Advances in Food and Nutrition Research*; Academic Press Inc.: Cambridge, MA, USA, 2011; Volume 63, pp. 17–40.
6. Consejería de Agricultura y Pesca. *Orden de 13 de Mayo de 2010, Por La Que se Aprueba el Reglamento de las Denominaciones de Origen «Jerez-Xérès-Sherry» y «Manzanilla de Sanlúcar de Barrameda», así Como sus Correspondientes Pliegos de Condiciones*; Boletín Oficial de la Junta de Andalucía (BOJA): Sevilla, Spain, 2010; Volume 103, pp. 53–74.
7. Council of the European Union. *Council Regulation (EEC) No 4252/88 of 21 December 1988 on the Preparation and Marketing of Liqueur Wines Produced in the Community*; European Union: Brussels, Belgium, 1988.
8. Ministerio de Agricultura, Pesca y Alimentación. *Orden de 26 de Diciembre de 2000 por la que se Ratifica el Reglamento de la Denominación de Origen «Vinagre de Jerez»*; Boletín Oficial del Estado (BOE): Madrid, Spain, 2000; p. 24477.
9. Council of the European Union. *Council Regulation (EC) No 510/2006 of 20 March 2006 on the Protection of Geographical Indications and Designations of Origin for Agricultural Products and Foodstuffs*; European Union: Brussels, Belgium, 2006.
10. Dodero, M.C.R.; Sánchez, D.A.G.; Rodríguez, M.S.; Barroso, C.G. Phenolic compounds and furanic derivatives in the characterization and quality control of Brandy de Jerez. *J. Agric. Food Chem.* **2010**, *58*, 990–997. [CrossRef] [PubMed]
11. Carpena, M.; Pereira, A.G.; Prieto, M.A.; Simal-Gandara, J. Wine aging technology: Fundamental role of wood barrels. *Foods* **2020**, *9*, 1160. [CrossRef]
12. Pérez-Prieto, L.J.; López-Roca, J.M.; Martínez-Cutillas, A.; Pardo Mínguez, F.; Gómez-Plaza, E. Maturing wines in oak barrels. Effects of origin, volume, and age of the barrel on the wine volatile composition. *J. Agric. Food Chem.* **2002**, *50*, 3272–3276. [CrossRef]
13. Sánchez-Guillén, M.M.; Schwarz-Rodríguez, M.; Rodríguez-Dodero, M.C.; García-Moreno, M.V.; Guillén-Sánchez, D.A.; García-Barroso, C. Discriminant ability of phenolic compounds and short chain organic acids profiles in the determination of quality parameters of Brandy de Jerez. *Food Chem.* **2019**, *286*, 275–281. [CrossRef]
14. Martínez Montero, C.; Rodríguez Dodero, M.D.C.; Guillén Sánchez, D.A.; García Barroso, C. Sugar contents of Brandy de Jerez during its aging. *J. Agric. Food Chem.* **2005**, *53*, 1058–1064. [CrossRef]
15. Ruiz-Muñoz, M.; Cordero-Bueso, G.; Benítez-Trujillo, F.; Martínez, S.; Pérez, F.; Cantoral, J.M. Rethinking about flor yeast diversity and its dynamic in the "criaderas and soleras" biological aging system. *Food Microbiol.* **2020**, *92*. [CrossRef]
16. Palacios, V.; Valcárcel, M.; Caro, I.; Pérez, L. Chemical and biochemical transformations during the industrial process of sherry vinegar aging. *J. Agric. Food Chem.* **2002**, *50*, 4221–4225. [CrossRef]
17. Schwarz, M.; Rodríguez, M.C.; Guillén, D.A.; Barroso, C.G. Analytical characterisation of a Brandy de Jerez during its ageing. *Eur. Food Res. Technol.* **2011**, *232*, 813–819. [CrossRef]
18. Cejudo Bastante, M.J.; Durán Guerrero, E.; Castro Mejías, R.; Natera Marín, R.; Rodríguez Dodero, M.C.; Barroso, C.G. Study of the polyphenolic composition and antioxidant activity of new sherry vinegar-derived products by maceration with fruits. *J. Agric. Food Chem.* **2010**, *58*, 11814–11820. [CrossRef]
19. Delgado-González, M.J.; García-Moreno, M.V.; Sánchez-Guillén, M.M.; García-Barroso, C.; Guillén-Sánchez, D.A. Colour evolution kinetics study of spirits in their ageing process in wood casks. *Food Control* **2021**, *119*. [CrossRef]
20. Recamales, A.F.; Hernanz, D.; Álvarez, C.; González-Miret, M.L.; Heredia, F.J. Colour of Amontillado wines aged in two oak barrel types. *Eur. Food Res. Technol.* **2007**, *224*, 321–327. [CrossRef]
21. Ding, Y.; Ma, Y.; Li, S.; Liang, J.; Xi, X.; Wang, J.; Sun, J.; Yu, H.; Guo, S. Characterization of biochemical compositions, volatile compounds and sensory profiles of cabernet sauvignon wine in successive vintages (2008–2017). *Am. J. Biochem. Biotechnol.* **2020**, *16*, 380–391. [CrossRef]
22. Pati, S.; Crupi, P.; Savastano, M.L.; Benucci, I.; Esti, M. Evolution of phenolic and volatile compounds during bottle storage of a white wine without added sulfite. *J. Sci. Food Agric.* **2020**, *100*, 775–784. [CrossRef] [PubMed]
23. Niimi, J.; Tomic, O.; Næs, T.; Bastian, S.E.P.; Jeffery, D.W.; Nicholson, E.L.; Maffei, S.M.; Boss, P.K. Objective measures of grape quality: From Cabernet Sauvignon grape composition to wine sensory characteristics. *LWT Food Sci. Technol.* **2020**, *123*. [CrossRef]

24. Zea, L.; Ruiz, M.J.; Moyano, L. Using Odorant Series as an Analytical Tool for the Study of the Biological Ageing of Sherry Wines. In *Gas Chromatography in Plant Science, Wine Technology, Toxicology and Some Specific Applications*; InTechOpen: Rijeka, Croatia, 2012.
25. Dunkel, A.; Steinhaus, M.; Kotthoff, M.; Nowak, B.; Krautwurst, D.; Schieberle, P.; Hofmann, T. Nature's chemical signatures in human olfaction: A foodborne perspective for future biotechnology. *Angew. Chem.* **2014**, *53*, 7124–7143. [CrossRef]
26. Souza Gonzaga, L.; Capone, D.L.; Bastian, S.E.P.; Jeffery, D.W. Defining wine typicity: Sensory characterisation and consumer perspectives. *Aust. J. Grape Wine Res.* **2020**, *27*, 246–256. [CrossRef]
27. Welke, J.E.; Hernandes, K.C.; Nicolli, K.P.; Barbará, J.A.; Biasoto, A.C.T.; Zini, C.A. Role of gas chromatography and olfactometry to understand the wine aroma: Achievements denoted by multidimensional analysis. *J. Sep. Sci.* **2021**, *44*, 135–168. [CrossRef]
28. Marín-San Román, S.; Rubio-Bretón, P.; Pérez-Álvarez, E.P.; Garde-Cerdán, T. Advancement in analytical techniques for the extraction of grape and wine volatile compounds. *Food Res. Int.* **2020**, *137*. [CrossRef]
29. Pérez-Elortondo, F.J.; Symoneaux, R.; Etaio, I.; Coulon-Leroy, C.; Maître, I.; Zannoni, M. Current status and perspectives of the official sensory control methods in protected designation of origin food products and wines. *Food Control* **2018**, *88*, 159–168. [CrossRef]
30. Cruces-Montes, S.J.; Merchán-Clavellino, A.; Romero-Moreno, A.; Paramio, A. Perception of the attributes of sherry wine and its consumption in young people in the South of Spain. *Foods* **2020**, *9*, 417. [CrossRef]
31. Cortes, M.B.; Moreno, J.J.; Zea, L.; Moyano, L.; Medina, M. Response of the aroma fraction in Sherry wines subjected to accelerated biological aging. *J. Agric. Food Chem.* **1999**, *47*, 3297–3302. [CrossRef] [PubMed]
32. Zea, L.; Moyano, L.; Moreno, J.; Cortes, B.; Medina, M. Discrimination of the aroma fraction of Sherry wines obtained by oxidative and biological ageing. *Food Chem.* **2001**, *75*, 79–84. [CrossRef]
33. Zea, L.; Moyano, L.; Moreno, J.A.; Medina, M. Aroma series as fingerprints for biological ageing in fino sherry-type wines. *J. Sci. Food Agric.* **2007**, *87*, 2319–2326. [CrossRef]
34. Zea, L.; Moyano, L.; Ruiz, M.J.; Medina, M. Chromatography-Olfactometry Study of the Aroma of Fino Sherry Wines. *Int. J. Anal. Chem.* **2010**, *2010*, 1–5. [CrossRef]
35. Moyano, L.; Zea, L.; Moreno, J.; Medina, M. Analytical study of aromatic series in sherry wines subjected to biological aging. *J. Agric. Food Chem.* **2002**, *50*, 7356–7361. [CrossRef] [PubMed]
36. Moreno, J.A.; Zea, L.; Moyano, L.; Medina, M. Aroma compounds as markers of the changes in sherry wines subjected to biological ageing. *Food Control* **2005**, *16*, 333–338. [CrossRef]
37. Moyano, L.; Zea, L.; Moreno, J.A.; Medina, M. Evaluation of the active odorants in Amontillado sherry wines during the aging process. *J. Agric. Food Chem.* **2010**, *58*, 6900–6904. [CrossRef]
38. Zea, L.; Moyano, L.; Ruiz, M.J.; Medina, M. Odor descriptors and aromatic series during the oxidative aging of oloroso sherry wines. *Int. J. Food Prop.* **2013**, *16*, 1534–1542. [CrossRef]
39. García-Moreno, M.V.; Sánchez-Guillén, M.M.; Delgado-González, M.J.; Durán-Guerrero, E.; Rodríguez-Dodero, M.C.; Guillén-Sánchez, D.A.; García-Barroso, C. Chemical content and sensory changes of Oloroso Sherry wine when aged with four different wood types. *LWT Food Sci. Technol.* **2021**, *140*, 110706. [CrossRef]
40. Garcia-Maiquez, E. Les levures de voile dans l'élaboration des vins de Xéres. In *Proceedings of the Application a l'Oenologie des Progres Recents en Microbiologie et en Fermentation*; OIV (Ed.): Paris, France, 1988; pp. 341–351.
41. Martínez, P.; Pérez Rodríguez, L.; Benítez, T. Velum formation by flor yeast isolated from Sherry wine. *Am. J. Enol. Vitic.* **1997**, *48*, 55–62.
42. Zea, L.; Serratosa, M.P.; Mérida, J.; Moyano, L. Acetaldehyde as Key Compound for the Authenticity of Sherry Wines: A Study Covering 5 Decades. *Compr. Rev. Food Sci. Food Saf.* **2015**, *14*, 681–693. [CrossRef]
43. Muñoz, D.; Peinado, R.A.; Medina, M.; Moreno, J. Biological aging of sherry wines under periodic and controlled microaerations with Saccharomyces cerevisiae var. capensis: Effect on odorant series. *Food Chem.* **2007**, *100*, 1188–1195. [CrossRef]
44. Hevia, K.; Castro, R.; Natera, R.; González-García, J.A.; Barroso, C.G.; Durán-Guerrero, E. Optimization of Head Space Sorptive Extraction to Determine Volatile Compounds from Oak Wood in Fortified Wines. *Chromatographia* **2016**, *79*, 763–771. [CrossRef]
45. Herrera, P.; Durán-Guerrero, E.; Sánchez-Guillén, M.M.; García-Moreno, M.V.; Guillén, D.A.; Barroso, C.G.; Castro, R. Effect of the type of wood used for ageing on the volatile composition of Pedro Ximénez sweet wine. *J. Sci. Food Agric.* **2020**, *100*, 2512–2521. [CrossRef]
46. Ruiz, M.J.; Moyano, L.; Zea, L. Changes in aroma profile of musts from grapes cv. Pedro Ximenez chamber-dried at controlled conditions destined to the production of sweet Sherry wine. *LWT Food Sci. Technol.* **2014**, *59*, 560–565. [CrossRef]
47. Ruiz, M.J.; Moyano, L.; Zea, L. Sweet wines produced by an innovative winemaking procedure: Colour, active odorants and sensory profile. *S. Afr. J. Enol. Vitic.* **2014**, *35*, 205–216. [CrossRef]
48. Franco, M.; Peinado, R.A.; Medina, M.; Moreno, J. Off-vine grape drying effect on volatile compounds and aromatic series in must from Pedro Ximénez grape variety. *J. Agric. Food Chem.* **2004**, *52*, 3905–3910. [CrossRef]
49. López De Lerma, N.; García-Martínez, T.; Moreno, J.; Mauricio, J.C.; Peinado, R.A. Volatile composition of partially fermented wines elaborated from sun dried Pedro Ximénez grapes. *Food Chem.* **2012**, *135*, 2445–2452. [CrossRef]
50. Ruiz, M.J.; Zea, L.; Moyano, L.; Medina, M. Aroma active compounds during the drying of grapes cv. Pedro Ximenez destined to the production of sweet Sherry wine. *Eur. Food Res. Technol.* **2010**, *230*, 429–435. [CrossRef]
51. Lopez De Lerma, N.; Bellincontro, A.; Mencarelli, F.; Moreno, J.; Peinado, R.A. Use of electronic nose, validated by GC-MS, to establish the optimum off-vine dehydration time of wine grapes. *Food Chem.* **2012**, *130*, 447–452. [CrossRef]

52. Márquez, R.; Castro, R.; Natera, R.; García-Barroso, C. Characterisation of the volatile fraction of Andalusian sweet wines. *Eur. Food Res. Technol.* **2008**, *226*, 1479–1484. [CrossRef]
53. García-Martínez, T.; Bellincontro, A.; De Lerma, M.D.L.N.L.; Peinado, R.A.; Mauricio, J.C.; Mencarelli, F.; Moreno, J.J. Discrimination of sweet wines partially fermented by two osmo-ethanol-tolerant yeasts by gas chromatographic analysis and electronic nose. *Food Chem.* **2011**, *127*, 1391–1396. [CrossRef] [PubMed]
54. Ruiz-Bejarano, M.J.; Castro-Mejías, R.; del Carmen Rodríguez-Dodero, M.; García-Barroso, C. Volatile composition of Pedro Ximénez and Muscat sweet Sherry wines from sun and chamber dried grapes: A feasible alternative to the traditional sun-drying. *J. Food Sci. Technol.* **2016**, *53*, 2519–2531. [CrossRef]
55. Ruiz-Bejarano, M.J.; Castro-Mejías, R.; Rodríguez-Dodero, M.d.C.; García-Barroso, C. Study of the content in volatile compounds during the aging of sweet Sherry wines obtained from grapes cv. Muscat and fermented under different conditions. *Eur. Food Res. Technol.* **2013**, *237*, 905–922. [CrossRef]
56. Campo, E.; Cacho, J.; Ferreira, V. The chemical characterization of the aroma of dessert and sparkling white wines (Pedro Ximénez, Fino, Sauternes, and Cava) by gas chromatography-olfactometry and chemical quantitative analysis. *J. Agric. Food Chem.* **2008**, *56*, 2477–2484. [CrossRef]
57. Chaves, M.; Zea, L.; Moyano, L.; Medina, M. Changes in color and odorant compounds during oxidative aging of Pedro Ximenez sweet wines. *J. Agric. Food Chem.* **2007**, *55*, 3592–3598. [CrossRef]
58. García-Martínez, T.; Moreno-García, J. Alternativas a la elaboración tradicional de vinos dulces de uva Pedro Ximénez pasificada. In Proceedings of the Grupos de Investigación Enológica, Tarragona, Spain, 9–12 June 2015; pp. 240–243.
59. Flanzy, C. Anaerobic grape metabolism. In *Enology: Scientific and Technological Base*; Mundi-Prensa: Madrid, Spain, 2000.
60. Cabañares, F.J.; Accensi, F.; Bragulat, M.R.; Abarca, M.L.; Castella, G.; Mínguez, S.; Pons, A. What is the source of ochratoxin A in wine? *Int. J. Food Microbiol.* **2002**, *79*, 213–215. [CrossRef]
61. Serratosa, M.P.; Marquez, A.; Lopez-Toledano, A.; Merida, J. Sensory analysis of sweet musts in pedro ximenez cv. grapes dried using different methods. *S. Afr. J. Enol. Vitic.* **2012**, *33*, 14–20. [CrossRef]
62. Ruiz-Bejarano, M.J.; Durán-Guerrero, E.; Castro, R.; Barroso, C.G.; Rodríguez-Dodero, M.C. Use of sensory analysis to investigate the influence of climate chambers and other process variables in the production of sweet wines. *Foods* **2020**, *9*, 424. [CrossRef] [PubMed]
63. Ruíz Bejarano, M.J.; Rodríguez Dodero, M.C.; Barroso, C.G. Optimizing the process of making sweet wines to minimize the content of ochratoxin A. *J. Agric. Food Chem.* **2010**, *58*, 13006–13012. [CrossRef]
64. Ruiz-Bejarano, M.J.; Castro-Mejías, R.; Rodríguez-Dodero, M.C.; García-Barroso, C. Effect of ageing of sweet Sherry wines obtained from cvs Muscat and Pedro Ximénez on ethyl carbamate concentration. *Aust. J. Grape Wine Res.* **2015**, *21*, 396–403. [CrossRef]
65. Espejo, F.; Armada, S. Effect of Enzyme Addition in the Making of Pedro Ximenez Sweet Wines Using Dynamic Pre-fermentative Maceration. *S. Afr. J. Enol. Vitic.* **2016**, *31*. [CrossRef]
66. García-Martínez, T.; De Lerma, N.L.; Moreno, J.; Peinado, R.A.; Millán, M.C.; Mauricio, J.C. Sweet Wine Production by Two Osmotolerant Saccharomyces cerevisiae Strains. *J. Food Sci.* **2013**, *78*, 874–879. [CrossRef]
67. Ruiz-Bejarano, M.J.; Rodríguez-Dodero, M.C.; Castro, R.; García-Barroso, C. Empleo de Torulaspora delbrueckii en la elaboración de vinos dulces andaluces. Caracterización química y sensorial. In Proceedings of the Grupos de Investigación Enológica, Tarragona, Spain, 9–12 June 2015; pp. 352–355.
68. Santiago Hurtado, J.I.; López de Lerma, N.; Moreno, J.; Peinado, R.A. Effect of thermal treatment and oak chips on the volatile composition of pedro ximénez sweet wines. *Am. J. Enol. Vitic.* **2010**, *61*, 91–95.
69. López de Lerma, N.; Peinado, J.; Moreno, J.; Peinado, R.A. Antioxidant activity, browning and volatile Maillard compounds in Pedro Ximénez sweet wines under accelerated oxidative aging. *LWT Food Sci. Technol.* **2010**, *43*, 1557–1563. [CrossRef]
70. Chinnici, F.; Guerrero, E.D.; Sonni, F.; Natali, N.; Marín, R.N.; Riponi, C. Gas chromatography-mass spectrometry (GC-MS) characterization of volatile compounds in quality vinegars with protected European geographical indication. *J. Agric. Food Chem.* **2009**, *57*, 4784–4792. [CrossRef]
71. Callejón, R.M.; González, A.G.; Troncoso, A.M.; Morales, M.L. Optimization and validation of headspace sorptive extraction for the analysis of volatile compounds in wine vinegars. *J. Chromatogr. A* **2008**, *1204*, 93–103. [CrossRef] [PubMed]
72. Ríos-Reina, R.; Morales, M.L.; García-González, D.L.; Amigo, J.M.; Callejón, R.M. Sampling methods for the study of volatile profile of PDO wine vinegars. A comparison using multivariate data analysis. *Food Res. Int.* **2018**, *105*, 880–896. [CrossRef]
73. Durán-Guerrero, E.; Schwarz, M.; Fernández-Recamales, M.Á.; Barroso, C.G.; Castro, R. Characterization and Differentiation of Spanish Vinegars from Jerez and Condado de Huelva Protected Designations of Origin. *Foods* **2019**, *8*, 341. [CrossRef]
74. Aceña, L.; Vera, L.; Guasch, J.; Busto, O.; Mestres, M. Chemical characterization of commercial sherry vinegar aroma by headspace solid-phase microextraction and gas chromatography-olfactometry. *J. Agric. Food Chem.* **2011**, *59*, 4062–4070. [CrossRef]
75. Callejón, R.M.; Morales, M.L.; Troncoso, A.M.; Silva Ferreira, A.C. Targeting key aromatic substances on the typical aroma of Sherry vinegar. *J. Agric. Food Chem.* **2008**, *56*, 6631–6639. [CrossRef] [PubMed]
76. Marrufo-Curtido, A.; Cejudo-Bastante, M.J.; Rodríguez-Dodero, M.C.; Natera-Marín, R.; Castro-Mejías, R.; García-Barroso, C.; Durán-Guerrero, E. Novel vinegar-derived product enriched with dietary fiber: Effect on polyphenolic profile, volatile composition and sensory analysis. *J. Food Sci. Technol.* **2015**, *52*, 7608–7624. [CrossRef]

77. Pizarro, C.; Esteban-Díez, I.; Sáenz-González, C.; González-Sáiz, J.M. Vinegar classification based on feature extraction and selection from headspace solid-phase microextraction/gas chromatography volatile analyses: A feasibility study. *Anal. Chim. Acta* **2008**, *608*, 38–47. [CrossRef]
78. Jiménez-Sánchez, M.; Durán-Guerrero, E.; Rodríguez-Dodero, M.C.; Barroso, C.G.; Castro, R. Use of ultrasound at a pilot scale to accelerate the ageing of sherry vinegar. *Ultrason. Sonochem.* **2020**, *69*. [CrossRef] [PubMed]
79. Marrufo-Curtido, A.; Cejudo-Bastante, M.J.; Durán-Guerrero, E.; Castro-Mejías, R.; Natera-Marín, R.; Chinnici, F.; García-Barroso, C. Characterization and differentiation of high quality vinegars by stir bar sorptive extraction coupled to gas chromatography-mass spectrometry (SBSE-GC-MS). *LWT Food Sci. Technol.* **2012**, *47*, 332–341. [CrossRef]
80. Ríos-Reina, R.; Segura-Borrego, M.P.; Morales, M.L.; Callejón, R.M. Characterization of the aroma profile and key odorants of the Spanish PDO wine vinegars. *Food Chem.* **2020**, *311*. [CrossRef]
81. Natera Marín, R.; Castro Mejías, R.; de Valme García Moreno, M.; García Rowe, F.; García Barroso, C. Headspace solid-phase microextraction analysis of aroma compounds in vinegar: Validation study. *J. Chromatogr. A* **2002**, *967*, 261–267. [CrossRef]
82. Callejón, R.M.; Morales, M.L.; Ferreira, A.C.S.; Troncoso, A.M. Defining the Typical Aroma of Sherry Vinegar: Sensory and Chemical Approach. *J. Agric. Food Chem.* **2008**, *56*, 8086–8095. [CrossRef]
83. Blanch, G.P.; Tabera, J.; Herraiz, M.; Sanz, J.; Reglero, G. Volatile Composition of Vinegars. Simultaneous Distillation-Extraction and Gas Chromatographic-Mass Spectrometric Analysis. *J. Agric. Food Chem.* **1992**, *40*, 1046–1049. [CrossRef]
84. Morales, M.L.L.; Troncoso, A.M.M. Note: Evaluation of Aroma Compounds in Wine Vinegars: Effect of Previous Neutralisation of Samples. *Food Sci. Technol. Int.* **2003**, *9*, 397–402. [CrossRef]
85. Morales, M.L.; González, G.A.; Casas, J.A.; Troncoso, A.M. Multivariate analysis of commercial and laboratory produced Sherry wine vinegars: Influence of acetification and aging. *Eur. Food Res. Technol.* **2001**, *212*, 676–682. [CrossRef]
86. Morales, M.L.; Tesfaye, W.; Carmen García-Parrilla, M.; Casas, J.A.; Troncoso, A.M. Evolution of the aroma profile of sherry wine vinegars during an experimental aging in wood. *J. Agric. Food Chem.* **2002**, *50*, 3173–3178. [CrossRef]
87. Bruna-Maynou, F.J.; Castro, R.; Rodríguez-Dodero, M.C.; Barroso, C.G.; Durán-Guerrero, E. Flavored Sherry vinegar with citric notes: Characterization and effect of ultrasound in the maceration of orange peels. *Food Res. Int.* **2020**, *133*. [CrossRef] [PubMed]
88. Morales, M.L.; Tesfaye, W.; García-Parrilla, M.C.; Casas, J.A.; Troncoso, A.M. Sherry wine vinegar: Physicochemical changes during the acetification process. *J. Sci. Food Agric.* **2001**, *81*, 611–619. [CrossRef]
89. Cejudo-Bastante, M.J.; Durán, E.; Castro, R.; Rodríguez-Dodero, M.C.; Natera, R.; García-Barroso, C. Study of the volatile composition and sensory characteristics of new Sherry vinegar-derived products by maceration with fruits. *LWT Food Sci. Technol.* **2013**, *50*, 469–479. [CrossRef]
90. Ríos-Reina, R.; Segura-Borrego, M.P.; García-González, D.L.; Morales, M.L.; Callejón, R.M. A comparative study of the volatile profile of wine vinegars with protected designation of origin by headspace stir bar sorptive extraction. *Food Res. Int.* **2019**, *123*, 298–310. [CrossRef]
91. Casale, M.; Armanino, C.; Casolino, C.; Oliveros, C.C.; Forina, M. A chemometrical approach for vinegar classification by headspace mass spectrometry of volatile compounds. *Food Sci. Technol. Res.* **2006**, *12*, 223–230. [CrossRef]
92. Chanivet, M.; Durán-Guerrero, E.; Barroso, C.G.; Castro, R. Suitability of alternative wood types other than American oak wood for the ageing of Sherry vinegar. *Food Chem.* **2020**, *316*. [CrossRef]
93. Chanivet, M.; Durán-Guerrero, E.; Rodríguez-Dodero, M.D.C.; Barroso, C.G.; Castro, R. Application of accelerating energies to the maceration of sherry vinegar with citrus fruits. *J. Sci. Food Agric.* **2020**, *101*, 2235–2246. [CrossRef]
94. Durán-Guerrero, E.; Chinnici, F.; Natali, N.; Riponi, C. Evaluation of volatile aldehydes as discriminating parameters in quality vinegars with protected European geographical indication. *J. Sci. Food Agric.* **2015**, *95*, 2395–2403. [CrossRef]
95. Troncoso-González, A.M.; Guzmán-Chozas, M. Volatile components in Andalusian vinegars. *Z. Lebensm.-Unters. Forsch.* **1987**, *185*, 130–133. [CrossRef]
96. Guerrero, M.I.; Heredia, F.J.; Troncoso, A.M. Characterisation and differentiation of wine vinegars by multivariate analysis. *J. Sci. Food Agric.* **1994**, *66*, 209–212. [CrossRef]
97. Guerrero, E.D.; Marín, R.N.; Mejías, R.C.; Barroso, C.G. Optimisation of stir bar sorptive extraction applied to the determination of volatile compounds in vinegars. *J. Chromatogr. A* **2006**, *1104*, 47–53. [CrossRef]
98. Guerrero, E.D.; Marín, R.N.; Mejías, R.C.; Barroso, C.G. Stir bar sorptive extraction of volatile compounds in vinegar: Validation study and comparison with solid phase microextraction. *J. Chromatogr. A* **2007**, *1167*, 18–26. [CrossRef]
99. Ríos-Reina, R.; Callejón, R.M.; Oliver-Pozo, C.; Amigo, J.M.; García-González, D.L. ATR-FTIR as a potential tool for controlling high quality vinegar categories. *Food Control* **2017**, *78*, 230–237. [CrossRef]
100. Guerrero, E.D.; Mejías, R.C.; Marín, R.N.; Lovillo, M.P.; Barroso, C.G. A new FT-IR method combined with multivariate analysis for the classification of vinegars from different raw materials and production processes. *J. Sci. Food Agric.* **2010**, *90*, 712–718. [CrossRef]
101. Benito, M.J.; Ortiz, M.C.; Sánchez, M.S.; Sarabia, L.A.; Iñiguez, M. Typification of vinegars from Jerez and Rioja using classical chemometric techniques and neural network methods. *Analyst* **1999**, *124*, 547–552. [CrossRef]
102. Durán, E.; Palma, M.; Natera, R.; Castro, R.; Barroso, C.G. New FT-IR method to control the evolution of the volatile constituents of vinegar during the acetic fermentation process. *Food Chem.* **2010**, *121*, 575–579. [CrossRef]
103. González-Viñas, M.A.; Salvador, M.D.; Cabezudo, M.D. Taste group thresholds and sensory evaluation of spanish wine vinegars. *J. Sens. Stud.* **1996**, *11*, 129–140. [CrossRef]

104. Tesfaye, W.; García-Parrilla, M.C.; Troncoso, A.M. Sensory evaluation of Sherry wine vinegar. *J. Sens. Stud.* **2002**, *17*, 133–144. [CrossRef]
105. Tesfaye, W.; Morales, M.L.; Callejón, R.M.; Cerezo, A.B.; González, A.G.; García-Parrilla, M.C.; Troncoso, A.M. Descriptive sensory analysis of wine vinegar: Tasting procedure and reliability of new attributes. *J. Sens. Stud.* **2010**, *25*, 216–230. [CrossRef]
106. Morales Gómez, M.L.; Bellido, B.B.; Tesfaye, W.; Callejón Fernandez, R.M.; Valencia, D.V.; Fernandez-Pachón, M.S.; García-Parrilla, M.D.C.; Troncoso González, A.M. Sensory evaluation of sherry vinegar: Traditional compared to accelerated aging with oak chips. *J. Food Sci.* **2006**, *71*. [CrossRef]
107. Durán Guerrero, E.; Mejías, R.C.; Marín, R.N.; Bejarano, M.J.R.; Dodero, M.C.R.; Barroso, C.G. Accelerated aging of a Sherry wine vinegar on an industrial scale employing microoxygenation and oak chips. *Eur. Food Res. Technol.* **2011**, *232*, 241–254. [CrossRef]
108. Cejudo-Bastante, M.J.; Rodríguez Dodero, M.C.; Durán Guerrero, E.; Castro Mejías, R.; Natera Marín, R.; García Barroso, C. Development and optimisation by means of sensory analysis of new beverages based on different fruit juices and sherry wine vinegar. *J. Sci. Food Agric.* **2013**, *93*, 741–748. [CrossRef]
109. Consejería de Agriculura, Pesca y Desarrollo Rural. *Orden de 28 de Junio de 2018, Por La Que Se Aprueba el Expediente Técnico de Indicación Geográfica "Brandy de Jerez."*; Boletin Oficial Junta Andalucía (BOJA): Sevilla, Spain, 2018; Volume 127, pp. 19–20.
110. Schwarz, M.; Rodríguez-Dodero, M.C.; Jurado, M.S.; Puertas, B.; Barroso, C.G.; Guillén, D.A. Analytical Characterization and Sensory Analysis of Distillates of Different Varieties of Grapes Aged by an Accelerated Method. *Foods* **2020**, *9*, 277. [CrossRef]
111. Trillo, L.M.; Álvarez, M.A.; Pinedo, J.M.; Arnedo, R.; Casas, J.A.; Sanz, J. Influencia de la selección varietal en la composición química de los aguardientes de vino destinados a la elaboración de Brandy de Jerez. In Proceedings of the XI Congreso Nacional de Enólogos, Toledo, Spain, 14–16 April 2005.
112. Álvarez, M.A.; Trillo, L.M.; Pinedo, J.M.; Casas, J.A.; García-Maíquez, E. Influencia de la temperatura de fermentación y el contenido de SO2 en la elaboración de vinos de "La Mancha", destinados a la obtención de aguardientes vínicos de bajo grado (Holandas). In Proceedings of the VI Congreso Nacional de Enólogos, Jerez de la Frontera, Spain, 7–9 April 2000.
113. Trillo, L.M.; Pinedo, J.M.; Arnedo, R.; Casas, J.A.; Sanz, J.; Roselló, A.; Cota, J.; Carrillo, F. Estudio comparativo de los procesos de destilación discontinuos en alambique charentais y en columna de platos para la obtención de aguardientes de vino destinados a la elaboración de Brandy de Jerez. In Proceedings of the XII Congreso Nacional de Enólogos, Santa Cruz de Tenerife, Spain, 5–7 May 2006.
114. Delgado-González, M.J.; Sánchez-Guillén, M.M.; García-Moreno, M.d.V.; Rodríguez-Dodero, M.C.; García-Barroso, C.; Guillén-Sánchez, D.A. Estudio del envejecimiento de destilados vínicos (holandas) monovarietales obtenidos mediante destilación de conos rotatorios. In Proceedings of the XIV Congreso Nacional de Investigación Enológica, Ciudad Real, Spain, 26–29 June 2019; pp. 204–207.
115. Guerrero-Chanivet, M.; Valcárcel-Muñoz, M.J.; García-Moreno, M.V.; Guillén-Sánchez, D.A. Characterization of the Aromatic and Phenolic Profile of Five Different Wood Chips Used for Ageing Spirits and Wines. *Foods* **2020**, *9*, 1613. [CrossRef]
116. Valme García-Moreno, M.; Sánchez-Guillén, M.M.; de Mier, M.R.; Delgado-González, M.J.; Carmen Rodríguez-Dodero, M.; García-Barroso, C.; Guillén-Sánchez, D.A. Use of alternative wood for the ageing of brandy de Jerez. *Foods* **2020**, *9*, 250. [CrossRef] [PubMed]
117. Valcárcel-Muñoz, M.J.; Guerrero-Chanivet, M.; García-Moreno, M.V.; Rodríguez-Dodero, M.C.; Guillén-Sánchez, D.A. Comparative Evaluation of Brandy de Jerez Aged in American Oak Barrels with Different Times of Use. *Foods* **2021**, *10*, 288. [CrossRef] [PubMed]
118. Delgado, R.; Durán, E.; Castro, R.; Natera, R.; Barroso, C.G. Development of a stir bar sorptive extraction method coupled to gas chromatography-mass spectrometry for the analysis of volatile compounds in Sherry brandy. *Anal. Chim. Acta* **2010**, *672*, 130–136. [CrossRef]
119. Durán Guerrero, E.; Cejudo Bastante, M.J.; Castro Mejías, R.; Natera Marín, R.; García Barroso, C. Characterization and differentiation of sherry brandies using their aromatic profile. *J. Agric. Food Chem.* **2011**, *59*, 2410–2415. [CrossRef]
120. Schwarz, M.; Rodríguez, M.C.; Guillén, D.A.; Barroso, C.G. Development and validation of UPLC for the determination of phenolic compounds and furanic derivatives in Brandy de Jerez. *J. Sep. Sci.* **2009**, *32*, 1782–1790. [CrossRef] [PubMed]
121. Rodríguez-Dodero, M.C.; García-Barroso, C.; Pérez-Bustamante-De Monasterio, J.A. Análisis de cumarinas en brandys de Jerez mediante el uso de HPLC con detección en fluorescencia. In Proceedings of the Jornadas de Viticultura y Enología en Tierra de Barros, Almendralejo, Spain, 6–10 May 1997.
122. Rodríguez-Dodero, M.C.; Guillén-Sánchez, D.A.; García-Barroso, C. La nariz electrónica como herramienta para la diferenciación de tipos de Brandy de Jerez. In Proceedings of the Congreso Nacional de Enólogos, Bilbao, Spain, 15–17 February 2003.
123. Consejo Regulador de la IGP Brandy de Jerez Bodegas y Marcas de Brandy de Jerez. Available online: https://www.brandydejerez.es/bodegas-y-marcas (accessed on 1 April 2021).
124. Delgado-González, M.J.; Sánchez-Guillén, M.M.; García-Moreno, M.V.; Rodríguez-Dodero, M.C.; García-Barroso, C.; Guillén-Sánchez, D.A. Study of a laboratory-scaled new method for the accelerated continuous ageing of wine spirits by applying ultrasound energy. *Ultrason. Sonochem.* **2017**, *36*, 226–235. [CrossRef] [PubMed]
125. Schwarz, M.; Rodríguez, M.C.; Sánchez, M.; Guillén, D.A.; Barroso, C.G. Development of an accelerated aging method for Brandy. *LWT Food Sci. Technol.* **2014**, *59*, 108–114. [CrossRef]

126. Sánchez-Guillén, M.M.; Rodríguez-Dodero, M.C.; Schwarz-Rodríguez, M.; Guillén-Sánchez, D.A.; García-Barroso, C. Aplicación del análisis sensorial durante la optimización del test patentado sobre envejecimiento acelerado de aguardientes. In Proceedings of the XII Congreso de los Grupos de Investigación Enológica GIENOL 2013, Madrid, Spain, 18–21 June 2013; pp. 555–558.
127. Trillo-Ollero, R.; García-Moreno, M.d.V.; Delgado-González, M.J.; Rodríguez-Dodero, M.C.; García-Barroso, C.; Guillén-Sánchez, D.A. Caracterización de cinco tipos de maderas para su uso en el envejecimiento de destilados. In Proceedings of the XIV Congreso Nacional de Investigación Enológica, Ciudad Real, Spain, 26–29 June 2018; pp. 200–203.

foods

*Review*

# Secondary Aroma: Influence of Wine Microorganisms in Their Aroma Profile

**Maria Carpena [1], Maria Fraga-Corral [1,2], Paz Otero [1,3], Raquel A. Nogueira [1], Paula Garcia-Oliveira [1,2], Miguel A. Prieto [1,2,*] and Jesus Simal-Gandara [1,*]**

1 Nutrition and Bromatology Group, Analytical and Food Chemistry Department, Faculty of Food Science and Technology, University of Vigo, Ourense Campus, E-32004 Ourense, Spain; maria.carpena.rodriguez@uvigo.es (M.C.); mfraga@uvigo.es (M.F.-C.); pazoterofuertes@gmail.com (P.O.); raquelnogueira36@gmail.com (R.A.N.); paula.garcia.oliveira@uvigo.es (P.G.-O.)
2 Centro de Investigação de Montanha (CIMO), Instituto Politécnico de Bragança, Campus de Santa Apolonia, 5300-253 Bragança, Portugal
3 Department of Pharmacology, Pharmacy and Pharmaceutical Technology, Faculty of Veterinary, University of Santiago of Compostela, 27002 Lugo, Spain
* Correspondence: mprieto@ipb.pt (M.A.P.); jsimal@uvigo.es (J.S.-G.)

**Abstract:** Aroma profile is one of the main features for the acceptance of wine. Yeasts and bacteria are the responsible organisms to carry out both, alcoholic and malolactic fermentation. Alcoholic fermentation is in turn, responsible for transforming grape juice into wine and providing secondary aromas. Secondary aroma can be influenced by different factors; however, the influence of the microorganisms is one of the main agents affecting final wine aroma profile. *Saccharomyces cerevisiae* has historically been the most used yeast for winemaking process for its specific characteristics: high fermentative metabolism and kinetics, low acetic acid production, resistance to high levels of sugar, ethanol, sulfur dioxide and also, the production of pleasant aromatic compounds. Nevertheless, in the last years, the use of non-saccharomyces yeasts has been progressively growing according to their capacity to enhance aroma complexity and interact with *S. cerevisiae*, especially in mixed cultures. Hence, this review article is aimed at associating the main secondary aroma compounds present in wine with the microorganisms involved in the spontaneous and guided fermentations, as well as an approach to the strain variability of species, the genetic modifications that can occur and their relevance to wine aroma construction.

**Keywords:** wine secondary aroma; fermentation; non-saccharomyces yeasts; lactic acid bacteria; volatile compounds; strain variability



**Citation:** Carpena, M.; Fraga-Corral, M.; Otero, P.; Nogueira, R.A.; Garcia-Oliveira, P.; Prieto, M.A.; Simal-Gandara, J. Secondary Aroma: Influence of Wine Microorganisms in Their Aroma Profile. *Foods* **2020**, *10*, 51. https://dx.doi.org/10.3390/foods10010051

Received: 29 November 2020
Accepted: 23 December 2020
Published: 27 December 2020



## 1. Introduction

### *1.1. Secondary Wine Aroma*

The combination of two modest substrates such as grapes and microorganisms (those belonging to the grape microbiota and/or those intentionally added) results in a huge variability and diversity of wines. However, this apparently simple conjugation hinders extremely specific chemical reactions that can be modified to obtain a stunning array of aromas and flavors. The wine aroma comprises a mixture of volatiles that can account up to 800 compounds, although just few of them are odor-active [1]. This complex chemical composition can be split in terms of aromas into three different categories that are mostly related with the three production steps: grape culture, fermentation stage and transformation process, respectively [2].

Primary or varietal aromas, as this second name indicates, are due to the grape variety. Primary aromas belonging to the same grape variety may present different features depending on natural factors derived from weather, type of soil, fertilization, presence/absence of plagues or even the geographical location, that prompt different cultivation conditions in each area and also each year. Besides, the ripening period and the care of the collector when selecting grapes may have influence in the final primary aroma of wine [1]. Grapes are known to contain free and sugar-glycosidically-linked terpenes, being the monoterpenes and sesquiterpenes the ones that contribute with aroma and flavor. Among them, the most odoriferous monoterpenes alcohols are linalool, geraniol, nerol, citronellol, 3,6-dimethyl-1,5-octa-dien-1,7-diol, hotrienol and α-terpineol, which provide floral, fruity and citrus aroma to the wine [3,4]. Even though just few aromas have been directly associated to specific varieties, each grape variety possesses an aroma fingerprint. Monoterpene glycosides or ethers do not show significant changes in their amount during yeast fermentations. Therefore, they can be used to classify different varieties, such as Muscat and Riesling wines, by the study of their analytical composition based on just 12 monoterpene compounds [5–7]. However, the low concentration of these aromas (e.g., hotrienol thresholds between 18 and 400 μg/L and linalool in 50 μg/L) does not permit their sensorial appreciation at least their potential gets boosted in later steps by enzymatic reactions thus, having a major impact in the final wine aroma [8,9]. Wine has been demonstrated to have more than 800 volatile compounds with wide range of concentration, from ng/L to hundreds of mg/L [10]. In addition, other precursors that do not possess odoriferous characteristics, are involved in the development of other aroma substances (e.g., monoterpenes, diols or terpene, polyols, fatty acids, carotenoids, glycosylated precursors of aroma and volatile phenols) [1].

In the next aromatic level, yeasts and bacteria carry out the fermentation, this is the chemical reactions chain responsible for transforming grape juice into wine and providing secondary aromas to wine. Secondary aromas can be divided into pre-fermentative, those arisen due to the mechanical treatment of grapes, and fermentative, those boosted during alcoholic or malolactic fermentation processes [1,2]. The most utilized species for the alcoholic fermentation is *Saccharomyces cerevisiae*, although there are about 20 yeast genera with the same capacity such as *Saccharomycodes*, *Candida*, *Issatchenkia*, *Pichia*, *Hanseniaspora* (*Kloeckera*) or *Brettanomyces* (*Dekkera*) [3]. These non-saccharomyces species drive the aroma release by the secretion of proteins, mainly enzymes, and the synthesis of new secondary metabolites. In addition, they contribute to color wine stability and they do not use up the available sugar in must. Thus, they are strategically utilized for creating multi-starter, mixed or sequential cultures in combination with *S. cerevisiae* [8,11,12]. Normally, after the alcoholic fermentation, wine is submitted to the malolactic fermentation by the inoculation of lactic acid bacteria (LAB). During this stage, malic acid responsible for the tart taste gets decarboxylated by the action of *Oenococcus oeni* or *Lactobacillus plantarum*, two common used LAB species [8]. After this fermentation process, wine is microbiologically stabilized. Along the fermentation, the main created aromas belong to the volatile fatty acids, higher alcohols, acetate and ethyl ester categories which make evolve the aroma profile of wine [3]. These molecules are usually present at high sensory thresholds (the oxidation products of linalool possesses a perception threshold of 6000 μg/L) and their combination creates the matrix of wine aroma [3,13]. In fact, by the end of this fermentative stage, the term aroma becomes more complex from a chemical and sensorial point of view and thus, it turns into the term bouquet. Therefore, even though the final wine aroma composition is highly dependent on fermentative species and strains, the grape microbiome is gaining attention, since different works point to its relevance in the final sensorial properties of wines [8,12].

Tertiary aromas are created during the last step, aging of wine, where the storage of the final product is the main responsible for the transference of aromas and flavors to wine. The typical aging method is the use of wood barrels mostly built with different oak species such as *Quercus alba*, *Q. robur* or *Q. petraea* [14]. Wine aged in these barrels may be transferred with volatiles such as guaiacol-oak lactones or vanillin and even with furfural, 5-methylfurfural, eugenol, guaiacol, 5-hydroxymethylfurfural, 4-methylguaiacol, guaiacol and syringol, when applying different toast treatments to wood [15]. The use of different wood provides different volatiles to aged wine, for instance, brandies aged in

*Quercus*-barrels were found to contain higher levels of ethyl-2-methylpropanoate, -butyrate and -octanoate and lower levels of butanoic acid, cis-β-methyl-γ-octalactone and syringol than when aged in *Castanea*-barrels [15]. In fact, apart from *Quercus*, other kind of woods such as *Acacia*, *Prunus* or *Castanea* are known to contain high concentrations of tannins, a kind of polyphenols, that are utilized to age wines since these non-volatile molecules can be transferred and may contribute to sensory properties such as color, astringency and bitterness [2,16].

Among the three classes of aroma, achieving an appropriate combination of secondary aromas represents the most intricate procedure. This stage implies the correct selection of yeasts and bacteria to perform the fermentation steps while avoiding wine spoiling due to cross contamination or due to the innate grape microbiota. Moreover, the high sensory threshold of the volatiles synthesized during this stage will define the wine aroma matrix. For obtaining a wine with well-defined secondary aromas and flavors, it is essential to understand how different microbial species interact with each other and which sensorial properties are capable to provide based on the metabolic pathways they develop.

### *1.2. Fermentation Implication on Wine Secondary Aroma*

As aforementioned, yeasts and bacteria are responsible for the production of the secondary aroma during the pre-fermentative and fermentation processes. Naturally, in traditional winemaking, fermentation of grape juice is carried out by different yeast species following an order. The fermentation is initialized by non-saccharomyces yeasts, which is called spontaneous fermentation. However, these yeasts do not resist the increase of ethanol and so, they are commonly replaced by the strongly winery fermentative yeast *Saccharomyces cerevisiae* [17]. Nevertheless, it has been suggested that some of the non-saccharomyces species could persist from one year to another in wine and become dominant during fermentation like *S. cerevisiae* [18]. In the past, non-saccharomyces wine yeasts were considered as undesired microorganisms but in recent years it is well known that they can enhance the analytical composition and aroma profile of wine [19]. Therefore, wine fermentation can be defined as a complex process where different microorganisms coexist and microbial interactions influence the final product [20]. Non-saccharomyces yeasts can influence both the primary and secondary aroma through the production of enzymes and metabolites, respectively [19]. In this context, and for the development of the following sections, it is important to differentiate between the three types of alcoholic fermentations that can occur. Besides, malolactic fermentation is a process that some type of wines can also undergo (i.e., wines with high acidity) and consequently, causes an improvement of the aromatic profile of wines [5]. This process will be explained in the following sections.

Firstly, spontaneous fermentation is a process that naturally takes place on grape must: at the initial stages, non-saccharomyces species (already present in grapes) dominate grape juice and are then replaced by winery yeasts, commonly *S. cerevisiae*, leading to wines with a complex profile but with lower microbiological control and submitted to variability and the risk of spoilage depending on the year and the exogenous microbiota of the grapes [19]. Next, the second type of fermentation is called guided, since wine is inoculated with selected cultures named as starters which compete and limit the growth of non-saccharomyces strains [21]. This way, industrial fermentations begun to use starters of selected wine yeast strains of *S. cerevisiae* for their fermentative behaviour, their ability to enhance secondary aroma but also to achieve more uniformity in the quality of wines [19]. However, it is currently accepted that those fermentations that use more than one yeast strain can produce wine with higher quality and complexity and less content in alcohol, while microbiological control is ensured. These are called mixed fermentations [20]. Mixed cultures have shown to exert additive or synergistic effect (e.g., by metabolites exchange between yeasts) resulting in the enhancement of the chemical and sensory profile of wines [22,23]. A representation of the types of fermentations is shown in Figure 1.

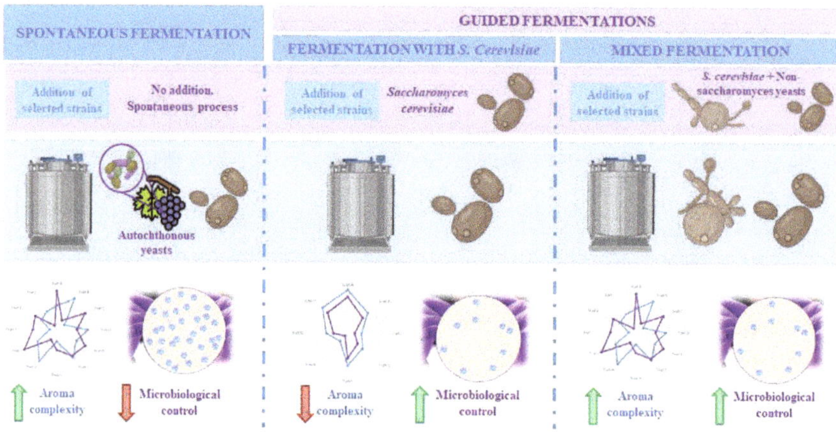


**Figure 1.** Schematic representation of the types of fermentation, the implied yeasts and the final characteristics of the obtained wine according to aroma complexity and microbiological control. Based on Reference [19].

It has been also highlighted that the selected inoculation strategy can modulate wine aroma profile in the case of mixed fermentations. If simultaneous inoculation is chosen, non-saccharomyces yeast and *S. cerevisiae* are added together, whereas in the case of sequential inoculation, non-saccharomyces starter is inoculated before *S. cerevisiae*, thus delaying the development of this last one [24,25]. In this sense, both strategies have shown aroma improvements depending on the utilized species. For instance, the aroma compounds resulting from the sequential fermentation of *Issatchenkia terricola* and *Pichia kudriavzevii* together with *S. cerevisiae* where higher than in the case of simultaneous fermentation [24]. In the other hand, a different experiment carried out with *Torulaspora delbrueckii* and *S. cerevisiae* showed an increase in the production of esters (fruity aroma) in the case of simultaneous fermentation when compared to sequential fermentation [26].

At last, it is worth to mention that other parameters during fermentation can also influence the wine aroma. These are temperature, molecular oxygen available during fermentation, maturation or ageing, the nitrogen source also known as yeast assimilable nitrogen (YAN) and the inoculation rate of yeasts as well as other post-fermentative parameters, such as yeasts final autolysis [27].

### *1.3. Microorganisms Implied in Wine Aroma*

Wine is a complex matrix where the development of alcoholic fermentation, leaded by different yeasts coupled to the volatile compounds released during malolactic fermentation, leaded by LAB and acetic acid bacteria (AAB), defines wine secondary aroma [8]. Yeasts are responsible for alcoholic fermentation, and particularly, the unicellular fungi *Saccharomyces cerevisiae* governs the process, which can occur spontaneously or guided by the use of starter cultures [17]. Yeast domain counts up to more than 2000 species, among which Saccharomyces has traditionally been the most studied and important genus for industrial fermentation [8]. Within Saccharomyces species, *S. cerevisiae* is the most known since the first inoculation processes with a pure yeast culture were carried out with this species. This trend continued for many decades and resulted in the generalized use of *S. cerevisiae* as starter yeasts inmost wine fermentations [28]. However, as aforementioned, non-saccharomyces species also play an important role during fermentation. Among this group, the genera most commonly present and studied are *Hanseniaspora*, *Hansenula*, *Metschnikowia*, *Candida*, *Pichia*, *Lachancea*, *Brettanomyces*, *Kluyveromyces*, *Schizosaccharomyces*, *Torulaspora*, *Zygosaccharomyces* and *Saccharomycodes* [5,8]. In respect of bacteria, most abundant LAB belongs to genera *Lactobacillus*, *Oenococcus*, *Pediococcus* and *Leuconostoc* whereas most predominant AAB during winemaking are *Acetobacter*, *Gluconobacter* or

*Gluconacetobacter* [29]. Figure 2 represents the main groups and taxonomy of the microorganisms implied in wine aroma. The challenge of winemakers and researches lies on the detection, characterization and quantification of all these microorganisms populations during fermentation to assess their participation on the development of wine secondary aroma [29].

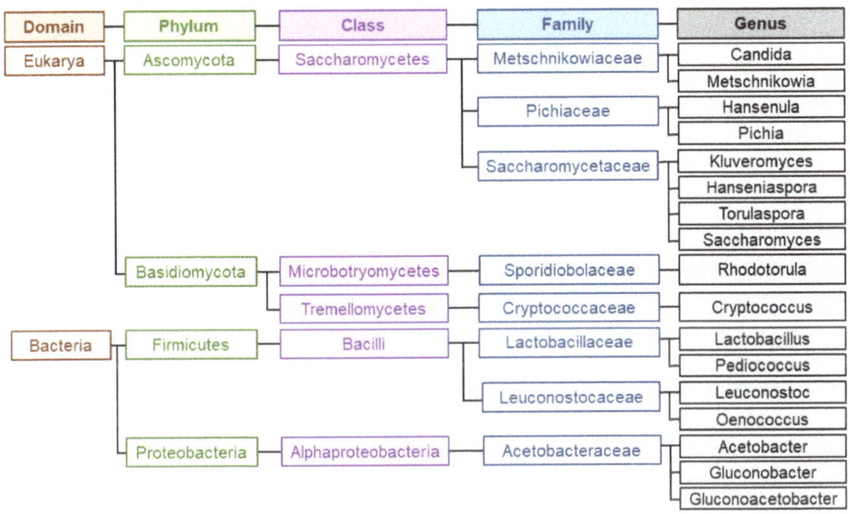


**Figure 2.** Main groups and taxonomy of the microorganisms implied in wine aroma production. Information extracted from Reference [30].

Therefore, this review presents an overview of the main secondary aromas present in wine, the microorganisms involved in the spontaneous and guided (simultaneous or mixed) fermentations as well as an approach to the aroma variation that wine can suffer when different strains and genetic modifications have occurred.

## 2. Compounds Involved in Secondary or Fermentative Aroma

The quality of wine is derived from its aroma which is, in turn, characterized by its volatile composition, mainly created during the fermentation stages. Fermentation is highly dependent on the species and strains selected but also on the components of the wine matrix. Among the main volatiles that define wine, higher alcohols, esters and fatty acids play a key role in the creation of secondary aromas (Table 1 and Figure 3).

### *2.1. Volatile Fatty Acids*

In the category of aliphatic fatty acids, apart from the most abundant volatile acid, i.e., the acetic acid, the major medium chain fatty acids are hexanoic, octanoic or decanoic. Besides, in the group of the unsaturated fatty acids is worthy to mention 9-decenoic acid which possesses preservative properties and is relevant from an aroma point of view when transformed into ethyl ester [31].

**Table 1.** Compounds involved in secondary aroma, classes of volatile aroma, main representative, desirable concentration, sensorial properties and producer microorganisms.

| Aromatic Class | Main Compounds | Desirable Concentration | Sensorial Properties | Producer Organism | Ref. |
|---|---|---|---|---|---|
| Fatty acids | Acetic acid, pentanoic acid, hexanoic acid, octanoic acid, decanoic acid, 9-Decenoic acid, 3-methylbutanoic acid, sobutyric acid | 200–700 mg/L | In excessive amount: rancid, greasy, and cheesy notes | *S. cerevisiae*, *P. fermentas*, *C. zemplinina*, *H. guilliermondii*, *H. vineae*, *H. uvarum* | [3,19,31] |
| Higher alcohols | 1-Propanol-isobutanol, isoamyl alcohol, 2-Phenylethanol, tyrosol, tryptophol, 2-methylbutanol-1, 3-methyl-1-butanol-1 | <300 mg/L | Floral, honey, and fruity notes (<300 mg/L). Pungent aroma (>400 mg/L) | *S. cerevisiae*, *C. zemplinin*, *H. uvarum*, *H. osmophila*, *H. guilliermondii*, *P. anomala*, *P. membranifaciens* | [2,3,5,8,19] |
| Esters | Ethyl hexanoate, ethyl octanoate, ethyl decanoate, ethyl acetate, isobutyl acetate, amyl acetate, hexyl acetate, 2PA, isoamyl acetate | 150–200 mg/L | Fruity aroma, including banana or apple, honey, and floral tones | *S. cerevisiae*, *Candida*, *Hansenula*, *Pichia* | [2,3,7,19] |
| Phenolics | 4-Vinyl guajacol, 4-Vinylphenol | - | Sweet vanillin aroma | LAB | [1,7,31] |

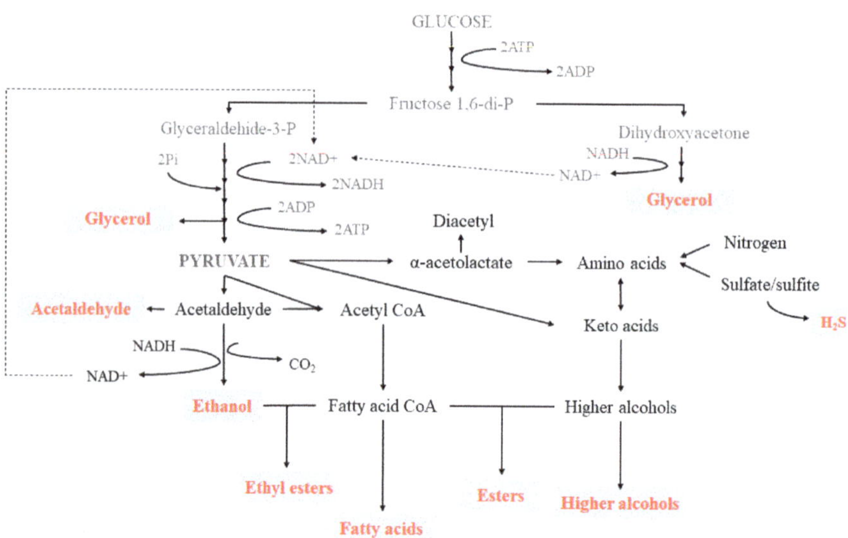


**Figure 3.** General representation of aroma compound metabolism.

Yeasts are the primary producers of these fatty acids which are the initial substrate for the final formation of ethyl esters. Among yeasts, *S. cerevisiae* is capable of synthesizing mainly hexanoic and octanoic acids in high amounts, but also pentanoic, decanoic and 3-methylbutanoic acids. Other non-saccharomyces species such as the genus *Hanseniaspora* has been described to produce acetic acid (in very variable ranges, from 0.6 up to 3.4 g/L) and species such as *Hanseniaspora vineae*, *H. uvarum*, *H. guilliermondii* or *Candida zemplinina* displayed higher synthesis rates for isobutyric acid [3,19]. However, it has been stated that this group of yeasts does not present a distinct biosynthesis of fatty acids. In fact, the use of mixed fermentations including *S. cerevisiae* and non-saccharomyces can modify the chemical profile of the single *S. cerevisiae* model. In general terms, this combination shows a reduction in the amount of medium-chain fatty acids, as it happens when inoculating *S. cerevisiae* with *H. osmophila*. Even though, the utilization of a mixture of *C. stellata* and *S. cerevisiae* could increase the amount of hexanoic and octanoic acids, followed *Pichia fermentas*. Similarly, the application of sequential inoculations based on *S. cerevisiae* and non-saccharomyces usually provides wines with lower concentrations of fatty acids [3]. The use of mixed or sequential fermentations can have benefits to regulate the content of these medium chain fatty acids, since their excessive presence may provide negative aromas with greasy, rancid and cheesy notes [3,32].

### 2.2. Higher Alcohols

The most abundant alcohols in wine, apart from ethanol and glycerol, are diols, higher alcohols and esters. Ethanol provides viscosity, balance taste and fix odors while higher alcohols and glycerol strongly contribute to the aroma complexity of wine and to the overall mouthfeel of wine. Higher alcohols are the result of the catabolism of amino acids by a process known as Ehrlich reaction, which affect directly or indirectly to the synthesis of aroma compounds. Higher alcohols are also involved as ester precursors which are important compounds in wine aroma [5,19]. Major higher alcohols are 1-propanol, isobutanol and isoamyl alcohol. Other important volatiles are the precursors of aromatic alcohols such as 2-phenylethanol, tyrosol or tryptophol and other higher alcohols but present in lower amounts, like 2-methylbutanol-1, 3-or methyl-1-butanol-1. Moderate concentrations of some of the volatiles considered to have high odor intensity, such as 3-methyl-1-butanol, 2-phenylethanol or isoamyl alcohol, can provide positive impact in the wine providing flower, honey and fruit aroma notes. However, the higher alcohol concentration plays a key role in the complexity of the aroma composition. Optimal alcohol values under 300 mg/L provide fruity and flowery notes, whereas alcohol values above 400 mg/L become negative by adding pungent and unpleasant aromas [2,3,8,19]. Among the different fermentation parameters that affect the final concentration of alcohol in wine, yeast strain is one of the key parameters followed by temperature, pH or oxygen, apart from grape ripeness and variety [33]. Higher alcohol synthesis has been widely studied and related to different species and/or inoculation protocols to obtain wines with an equilibrated higher alcohol composition. Different works have evaluated the efficiency of *S. cerevisiae* in terms of higher alcohol production [34,35]. Generally, no significant differences have been observed for 1-propanol while isobutanol, isoamyl alcohol, 3-methyl-1-butanol or 2-phenylethanol production seems to be strain-dependent and related to the presence of *S. cerevisiae*, both as pure or mixed cultures. In general terms, *H. uvarum*, *C. zemplinina* or *P. anomala* are considered as high alcohol producers, used both as single and mixed (with *S. cerevisiae*) fermentation agents [3,8,19]. Nevertheless, the single application of non-saccharomyces yeasts has been stated to produce lower amounts of total alcohols than *S. cerevisiae* and so, a reduction of the final amount of higher alcohols when using mixed cultures [36]. Indeed, *H. osmophila*, *H. guilliermondii* and *P. membranifaciens* were demonstrated to produce lower amounts of higher alcohols when tested against *S. cerevisiae*, even though *H. osmophila* provided high levels of 2-phenylethanol and isoamyl alcohol. Similarly, for the genus *Candida*, *C. zemplinina* has been described to synthesize 2-phenylethyl, glycerol and low amounts of ethanol and acetic acid. This combination has prompted its classification as fructophilic species, whereas *C. stellata* is classified as low producer. Another study with *H. uvarum* strains displayed variability in all produced higher alcohols except for isobutanol whose production seems to be boosted by *Hanseniaspora*. Indeed, another species, *H. guilliermondii*, also has a higher production rate of isobutanol than *S. cerevisiae*. Besides, same species synthesized very low amounts of 1-propanol [3,8,19].

### 2.3. Esters

Esters are another relevant group, also responsible for the aroma complexity of wines with more than 160 representatives already identified. From a chemical point of view, they can be classified into ethyl fatty acid esters or acetate esters. In the first category, ethyl hexanoate, ethyl octanoate, and ethyl decanoate are the most abundant ones. In these molecules, ethanol represents an important contribution to their structure. In the second class, higher alcohols are essential for the formation of these esters. The major acetate esters are isobutyl acetate, amyl acetate, hexyl acetate, ethyl acetate (fruity aroma), isoamyl acetate (banana aroma) and 2-phenylethyl acetate (2PA), which has been described to provide honey, fruity and floral aromas to the wine [2,3,7]. In white wine, the main fatty acid ethyl esters include ethyl butanoate, caproate, caprylate, caprate and laurate. As other esters, they can also provide fruity tones that become softer with the increasing number of carbons in their chemical structure of the formation of these esters depends on the selection of yeast species and other fermentation parameters such as low temperatures, are [7]. Different yeasts have been used to give complexity to wines through ester production including *S. cerevisiae* but also non-saccharomyces species such as *Candida*, *Hansenula* and *Pichia* since their differential enzymatic mechanisms allow the introduction of novel aromas in wines [3]. In general terms, esters have positive effects on the aroma of young wines, especially in those with neutral flavors. Nevertheless, as it happens in the case of higher alcohols, excessive amounts of esters may induce negative effects on the quality of wine. A high concentration of esters can hidden varietal aromas and simplify the composition of aroma of the final product or spoil wine, for instance, if ethyl acetate exceeds 150–200 mg/L [2,19].

### 2.4. Volatile Phenols

The positive aroma notes of this group of molecules have been mainly related to the aging process where the main volatile phenols are guaiacol, 4-methylguaiacol, 4-ethylguaiacol, phenol, o-cresol or vanillin. The enzymes involved in these metabolic steps are mainly associated with LAB, such as β-glucosidases, proteases, esterases, citrate lyases and phenolic acid decarboxylases. In fact, many malolactic fermentations take place in oak barrels even though LAB can synthesize oak-like derived compounds from non-volatile phenols present in wine. Among the non-volatile phenols present in grapes it is common to find phenolic acids (caffeic, ferulic and *p*-coumaric) or their tartaric esters (caftaric acid, feruloyl tartaric acid, *p*-coumaroyl tartaric). LAB have the capacity to metabolize cinnamic acids, such as *p*-coumaric or ferulic, that through a decarboxylation step can be transformed into 4-vinyl guajacol and 4-vinylphenol. Thus, the use of LAB to obtain these compounds before the aging step has gained attention since it can help to modify the aroma complexity of wine. LAB can transform non-volatile phenols that contribute with unpleasant aromas such as pharmacy, smoke, forest, leather or pepper, into volatile pleasant ones, such as those related to vanillin, methyl vanilla or homovainyl alcohol. Apart from those that can be synthesized during fermentation stages due to their presence in grapes, another volatile phenols not present in grapes can be found in wines, i.e., acetovanillone [1,7,31].

## 3. Saccharomyces Cerevisiae

*Saccharomyces cerevisiae* is the most known yeast regarding the winemaking process. The historical importance of this yeast comes from far below as it was the first yeast observed by Antoine van Leeuwenhoek using a primitive microscope and it was then described as a living agent of transformation by Louis Pasteur [37]. As "agent of transformation", *S. cerevisiae* was domesticated from the production of food and beverages such as bread and wine or beer, respectively [38,39]. Apart from its traditional application in food and alcoholic beverages, *S. cerevisiae* has been also used for fuel production, for the expression of engineered designed proteins and as genetic model organism [40]. Particularly, in wine production, *S. cerevisiae* was selected and has been used for centuries due to its specific characteristics: high fermentative metabolism, suitable fermentation kinetics, low acetic acid production, resistant character against higher concentrations of sugar, ethanol and sulfur dioxide and also, the production of pleasant aromatic compounds [38,41]. Therefore, in 1890, *S. cerevisiae* cultures started to be inoculated to wine and commercial starters were introduced into the market [39]. Since this moment, different approaches have been followed up such as guided or mixed fermentations for optimizing wine production and its organoleptic characteristics [19].

*S. cerevisiae* possesses a specific metabolism that regulates the production of volatile and aroma molecules. As it can be seen in Table 1, this yeast contributes to many of the aroma compounds classes present in wine (fatty acids, higher alcohols and esters), although varietal compounds and pre-fermentative compounds also contribute to the final wine complex aroma [2,3]. Some of these groups have been intensively studied using *S. cerevisiae* fermentations and also, different enzymes have demonstrated a key role in their formation, such as alcohol acetyltransferases (Atf1p and Atf2p), isoamyl alcohol acetyltransferase or ethanol acetyltransferase (implied in the formation of acetate esters) or the acyl-CoA:ethanol O-acyltransferase, related with the production of the ethyl esters [5]. In general terms, *S. cerevisiae* produce lower amounts of higher alcohols and poorer extracellular enzymes involved in the hydrolysis of structural components when compared to other non-saccharomyces species. However, it produced higher quantities of esters or acetaldehyde [42]. Ethanol content also influences sensory characteristics, providing fruity, flowery or acid aromas to wine, in specific concentrations [5]. On the other hand, sulfur compounds have been associated with negative or unpleasant aromas thus, poorer producers of sulfur dioxide *S. cerevisiae* strains are frequently selected. Besides, terpenoids, can be *de novo* produced by *S. cerevisiae* through the mevalonic acid pathway, constituting an alternative pathway [43].

Apart from those desirable characteristics of yeasts, there are other variables that can affect or have consequences on the aroma profile [43]. For instance, in the case of sparkling wines, a recent study showed that depending on the employed strain of *S. cerevisiae* and the period of aging, different aroma profiles were obtained. The study evaluated the production of ethyl esters (sour and apple aromas) and alcohols (herbaceous, rose, sweet aromas). It was demonstrated that flocculent yeasts produced higher amounts of these volatile compounds after 3 months whereas yeasts with higher autolytic ability produced more elevated amount of esters and alcohols after 6 months [44]. Other aspect that influences different aromatic profiles is geographical origin of indigenous yeasts. Some authors have pointed out that aroma or *terroir* includes a microbial aspect, since its sensory profile varies depending on the microorganisms implied. Particularly, the specific "signature" of some *S. cerevisiae* indigenous populations is linked to certain regions and environment conditions [45,46]. In this sense, it has been found that different genotypes (original from a specific region) are related to changes in the released compounds and thus, in the aroma profile. For instance, different genotypes from New Zealand were compared and it was found that depending on the area, some genotypes produced higher amounts of β-damascenone (apple, honey and floral aromas), higher concentrations of ethyl isobutyrate and ethyl-2-methyl butanoate (apple and sweet fruit aromas) or ethyl butanoate (peach, apple and sweet aromas) [47]. Another work reported that, independently from the substrate characteristics, the production of specific aromatic compounds is related with yeast origin, showing differences in the amounts of acetic acid, acetoin, acetaldehyde, n-butanol and 2,3-butanediols, 2-methyl-1-butanol and 3-methyl-1-butanol, among others [45]. Therefore, the current thinking is that origin, genotype and phenotype of *S. cerevisiae* strains affect quality parameters of wine and has prompted the interest on selecting autochthonous yeasts over commercial ones [46].

Nutrients (e.g., initial nitrogen and lipids) concentration and temperature are other parameters that can influence aroma. A recent study evaluated how specific environmental conditions affect the production of volatile compounds and found that their effects depended on the target compounds. However, authors found that the strain was determinant for the effects of environmental parameters [48]. Regarding higher alcohols, initial nitrogen content played a fundamental role exerting a negative quadratic effect for 2-phenylethanol and it positively affected propanol production. In general, temperature and lipid content were positively correlated with the synthesis of isobutanol and isoamyl alcohol. Interactive effects were also found between parameters. Therefore, it has been suggested that the disposal of nitrogen sources (e.g., amino acids) and the production of aroma compounds by *S. cerevisiae* do not follows linear relationships [48,49].

## 4. Non-Saccharomyces Species

### *4.1. Yeasts*

#### 4.1.1. Major Yeasts

Hanseniaspora/Kloeckera

*Hanseniaspora* is a genus of apiculate yeasts whereas the name *Kloeckera* is applied to its anamorph form. Nowadays, the *Hanseniaspora/Kloeckera* group, which is naturally present in grapes, comprises ten species: *H. valbyensis, H. guilliermondii, H. uvarum, H. opuntiae, H. thailandica, H. meyeri, H. clermontiae; H. vineae, H. osmophila* and *H. occidentalis* [50]. This genus is widely found in grape must and is characterized by its low fermentative power but also for its production of wine volatile compounds and its contribution to wine complexity [51]. Although several groups of volatile molecules are produced in wines during their fermentation with *Hanseniaspora* spp., this genus has been characterized as high producer of volatile fatty acids, esters, aldehydes and sulfur compounds but low producer of higher alcohols [19]. The most characteristic compounds produced by *Hanseniaspora* spp. that confer positive aroma to wines are acetate esters (isoamyl acetate, ethyl hexanoate, ethyl caprylate, phenylethyl propionate, ethyl caprate, ethyl 9-decenoeate, ethyl acetate, phenethyl acetate, beta-phenylethyl acetate, benzyl acetate and 2PA) [50,52–54] and aldehydes, such as acetaldehyde, benzaldehyde, 4-ethylbenzaldehyde and benzene acetaldehyde [50]. In addition, some alcohols (glycerol, 1-pentanol, phenethyl alcohol, benzyl alcohol), carboxylic acids (hexanoic acid, octanoic acid) and terpenes (limonene) are implicated in the wine flavors [50,52].

At industrial scale, *H uvarum, H. vineae* and *H. guilliermondii* are the most appropriate species to achieve an intense wine flavor and aroma complexity [55]. As previously stated, these microorganisms can naturally appear and develop spontaneous fermentation or they can be inoculated in mixed fermentations with *S. cerevisiae*. These strains contribute with positive oenological properties to wines conferring mainly floral and fruity notes such as chocolate, fig and tobacco (*H. uvarum*) [56], fruity-sweet coconut and woody or vanilla aromas (*H. vineae*) [57] and rose and honey sensory markers (*H. guilliermondii*) [53]. In mixed cultures, they can contribute with an enhancement of the production of volatile compounds. For instance, *H. guilliermondii* contributed with higher levels of 2PA to wine and *H. uvarum* could increase their isoamyl acetate content, when inoculated respectively alone. On the other side, an increase of the content of other compounds such as methionol, acetic acid-3-(methylthio) propyl ester or 4-(methylthio)-1-butanol, among others, occurred when they were inoculated with *S. cerevisiae* [58]. Further, 2PA is one of the compounds more studied in terms of aroma implications within species of *Hanseniaspora*. Different researchers have found that the mixed culture of both, *H. vineae* and *H. uvarum* with *S. cerevisiae*, provoke a synergistic effect on the production of 2PA, enhancing their floral, fruity (banana, pear, apple or citric fruits, among others) and honey aromas [23,59]. In addition, mixed cultures of *H. guilliermondii* with *S. cerevisiae*, have shown an increase of higher alcohols, acetate esters and acetaldehyde, while a reduction of ethanol, hydrogen sulphide and ethyl esters, when *S. cerevisiae* was used alone [60]. Another species of this genus, in this case, *H. opuntiae* was evaluated in mixed culture and the sensory analysis of the resulting wine, showed a higher floral and sweet aroma. This increment was related to the major production of some compounds such as phenylethanol or 3-methyl-1-bu-tanol and minor levels of decanoic and octanoic acid [61]. Apiculate yeasts have sometimes been related with the release of unpleasant flavour compounds but as previously stated, they can positively influence aroma profile in certain cases [51]. Therefore, authors are cautious when considering this genus as high intra-strain variability is found regarding their production of aroma compounds [21,51].

Likewise, other parameters can influence aroma. Regarding the time of inoculation, a recent study tried to elucidate the differences in the volatile aroma compounds when occurring a sequential or a simultaneous fermentation of *H. uvarum* and *S. cerevisiae*. Kai et al. (2018) observed that volatile phenols and acetate ester levels were higher in sequential fermentation, suggesting that this could be linked to a population ratio *H. uvarum/S. cerevisiae* higher than 1 [62]. Another study proved that the simultaneous inoculation of *H. uvarum* and *S. cerevisiae* caused an increase of medium-chain fatty acid ethyl ester content, improving floral, berry, tropical and temperate fruity aromas whereas sequential inoculation also improved floral and tropical fruity traits but produced an unpleasant "nail polish" odor [63]. Furthermore, must or wine composition also influences aroma profile. In this sense, low initial levels of YAN in the case of Ecolly wine were related to higher levels of ethyl esters and fatty acids whereas elevated content of YAN in the case of Cabernet Sauvignon must, motivated the expression of ATF1 gene and thus, an increment of the acetate ester production [62].

Table 2 describes the influence of different *Hanseniaspora* strains in the wine aroma profile. In general and from a chemistry point of view, the aroma improvements are explained by the production of higher concentrations of acetate esters like 2PA and isoamyl acetate, terpenes, medium-chain fatty acid-ethyl esters, benzenoids and decrease of higher alcohols [55].

**Table 2.** Different aroma compounds produced by yeast which confers good characteristics and pleasant aromatic properties to wine.

| Yeast | Compounds | Matrix | Aroma (Odour Descriptor) | Ref. |
|---|---|---|---|---|
| *H. uvarum* and *C. stellata* | Benzyl alcohol | Cabernet sauvignon wine | Chocolate, fig and tobacco | [56] |
| *H. vineae* | Beta-phenylethyl acetate | Red wine from Uruguay (Tannat cultivar) | Intense fruity | [54,56] |
| *H. vineae* | P-hydroxybenzyl | Wine | Fruity, coconut, woody, vanilla | [56] |
| *H. guilliermondii* | Beta-phenylethyl acetate ester, 2PA | Wine | Rose, honey, fruity and flowery | [53] |
| *H. uvarum* and *H. guilliermondii* | 2-phenylethanol | Grape must from Douro, Portugal | Fruity and flowery | [58] |
| *H. uvarum* | Ethyl acetate | Wine | Fruity | [64] |
| *H. uvaum* | Terpenes, C13-norisoprenoids, volatile phenols, terpineol and linalool oxide | Ecolly and Cabernet Sauvignon wine | Tropical fruity and floral | [62] |
| *H. vinae* | 2PA, isoamyl acetate and esters | Chardonnay wine | Banana, pear, apple, citric fruits, guava | [65] |
| *H. vinae* | Phenyl ethyl acetate | Macabeo must | Fruity, floral and honey | [66] |
| *C. pulcherrima* | Ethyl acetate, Iso-amyl acetate | Wine | Fruity, sweet and banana-like | [17] |
| *C. zemplinina* | Hexyl acetate, ethyl hexanoate, ethyl heptanoate, ethyl dodecanoate and ethyl butanoate | Barbera wines | Apple, fruit, herb, sweet or waxy | [67] |
| *M. pulcherrima* | Phenol,2,6-dimethoxy | White wine | Smoky notes | [68] |

**Table 2.** *Cont.*

| Yeast | Compounds | Matrix | Aroma (Odour Descriptor) | Ref. |
|---|---|---|---|---|
| *R. mucillaginosa* | Terpenic compounds (b-damascenone, geraniol, citronellol, linalool, b-terpineol) | Irpinian wines (Aglianico and Fiano wines) | Floral, sweet and ripened fruit | [69] |
| *R. mucillaginosa* | Terpenols | Chinese wine | Fruity and floral | [70] |
| *R. mucillaginosa* | C6 compounds (1-hexanol) and faty acids | Chinese wine | Grass and unpleasant fatty | [70] |
| *R. mucillaginosa* | 3-hexene-1-ol, neroloxide, acetates and ethyl groups | Ecolly dry white wine | Citrus, sweet/acid fruit, berry, floral | [71] |
| *P. anomala* | Isoamyl acetate | Wine | Banana | [53] |
| *P. kluyveri* | 2PA, ethyl octanoate | Sparkling wine | Fruity, rose, sweet, honey flavors and pineapple, pear, soapy | [55,72] |
| *T. delbrueckii* | Ethyl butyrate, ethyl acetate, ethyl hexanoate and ethyl hexanoate | Sparkling wine | Fruity, sweet, pineapple, green apple, brandy, wine-like, strawberry | [22] |
| *T. delbrueckii* | Ethyl propanoate, ethyl isobutanoate, ethyl dihydrocinnamate and isobutyl acetate | Sauvignon blanc and Merlot must | Fruitiness and complexity | [26] |
| *T. delbrueckii* | Isoamyl acetate, hexyl acetate, ethyl hexanoate and ethyl octanoate | Juice from Syrah grapes | Fresh and fruity | [73] |
| *T. delbrueckii* | 3-sulfanylhexan-1-ol | Sauvignon Blanc grape must | Grapefruit/passion fruit | [74] |

Candida

The genus *Candida* is a collection of approximately 150 asporogenus yeast species from which 11% are agents of human infection since when they are ingested, they can enter to the bloodstream and cause fungaemia [75]. Besides, several studies show the significant impact of *Candida* spp. on the production of metabolites that affects the flavor and aroma of wines during fermentation on its own and together with *S. cerevisiae* [17,76]. Generally, it has been related to high production of esters, sulfur compounds and higher alcohols and low production of volatile fatty acids, aldehydes and volatile phenols [19].

Among this genus, the most known and studied Candida species could be *C. albicans*, *C. stellata* (reclassified as *C. zemplinina*) or *C. pulcherrima* (telemorphic form of *Metschnikowia pulcherrima*), among others [21,51]. All these species have been studied in different times. *C. albicans* was able to produce higher levels of farnesol and farnesene (gardenia/perfume aroma) [21]. *C. stellata* was found to intensify the apricot, honey and sauerkraut aromas when used alone in monoculture and increase the production of ethyl-acetate in sequential fermentation with *S. cerevisiae* on Chardonnay wine [77]. In addition, in a recent experiment, *C. zemplinina* was used in mixed cultures with *S. cerevisiae* and produce more aliphatic alcohols, certain aldehydes and ketones and esters (hexyl acetate, ethyl hexanoate, ethyl heptanoate, ethyl dodecanoate and ethyl butanoate) providing apple, fruit, herb, sweet or waxy aromas to wine [67].

*C. pulcherrima/M. pulcherrima* are commercial starters able to induce changes in wine's profile, especially in terpenes, volatile thiols and esters [51]. For example, it was observed that *C. pulcherrima* in mixed cultures produced higher levels of ethyl acetate and less undesirable volatile compounds [17], being ethyl acetate strongly linked to a fruity flavor in wines at levels of 0.2 g/L [78]. The quantity of isoamyl acetate formed by *C. pulcherrima* was significantly higher than that produced by non-saccharomyces yeasts in pure cultures exhibiting a sweet, fruity and banana-like aroma at levels upper 0.001 g/L [17]. In another study, *M. pulcherrima* non-flocculant strain AWRI305 was tested in mixed culture with *S. cerevisiae*. The study showed an increment in the concentration of esters (especially, ethyl acetate and 2-methylbutyl acetate) and sulfur compounds. In this case, compared with *S. cerevisiae* alone, these wines showed lower content in brown tint and higher in red fruit aroma [79]. Likewise, another study showed that *M. pulcherrima* (sequential culture with *S. cerevisiae*) produced higher content in higher alcohols (specially 3-methyl-1-butanol and 2-methylpropanol), lower amounts of acetaldehyde, a severe decrease of butyl acetate and quite higher production of volatile phenols. These changes motivated the perception of smoky and flowery notes by tasters [68]. In addition, *M. pulcherrima* has been related to the production of low-alcohol wines, a desirable characteristic for the wine industry [80]. In this context, other authors have proposed this species and *C. zeylanoides* for this purpose, as they were poorer sugar consumers and effectively reduced ethanol content. *M. pulcherrima* has shown high production of higher alcohols (isobutanol and 2-phenylathanol), ethyl propionate, ethyl acetate and diacetyle, when compared to other species thus, being potentially suitable as inoculum [81]. Other strains like *C. molischiana* could produce terpenols and alcohols from a glycoside matrix. It has been also described the production of aldehydes by *Candida krusei* and volatile phenols and sulfur compounds by other species of *Candida* genus [82].

#### 4.1.2. Minor Yeasts

Spontaneous grape-must fermentation can also begin with the growth of other minor species belonging to genera such as Rhodotorula or Pichia, among others. These yeasts with low fermentative capacity can confer wine flavor and aroma complexity by increasing the amounts of the volatile compounds responsible for the fruity aroma, through hydrolysis of aroma precursors caused by enzymatic activity [83]. Studies reported that glycosidases from minor yeast have also remarkable potential to improve aroma complexity and regional characteristics of wine [70]. Table 2 shows various examples of non-saccharomyces yeasts and their implication in wine aroma.

##### Rhodotorula

*Rhodotorula* spp. has been referred by some authors as high producer of esters and isoamyl acetate [19]. One of the most studied species is *Rhodotorula mucillaginosa* [21]. *R. mucillaginosa* possesses high extra-cellular glycosidase activity able to convert the glycosylated form of terpenes into aromatic compounds. For example, a general increase of terpene compounds (β-damascenone, geraniol, citronellol, linalool, β-terpineol) was observed in Aglianico and Fiano wines from Italy (Irpinian wines) [69]. The application potential of a Chinese strain of *R. mucillaginosa* to wine aroma enhancement was also reported [70]. In other case, an increase in the concentration of volatile compounds (neroloxide, alpha-terpineol, farnesol, limonene, linalool, citronellol, geraniol, geranyl acetone and nerolidol) was observed in samples treated with glycosidase extracts from *R. mucilaginosa*. Moreover, the enzyme treatments improved the content of volatile phenols, $C^6$ compounds (1-hexanol) and fatty acids. Terpenic compounds and benzene compounds are positive aromatic compounds, while $C^6$ compound, volatile phenols and fatty acids could release unpleasant aromas, depending on their concentration on the final wines [70]. A recent study also assessed the glycosidase activity and the main compounds related with the fermentative aroma produced by *R. mucillaginosa* which were 1-butanol, isoamyl alcohol, ethyl acetate, ethyl lactate and phenyl ethyl alcohol [71].

##### Pichia

The genus Pichia has been described as producer of esters, especially ethyl acetate and isoamyl acetate [19]. The yeast *Pichia kluyveri* is usually co-inoculated together with *S. cerevisiae* since it is unable to ferment to dryness on their own [84]. The use of *P. kluyveri* to increase the levels of terpenic compounds in sequential fermentations with *S. cerevisiae* had been previously reported [55]. This study also described that this yeast was able to produce high levels of esters, specially 2PA and ethyl octanoate [55]. The characteristic fruity, rose, sweet, honey flavors of wine and other grape-derived alcoholic beverages are primarily due to 2PA and ethyl octanoate, thus they provide pineapple, pear, soapy odors [72]. It has been also reported that *P. anomala* wine yeasts produce increased concentrations of esters with a fruity aroma. The yeast was found to be a potent isoamyl acetate producer and the characteristic banana-like aroma of wine was primarily due to this compound [53]. Other study showed that *P. kudriavzevii* in mixed cultures exhibit a chemical profile with higher levels of glycerol, ethyl acetate and isoamyl acetate and less content in fatty acids, higher alcohols and phenyl ethanol. These profile resulted in a floral, sweet and fruity aromas [61]. Similarly, a recent study showed that wines produced in simultaneous fermentation with *P. kudriavzevii* and *S. cerevisiae* had lower volatile acidity, higher amounts of esters and lower higher alcohols, fatty acids, benzene derivatives and $C^6$ compounds concentration. In addition, the aroma profile and whole flavor and quality were improved and wines obtained higher scores in fruity and floral aromas, appearance and mouthfeel [24].

Torulaspora

As aforementioned, other genera and species have been also widely studied for their implication in the aroma profile of wine. This is the case of *Torulaspora delbrueckii* which has shown positive impact on wine's aroma and increasingly importance [85]. *T. delbrueckii* (anamorph *Candida colliculosa*, synonym *Saccharomyces rosei*) [19] has shown strain variability regarding aroma profile and some of them exhibited low production of acetaldehyde and acetoin, both positive attributes. However, it produced small amounts of higher alcohols (being isoamyl alcohol and β-phenylethanol, the major compounds), acetate esters and ethyl esters of fatty acids. This way, it has been suggested that they slightly contribute to aroma complexity when compared to other non-saccharomyces organisms [85]. Nevertheless, a recent study investigated the effects of *T. delbrueckii* alone or in mixed culture with *S. cerevisiae* and their volatile compounds profile. In general, the presence of *T. delbrueckii* was associated with an increase of the fruity, sweet, pineapple, green apple, brandy, wine-like and strawberry sensory descriptors [22]. Other studies have related it to high production of isovaleric acid, ethyl propionate, 1-butanol and low production of acetic acid [81]. Also other authors confirm the low acetic acid production, while an increase in higher alcohols concentration (1-butanol) was observed [68]. Another study demonstrated that *T. delbrueckii* was related with higher concentrations of esters and differences were observed between mixed and sequential fermentation, promoting polyols synthesis (2,3-butanediol and 1,2-propanediol) and 1-butanol, 3-ethoxy-1-propanol and furaneol production, respectively [73]. At last, combinations of more than one species of non-saccharomyces yeasts have been also researched [19].

### 4.2. *Bacteria*

#### 4.2.1. Lactic Acid Bacteria

Given the specific fermentation conditions, high ethanol production, presence of sulfur dioxide and low pH and nutrients concentrations, the environment turns out to be inhospitable for most bacteria genera. Nevertheless, lactic acid bacteria (LAB) and acetic acid bacteria (AAB) have managed to survive. LAB are in charge of the malolactic fermentation (MLF), also known as secondary fermentation which takes place in most of red wines and some white wines, performing the enzymatic decarboxylation of L-malic acid to L-lactic acid and carbon dioxide. This deacidification of wine reduces the sour taste that an excess of malic acid could give [5,86]. Simultaneously, as a result of LAB activity, volatile compounds are released enhancing aroma complexity with fruity or buttery notes, and reducing others such as vegetal or grass aroma. MLF also contributes to the microbiological stability of wine, decreasing the possibilities of spoilage by unwanted microbiota [5,87]. Within the LAB group, researchers have identified four main genera, the bacilli *Lactobacillus* and three cocci, *Oenococcus*, *Pediococcus* and *Leuconostoc* [87,88].

Among these genera, *Oenococcus oeni* is species most usually linked to MLF due to its resistance to fermentation conditions in red and white wines. [89]. It shows a well-adapted response to highly acidic wine conditions and better enzymatic activity than other selected starters [90]. Within this genus barely three have been isolated in must, *O. oenis*, *O. alcoholitolerans* [91] and *O. kitaharae* [92]. Several examples of the implications of *Oenococcus* and other LAB species is compiled in Table 3.

**Table 3.** Different aroma compounds produced by lactic and acetic acid bacteria strains.

| Bacteria | Compounds | Matrix | Aroma | Ref. |
|---|---|---|---|---|
| **Lactic Acid Bacteria** | | | | |
| *Lactobacillus brevis* | Methanethiol | Merlot wine | Unpleasant sulfur aroma | [93] |
| | 3-(methylsulfanyl) propan-1-ol | | Meaty aroma (<10 μM) | |
| *Lactobacillus plantarum* | Linalool, 2 phenyl-ethanol, 2,3-butanediol, 4-terpineol and geraniol | Fiano wine | Floral, fruity and spicy aroma | [94] |
| *Lactobacillus plantarum* | Terpenes, limonene and linalool | Synthetic wine | Flowery-citric aroma | [95] |
| | Benzyl alcohol and b-phenyl-ethyl-alcohol | | Rose-like odor | |
| *Oenococcus oeni* and *Lactobacillus* spp. | Terpenes, norisoprenoids, phenols and vanillins | Synthetic wine | Alcohol and dried sensory descriptors. Fruity aroma | [96] |
| *Oenococcus oeni* and *Lactobacillus plantarum* | 2 phenyl-ethanol, 2,3-butanediol, ethyl-lactate, terpenes and vanillate derivatives | Shiraz wine | Fruity, floral, earthy/nutty aromas | [97] |
| *Oenococcus oeni* | Ethyl esters | Wine | Fruit-like | [98] |
| *Oenococcus oeni* | 2-phenylethanol, terpenes, lactic acid ethyl-ester and succinic acid, diethyl-ester | Riesling wine | Rose notes, fruity and floral notes | [99] |
| *Oenococcus oeni* | Hexanol, 3-methylbutylester, acid esters | Chardonnay wine | Green and herbaceous, banana notes and fruity aroma | |
| *Oenococcus oeni* | Substituted ethyl esters: i.e., (2S)-2-hydroxy-n-me-thylpentanoic acid | Merlot wine | Black-berry and jammy-fruit notes | [100] |
| *Oenococcus oeni* | Fruity esters and lower production of alcohols and terpenes | Black raspberry wine | Strong fruity and slight notes of solvent and herbaceous | [101] |
| *Leuconostoc* | Phenyl-ethyl acetate | Black glutinous rice wine | Sweet, floral aroma | [102] |
| *Leuconostoc* | 2,3-butanediol | | Buttery aroma | |
| LAB commercial starter | Diacetyl, ethyl acetate, ethyl lactate, mono-ethyl and diethyl succinate | Single-varietal red wines | Fruity, smoked/toasted. | [103] |
| **Acetic Acid Bacteria** | | | | |
| *Acetobacter* | Ethyl esters | Highland barley wine | Fruity, grape-like aroma | [104] |
| | Acetic acid | | Vinegar | |
| *Acetobacter aceti* | 2PA, 3-methyl butanol, ethyl acetate | Pineapple wine | Floral-fruity aroma | [105] |
| *Gluconobacter* | Tartaric and citric acid, ethyl esters | Black glutinous rice wine | Acid and fruity aroma | [106] |

*Lactobacillus* genus is represented by approximately 30 species. Among them, *L. plantarum*, *L. brevis*, *L. buchneri*, *L. hilgardii* and *L. fructivorans* are the most found in must and wine. Also, other species such as *L. bobalius*, *L. casei*, *L. collinoides*, *L. fermentum*, *L. kunkeei*, *L. lindneri*, *L. mali*, *L. nagelii*, *L. oeni*, *L. paracasei*, *L. paraplantarum*, *L. uvarum* and *L. vini* have been also found [88,107]. Among them, *L. plantarum* is the most-liked by winemakers, due to its particular qualities: less nutrition requirements, lower inoculum concentration, tolerance to ethanol, high pH and sulfur dioxide and also, their diverse collection of enzymes able to enhance aroma profile in wines, such as glycosidases or esterases, among others [107–109]. For instance, the esterase activity of some LAB strains has been related to the increase of red- and black-berry fruit-like aroma and jam-fruit aroma as well [100].

In addition, research has been focused on the leverage of co-culture between *O. oeni* and *L. plantarum*, reporting more aroma complexity in mixed cultures [97]. For instance, a work evaluated the use of *O. oeni* and *Lactobacillus* strains and found that terpenes, norisoprenoids, phenols and vanillins were released in association, in general, with alcohol and dried sensory descriptors whereas oxidize notes were linked to phenyl-acetaldehyde and phenyl-acetic acid concentrations [96]. Furthermore, other studies have researched the application of LAB (i.e., *O. oeni*) together with mixed cultures (i.e., *S. cerevisiae* with *T. delbrueckii*). Results showed that volatile composition was quite different and scored better for global aroma than spontaneous MLF, which enhanced both pleasant and off-flavors [101]. Another study stated that *T. delbrueckii* together with *S. cerevisiae* (sequential fermentation) created more MLF favorable conditions, since lower levels of sulfur dioxide and medium chain fatty acids, promoting the development of *O. oeni* [110].

Regarding the aroma compounds related to LAB, diacetyl is the most important one. This compound is produced as a result of citric acid metabolism and can be further metabolized to 2,3-butanediol [5,88]. Therefore, citrate lyase enzyme plays an important role regulating the production of diacetyl. At low concentrations, it is related to yeasty, nutty and toasty aromas, whereas at high levels it produces sweet, buttery, creamy or milky aroma, sometimes linked to off-flavors. In this regard, some strains of *L. plantarum* do not present citrate lyase complex genes, and thus it is feasible to obtain wines with low diacetyl concentrations [5,109]. In addition, the perception of diacetyl is influenced by several factors such as the chemical composition of wines, the strain of LAB and the presence of sulfur dioxide, which can interact with diacetyl, decreasing wine's volatility [103,109].

Apart from Oenococcus and Lactobacillus, different pediococci species including *P. damnosus*, *P. inopinatus*, *P. parvulus* and *P. pentosaceus* have been isolated from wines. Among them, *P. parvulus* and *P. damnosus* are more commonly found in must and wine, due to its undesirable effects in wine, being associated with unpleasant aromas, bitterness and ropiness [88]. For instance, a study performed in pinot noir wine with *P. inopinatus*, *P. parvulus* and *P. pentosaceus* reported floral and fruit-like aromas [111]. It has also been reported the existence of other species, such as the recombinant strain, *P. acidiltactici* BD16, which could improve the aroma of wine due to the production of phenolics derived from MLF [112].

*Leuconostoc* strains are known by dominating the initial fermentation stages, conducting the MLF alongside with *Lactobacillus* and *Oenococcus*. Nonetheless, as acids levels raise, *Leuconostoc* is overcome by more acid tolerant *Lactobacillus* [113]. Further, its enzymatic activity involved in flavor and aroma has been barely studied; it has been reported protease and also citrate lyase activity [114,115]. In addition, *L. paramesenteroides* was renamed as *Weissella paramesenteroides* [88]. *L. mesenteroides* is the current dominant *Leuconostoc* strain in grape juice and must. More recent studies have positively correlated the presence of *Leuconostoc* sp. with floral and buttery-like aromas [102].

#### 4.2.2. Acetic Acid Bacteria

Acetic acid bacteria (AAB) belong to the family Acetobacteraceae. They are classified as aerobic strict gram-negative bacteria, well adapted to sugar and high ethanol environments and able to oxidize ethanol to acetic acid. AAB have been spotted on grapes and red wine; being notably higher in damage and rotten grapes and they are mainly classified in *Acetobacter* or *Gluconobacter* [116]. In contrast to LAB, AAB presence is less desirable in winemaking; they are considered as spoilage organisms due to the formation of acetaldehyde and acetic acid, among other spoilage compounds [117]. In general, low acetic acid concentration provides vinegar-like sourness, nutty and sherry-like aroma to wine, associated with a reduction in fruity characters, but as the concentration raises, the effect is replaced by an unpleasant smell [117]. The sensory threshold for acetic acid becoming undesirable depends on the wine type, e.g., in Canadian sweet wines is 1.0–1.5 g/L with an allowed maximum of 2.1 g/L [118], while in dry wine, the concentration must not exceed 1.0 g/L [119]. Another compound related to AAB metabolism that could affect wine quality is ethyl acetate, which can positively contribute to wine with floral or fruity-like aroma, though when it exceeds a specific threshold, it is also considered an undesirable compound. However, researchers have not reach an agreement about the establishment of that threshold [8].

*Acetobacter* are gram negative rods with an over-oxidative capacity, being able to oxidize ethanol to acetic acid to $CO_2$ and water [119]. Generally, *A. aceti* and *A. pasteurianus* are the most often species isolated in wine [120,121]. Further, other related species are *A. cerevisiae*, *A. malorum*, *A. tropicalis* [121] or *A. rancens* and *A. suboxydans* found in Indian palm wine [122]. *Acetobacter* appearance is related to a bitter aroma and acid flavor, due to the excessive acetic acid concentration. Other aroma compounds as hexa- and octadecanoic acid ethyl esters, acetaldehyde, propionic and succinic acid have been correlated with *Acetobacter* metabolism [104].

*Gluconobacter* are also gram-negative rods, strictly aerobes, but unlike *Acetobacter* they are not able to oxidize acetate and lactate to carbon dioxide. *Gluconobacter* strains are frequently detected in grapes and can persist during the fermentation despite being relatively weak acetic acid resistant and less ethanol tolerant, since, they can be even inhibited by high levels of alcohol [106,123,124]. The presence of *Gluconobacter* has been positively correlated with the appearance of butanoic, lactic, citric and tartaric acids and other compounds such as benzyl alcohol, octanoic acid ethyl ester or ethyl 9-decenoate, among others [106]. However, it is worth mentioned that all these species together with *Acetobacter* species are majorly considered as spoilage microorganisms in wine, not used as starters and so, few investigation has been developed on their aroma implications beyond their acetic acid production.

## 5. Strain Dependent Variability and Genetics Influence on Aroma Profile

Variability on aroma profile has been related to different factors such as soil, grapes, climate, type of fermentation, medium and involved microorganisms, among others [1]. Focusing on the microorganisms and their implication in secondary aroma, genetics helps to understand the origin of these changes based on strains genotypes and phenotypes. In this regard, it is essential to mentioned *S. cerevisiae*, as it is considered as the best understood genetic model organism and the first eukaryote genome completely sequenced, the best annotated and also likely to be genetically manipulated and analyzed [40]. *S. cerevisiae* has shown genetic divergence, as the phenotypes that are currently used have demonstrated different characteristics related to wine production, such as resistance to sulfites [125]. This fact is also explained in aroma terms, as it has been shown that wild strains of *S. cerevisiae* and other Saccharomyces species produced earthy and sulfurous characteristics, whereas wine domesticated strains produced fruity and floral notes [125]. In this sense, the development of "omics" technologies and the improvement of high throughput sequencing has deeply contributed to further study the microbial community of wines and thus, its implication in wine aroma [126]. These advances have been mostly directed towards different objectives: (1) mapping yeasts and bacteria genomes to identify new genetic variants that are responsible for desirable aroma characteristics, (2) using analytical techniques to isolate, identify and quantify volatile compounds involved in aroma profile and (3) modifying yeast and bacteria strains to obtain a specific character or ability. Further, progress is aimed at transcriptomics, proteomics, exometabolomics, etc. studies [126,127].

As previously stated, yeasts have been historically selected according to different characteristics: ethanol tolerance, low residual sugars levels, low volatile acids production, low nitrogen consumption or high growth rate. Frequently, these features come determined by multiple quantitative trait loci (QTL), i.e., regions linked to certain phenotypic traits [128,129]. For instance, a recent study, found in *S. cerevisiae* 51 potential QTLs related to the production of monoterpenes and found that three of them (UDV060, VLG19-I-1 and VLG3-A-1) on three different chromosomes were placed closely to genes connected to the production of aroma compounds [130]. In addition, other studies have investigated different genetic mechanisms that affect wine aroma: changes in transcription levels of ALD6 gene (involved in the conversion of acetaldehyde into acetic acid), haploinsufficiency effects on YFL040W related to acetic acid and glycerol and succinic acid production or the epistatic gene-gene interactions resulting in heterosis of FLX1 and MDH2, two genes associated to succinic acid production [128]. Other studies have tried to integrate different omics, thus identifying new genes related to wine aroma and flavor in different strains of *S. cerevisiae* and confirming the production of fatty acids and ethyl and acetate esters by using high performance liquid chromatography (HPLC), gas chromatography-mass spectrometry (GC-MS) and microarray techniques [131]. Besides, research has been performed on the influence of nitrogen availability and related gene expression of *S. cerevisiae*. It was found that depending on nitrogen levels, a total of 46 genes were up-or down-regulated, proposing some of them to be used as molecular markers. In this sense, potentially different strains could be used to obtain different aromas [27,132]. In the same context, other authors have investigated the effects on volatile compounds production and gene expression of *S. cerevisiae* when adding branched-chain amino acids to must. Different genes were identified and associated to yeast growth and amino acids transport; also 25 metabolites (higher alcohols, esters, fatty acids and branched-chain amino acid) were detected, among which 2,3-butanediol and ethyl lactate levels were highly increased. Therefore, it was suggested that the addition of branched chain amino acids was able to enhance aroma complexity [133]. All these techniques and studies have been developed to characterise different strains of *S. cerevisiae* that are tightly connected to differences in aroma compounds production, such in the case of a Gewürztraminer wine where a specific strain was able to produce increased amounts of 2-phenylethanol and cis-rose oxide and the most complex aroma profile [134].

On the other hand, transcriptional analyses have been also carried out on non-saccharomyces species mixed cultures with *S. cerevisiae*. It has been shown that culture in consortium of *S. cerevisiae* with other species can modify the genome transcriptional response of *S. cerevisiae* and differently express specific genes that encode for enzymes linked to the production of aroma compounds [60]. For instance, several studies have investigated *T. delbrueckii*. It has been demonstrated that the mixed cultures of these two species stimulates the transcription of some genes, such as those implied in the glucofermentative pathway, thus producing higher amounts of $CO_2$ [135]. More recently, a transcriptome analysis of the same species revealed that the *T. delbrueckii* lower production of higher alcohols and acetate esters was explained by the absence of transcripts of key enzymes in those pathways whereas low levels of ethyl esters were related to down-regulation of fatty acids biosynthesis genes [136]. At last, a study bared that the presence of *T. delbrueckii* affect the transcriptional and phenotypic response of *S. cerevisiae* to ammonium nutrition by reducing its global effects. This way, mixed cultures produced higher concentrations of esters (i.e., acetic acid ethyl ester and lactic acid ethyl ester), providing fruity aroma to wine [137].

At last, it is also worth mention that according to the previously mention objective 3, other approaches have been explored for modifying yeasts to achieve a specific characteristic. In this sense, genetic modified organisms (GMOs) have been developed to fulfil those requirements but also other methodologies have been used to generate enhanced wine organisms not considered as GMO, such as clonal selection, random mutagenesis or sexual hybridization [129]. Further, research has been focused on grapes and its genetic base regarding the synthesis of aroma compounds during fermentation [138]. Therefore, the development of new "omics" technologies and related sciences is necessary for the elucidation of the transcriptional and genetic mechanisms involved in wine aroma formation.

## 6. Future Perspectives and New Approaches

The fermentation of grape must and the production of premium quality wines is a complex biochemical process that involves the interactions of enzymes from many different microbial species, but mainly yeasts and LAB [139]. In recent years, the oenological industry has undergone an important transformation, becoming a sector with constant changes and innovations. As it is described before, non-saccharomyces yeasts can positively influence aroma [17,55,76]. This quality improvement allows the production of innovative and differentiated wines. These yeasts can be introduced into the winemaking process to obtain differentiated wines that reflect the characteristics of a specific region. In this context, the study of the use of non-saccharomyces autochthonous cultures to produce wines with particular oenological and sensory characteristics, would allow to choose suitable candidates to be included in commercial mixed cultures [82]. The presence of the non-saccharomyces species during the alcoholic fermentation might be of technological interest, but further studies on these yeasts for their biotechnological applications in winemaking are needed since the commercial assortment of non-saccharomyces cultures is still reduced [82]. It is yet possible to acquire some interesting species like *Torulaspora delbrueckii*, *Lachancea thermotolerans*, *Metschnikowia pulcherrima*, *Schizosaccharomyces pombe* and *Pichia kluyveri*. Other strains like *Starmerella bacillaris*, *Meyerozyma guilliermondii* and *Hanseniospora* spp., will probably be on market in coming years [55]. The use of non-saccharomyces yeasts can be also remarkable in regions where grape harvesting is put forward due to excessive rainfall and where the grape may contain insufficient amounts of aromatic compounds [70]. Bibliography shows big differences, depending on the non-saccharomyces strain employed due to the genomic diversity of those species and the importance of performing selective processes, such as those that were conducted for *S. cerevisiae* strains in the past [55]. In this sense, one parameter to consider is to identify yeast strains with a high level of β-glucosidase activity and to evaluate the hydrolysis characteristics of its enzyme extract.

Futures perspectives in the use of non-saccharomyces yeasts also aim to produce wines with lower alcohol content than those from pure *Saccharomyces* spp. starters [140]. Nowadays, consumers demand wine with low level of alcohol. Following this trend, winemakers search alternative methodologies to reduce the final content of ethanol in wine, especially in vineyards from warm climates where the grape over-ripening can occur giving an increase of sugar levels [140]. Inoculation of different non-saccharomyces yeast strains have been proposed for lowering alcohol levels in wine (<2%, depending on the yeast species and fermentation conditions) [55,141]. Different yeast species like *Hanseniaspora/Kloeckera* spp., *Pichia* spp. or *Candida* spp., which are predominant in the first stages of fermentation (up to 6% of alcohol content), consume sugars by respiration rather than fermentation. In this sense, non-saccharomyces species allow to reduce the initial ethanol content and would produce desirable levels of secondary metabolites, which will affect aroma profile [141]. Enzyme or osmotic filtration is another alternative strategy which can be used to reduce the content of ethanol in wine [140].

Nowadays, the production of efficient malolactic starter cultures has become another main challenge for oenological research [142]. There are several parameters to address when selecting LAB for possible use in a starter culture, such us their tolerance to acid conditions, high ethanol and $SO_2$ concentrations, their compatibility with the selected yeast strains, adequate growth characteristics under winemaking conditions, the inability to produce biogenic amines and the lack of off-flavor or off-odor production [142]. Recent research highlights the importance of choosing specific LAB strains to obtain the desired wine, as specific flavors such as ethyl ester, volatile sulfides and glycosidic aroma compounds have been associated with specific strains. Since GMOs are not widely accepted by consumers, research is focused on identifying strains that can be used to modulate the aroma and flavor compounds of wine [88]. Finally, it is important to highlight that wine aroma is complex and contain an enormous chemical diversity. In this context, research should be also focused in developing simpler and non-targeted LC-MS methodology to study the entire volatile fraction of wine metabolome as well as other approaches such 2D GC, which has previously helped to solve the aroma on many complex matrices, including wine [8,143].

## 7. Conclusions

Wine secondary aroma is complex and comes determined from the diversity of different chemical compounds. In the process of aroma formation, different factors play an important role. Different types of fermentation (single or mixed culture) and different strategies of inoculation (simultaneous or sequential) have shown differences on the final aroma profile. Further, other factors such as YAN, temperature, oxygen or time affect the sensory characteristics of wine. Regarding the whole chemical diversity found in wine and in particular, in volatile aroma compounds, those related with secondary or fermentative aroma are mainly higher alcohols and esters, together with volatile fatty acid and volatile phenols, and thus, they are the compounds mostly studied in research articles regarding wine aroma. In the last years, winemaking industry has undergone important transformations and despite *S. cerevisiae* is still used in production purposes for its desirable characteristics, non-saccharomyces yeasts have been highlighted as organisms that can positively influence aroma profile. According to provided data, there is a huge diversity of non-saccharomyces yeasts that can enhance or decrease the production of some aroma compounds, resulting in specific aroma attributes that are evaluated from a sensory point of view. Besides, MLF can affect wine aroma since LAB are tightly connected to the production of higher alcohols, esters and terpenes, together with norisoprenoids, phenols and vanillate derivatives in minor quantities. The main challenge still is to characterize their enzymatic activities and related genes. AAB are also revised since as spoilage microorganisms, they can negatively alter the aroma profile, although it has also been suggested that they can contribute with positive traits such as floral or fruity. Therefore, taking into account the diversity of yeasts and bacteria species and the necessities of the winemaking sector, genetics, transcriptomics and other sciences, aimed at decoding the strain dependent variability of species and its implications on wine aroma, are fundamental for the focused use of microorganisms and the achievement of wines with higher aroma complexity and pleasant characteristics that can fulfil the requirements of the consumers.

**Author Contributions:** Conceptualization, M.C., M.A.P. and J.S.-G.; methodology, M.C., M.F.-C., P.O., R.A.N., P.G.-O. and M.A.P.; formal analysis, M.C., M.F.-C., P.O., R.A.N., P.G.-O. and M.A.P.; investigation, M.C., M.F.-C., P.O., R.A.N. and P.G.-O.; writing—original draft preparation, M.C., M.F.-C., P.O., R.A.N., P.G.-O. and M.A.P.; writing—review and editing, M.C., M.A.P. and J.S.-G.; supervision, M.C., M.A.P. and J.S.-G.; project administration, M.C., M.A.P. and J.S.-G. All authors have read and agreed to the published version of the manuscript.

**Funding:** The research leading to these results was supported by FEDER under the program Interreg V-A Spain-Portugal (POPTEC) 2014–2020 ref. 0377_IBERPHENOL_6_E and ref. 0181_NANOEATERS_01_E; by Xunta de Galicia supporting with the Axudas Conecta Peme the IN852A 2018/58 NeuroFood Project and the program EXCELENCIA-ED431F 2020/12; by EcoChestnut Project (Erasmus+ KA202) that supports the work of M. Carpena; by Ibero-American Program on Science and Technology (CYTED—AQUA-CIBUS, P317RT0003) and by the Bio Based Industries Joint Undertaking (JU) under grant agreement No 888003 UP4HEALTH Project (H2020-BBI-JTI-2019), the JU receives support from the European Union's Horizon 2020 research and innovation program and the Bio Based Industries Consortium.

**Institutional Review Board Statement:** Not applicable.

**Informed Consent Statement:** Not applicable.

**Data Availability Statement:** No new data were created or analyzed in this study. Data sharing is not applicable to this article.

**Acknowledgments:** The research leading to these results was supported by MICINN with the Ramón&Cajal grant for M. A. Prieto (RYC-2017-22891) and the Juan de la Cierva_incorporación grant for P. Otero (IJCI-2016-27774); by Xunta de Galicia and University of Vigo supporting the post-doctoral grant of M. Fraga-Corral (ED481B-2019/096) and the pre-doctoral grant for P. García-Oliveira (ED481A-2019/295).

**Conflicts of Interest:** The authors declare no conflict of interest.

## Abbreviations

| | |
|---|---|
| LAB | Lactic Acid Bacteria |
| YAN | Yeast Assimilable Nitrogen |
| AAB | Acetic Acid Bacteria |
| 2PA | 2-Phenylethyl Acetate |
| MLF | Malolactic Fermentation |
| GC/MS | Gas Chromatography-Mass Spectrometry |
| QTL | Quantitative Trait Loci |
| HPLC | High Performance Liquid Chromatography |
| GMOs | Genetic Modified Organisms |

## References

1. Pereira, A.G.; Fraga, M.; Garcia-Oliveira, P.; Carpena, M.; Jimenez-Lopez, C.; Lourenço-Lopes, C.; Barros, L.; Ferreira, I.C.F.R.; Prieto, M.A.; Simal-Gandara, J. Management of Wine Aroma Compounds: Principal Basis and Future Perspectives. In *Winemaking-Stabilization, Aging Chemistry and Biochemistry*; IntechOpen: London, UK, 2020.
2. Perestrelo, R.; Silva, C.; Gonçalves, C.; Castillo, M.; Câmara, J.S. An Approach of the Madeira Wine Chemistry. *Beverages* **2020**, *6*, 12. [CrossRef]
3. Jeromel, A.; Korenika, A.-M.J.; Tomaz, I. 6—An Influence of Different Yeast Species on Wine Aroma Composition. In *Fermented Beverages*; Grumezescu, A.M., Holban, A.M.B.T.-F.B., Eds.; Woodhead Publishing: Cambridge, UK, 2019; pp. 171–285, ISBN 978-0-12-815271-3.
4. Ruiz, J.; Kiene, F.; Belda, I.; Fracassetti, D.; Marquina, D.; Navascués, E.; Calderón, F.; Benito, A.; Rauhut, D.; Santos, A.; et al. Effects on varietal aromas during wine making: A review of the impact of varietal aromas on the flavor of wine. *Appl. Microbiol. Biotechnol.* **2019**, *103*, 7425–7450. [CrossRef] [PubMed]
5. Styger, G.; Prior, B.; Bauer, F.F. Wine flavor and aroma. *J. Ind. Microbiol. Biotechnol.* **2011**, *38*, 1145–1159. [CrossRef] [PubMed]
6. Rapp, A. Volatile flavour of wine: Correlation between instrumental analysis and sensory perception. *Food Nahr.* **1998**, *42*, 351–363. [CrossRef]
7. Rapp, A.; Mandery, H. Wine aroma. *Experientia* **1986**, *42*, 873–884. [CrossRef]
8. Belda, I.; Ruiz, J.; Esteban-Fernández, A.; Navascués, E.; Marquina, D.; Santos, A.; Moreno-Arribas, M.V. Microbial Contribution to Wine Aroma and Its Intended Use for Wine Quality Improvement. *Molecules* **2017**, *22*, 189. [CrossRef]
9. Ribâereau-Gayon, P.; Dubourdieu, D.; Donáeche, B. *Handbook of Enology: Volume 1*; John Wiley & Sons: Hoboken, NJ, USA, 2006.
10. Marín-San Román, S.; Rubio-Bretón, P.; Pérez-Álvarez, E.P.; Garde-Cerdán, T. Advancement in analytical techniques for the extraction of grape and wine volatile compounds. *Food Res. Int.* **2020**, *137*, 109712. [CrossRef]
11. Blanco-Padilla, A.; Soto, K.M.; Hernández Iturriaga, M.; Mendoza, S. Food antimicrobials nanocarriers. *Sci. World J.* **2014**, *2014*, 1–11. [CrossRef]
12. Van Wyk, N.; Grossmann, M.; Wendland, J.; Von Wallbrunn, C.; Pretorius, I.S. The Whiff of Wine Yeast Innovation: Strategies for Enhancing Aroma Production by Yeast during Wine Fermentation. *J. Agric. Food Chem.* **2019**, *67*, 13496–13505. [CrossRef]
13. Fariña, L.; Villar, V.; Ares, G.; Carrau, F.; Dellacassa, E.; Boido, E. Volatile composition and aroma profile of Uruguayan Tannat wines. *Food Res. Int.* **2015**, *69*, 244–255. [CrossRef]
14. Carpena, M.; Pereira, A.G.; Prieto, M.A.; Simal-Gandara, J. Wine aging technology: Fundamental role of wood barrels. *Foods* **2020**, *9*, 1160. [CrossRef] [PubMed]
15. Tao, Y.; García, J.F.; Sun, D.W. Advances in Wine Aging Technologies for Enhancing Wine Quality and Accelerating Wine Aging Process. *Crit. Rev. Food Sci. Nutr.* **2014**, *54*, 817–835. [CrossRef] [PubMed]
16. Fraga-Corral, M.; García-Oliveira, P.; Pereira, A.G.; Lourenço-Lopes, C.; Jimenez-Lopez, C.; Prieto, M.A.; Simal-Gandara, J. Technological application of tannin-based extracts. *Molecules* **2020**, *25*, 614. [CrossRef] [PubMed]
17. Zohre, D.E.; Erten, H. The influence of *Kloeckera apiculata* and *Candida pulcherrima* yeasts on wine fermentation. *Process. Biochem.* **2002**, *38*, 319–324. [CrossRef]
18. Tenorio, C.; Gutie, A.R. Analysis of yeast population during spontaneous alcoholic fermentation: Effect of the age of the cellar and the practice of inoculation. *Int. J. Food Microbiol.* **2005**, *103*, 49–56. [CrossRef]
19. Padilla, B.; Gil, J.V.; Manzanares, P. Past and future of non-Saccharomyces yeasts: From spoilage microorganisms to biotechnological tools for improving wine aroma complexity. *Front. Microbiol.* **2016**, *7*, 411. [CrossRef]
20. Ciani, M.; Comitini, F. Yeast interactions in multi-starter wine fermentation. *Curr. Opin. Food Sci.* **2015**, *1*, 1–6. [CrossRef]

21. Rossouw, D.; Bauer, F.F. Exploring the phenotypic space of non-*Saccharomyces* wine yeast biodiversity. *Food Microbiol.* **2016**, *55*, 32–46. [CrossRef]
22. Canonico, L.; Comitini, F.; Ciani, M. *Torulaspora delbrueckii* for secondary fermentation in sparkling wine production. *Food Microbiol.* **2018**, *74*, 100–106. [CrossRef]
23. Zhang, B.; Xu, D.; Duan, C.; Yan, G. Synergistic effect enhances 2-phenylethyl acetate production in the mixed fermentation of *Hanseniaspora vineae* and *Saccharomyces cerevisiae*. *Process. Biochem.* **2020**, *90*, 44–49. [CrossRef]
24. Shi, W.K.; Wang, J.; Chen, F.S.; Zhang, X.Y. Effect of *Issatchenkia terricola* and *Pichia kudriavzevii* on wine flavor and quality through simultaneous and sequential co-fermentation with *Saccharomyces cerevisiae*. *LWT* **2019**, *116*, 1–9. [CrossRef]
25. Maturano, Y.P.; Mestre, M.V.; Kuchen, B.; Toro, M.E.; Mercado, L.A.; Vazquez, F.; Combina, M. Optimization of fermentation-relevant factors: A strategy to reduce ethanol in red wine by sequential culture of native yeasts. *Int. J. Food Microbiol.* **2019**, *289*, 40–48. [CrossRef] [PubMed]
26. Renault, P.; Coulon, J.; de Revel, G.; Barbe, J.C.; Bely, M. Increase of fruity aroma during mixed *T. delbrueckii/S. cerevisiae* wine fermentation is linked to specific esters enhancement. *Int. J. Food Microbiol.* **2015**, *207*, 40–48. [CrossRef] [PubMed]
27. Mina, M.; Tsaltas, D. Contribution of Yeast in Wine Aroma and Flavour. In *Yeast—Industrial Applications*; IntechOpen: London, UK, 2017; pp. 117–134.
28. Marsit, S.; Dequin, S. Diversity and adaptive evolution of Saccharomyces wine yeast: A review. *FEMS Yeast Res.* **2015**, *15*, 1–12. [CrossRef]
29. Longin, C.; Petitgonnet, C.; Guilloux-Benatier, M.; Rousseaux, S.; Alexandre, H. Application of flow cytometry to wine microorganisms. *Food Microbiol.* **2017**, *62*, 221–231. [CrossRef]
30. National Center for Biotechnology NCBI Taxonomy Browser. Available online: https://www.ncbi.nlm.nih.gov/Taxonomy/Browser/wwwtax.cgi (accessed on 15 November 2020).
31. Schreier, P.; Jennings, W.G. Flavor composition of wines: A review. *Crit. Rev. Food Sci. Nutr.* **1979**, *12*, 59–111. [CrossRef]
32. Louw, L.; Tredoux, A.G.J.; Van Rensburg, P.; Kidd, M.; Naes, T.; Nieuwoudt, H.H. Fermentation-derived aroma compounds in varietal young wines from South Africa. *S. Afr. J. Enol. Vitic.* **2010**, *31*, 213–225. [CrossRef]
33. Muñoz, D.; Peinado, R.A.; Medina, M.; Moreno, J. Higher alcohols concentration and its relation with the biological aging evolution. *Eur. Food Res. Technol.* **2006**, *222*, 629–635. [CrossRef]
34. Tokpohozin, S.E.; Fischer, S.; Becker, T. Selection of a new Saccharomyces yeast to enhance relevant sorghum beer aroma components, higher alcohols and esters. *Food Microbiol.* **2019**, *83*, 181–186. [CrossRef]
35. Stribny, J.; Gamero, A.; Pérez-Torrado, R.; Querol, A. Saccharomyces kudriavzevii and Saccharomyces uvarum differ from Saccharomyces cerevisiae during the production of aroma-active higher alcohols and acetate esters using their amino acidic precursors. *Int. J. Food Microbiol.* **2015**, *205*, 41–46. [CrossRef]
36. Canonico, L.; Solomon, M.; Comitini, F.; Ciani, M.; Varela, C. Volatile profile of reduced alcohol wines fermented with selected non-Saccharomyces yeasts under different aeration conditions. *Food Microbiol.* **2019**, *84*, 103247. [CrossRef] [PubMed]
37. Pretorius, I.S.; Curtin, C.D.; Chambers, P.J. Designing wine yeast for the future. In *Advances in Fermented Foods and Beverages: Improving Quality, Technologies and Health Benefits*; Woodhead Publishing: Cambridge, UK, 2015; pp. 197–226, ISBN 9781782420248.
38. Jouhten, P.; Ponomarova, O.; Gonzalez, R.; Patil, K.R. Saccharomyces cerevisiae metabolism in ecological context. *FEMS Yeast Res.* **2016**, *16*, 1–8. [CrossRef] [PubMed]
39. Jagtap, U.B.; Jadhav, J.P.; Bapat, V.A.; Pretorius, I.S. Synthetic biology stretching the realms of possibility in wine yeast research. *Int. J. Food Microbiol.* **2017**, *252*, 24–34. [CrossRef] [PubMed]
40. Goddard, M.R.; Greig, D. Saccharomyces cerevisiae: A nomadic yeast with no niche? *FEMS Yeast Res.* **2015**, *15*, 1–6. [CrossRef]
41. Suárez-Lepe, J.A.; Morata, A. New trends in yeast selection for winemaking. *Trends Food Sci. Technol.* **2012**, *23*, 39–50. [CrossRef]
42. Capozzi, V.; Garofalo, C.; Chiriatti, M.A.; Grieco, F.; Spano, G. Microbial terroir and food innovation: The case of yeast biodiversity in wine. *Microbiol. Res.* **2015**, *181*, 75–83. [CrossRef]
43. Hirst, M.B.; Richter, C.L. Review of aroma formation through metabolic pathways of Saccharomyces cerevisiae in beverage fermentations. *Am. J. Enol. Vitic.* **2016**, *67*, 361–370. [CrossRef]
44. Di Gianvito, P.; Perpetuini, G.; Tittarelli, F.; Schirone, M.; Arfelli, G.; Piva, A.; Patrignani, F.; Lanciotti, R.; Olivastri, L.; Suzzi, G.; et al. Impact of Saccharomyces cerevisiae strains on traditional sparkling wines production. *Food Res. Int.* **2018**, *109*, 552–560. [CrossRef]
45. Capece, A.; Granchi, L.; Guerrini, S.; Mangani, S.; Romaniello, R.; Vincenzini, M.; Romano, P. Diversity of Saccharomyces cerevisiae strains isolated from two Italian wine-producing regions. *Front. Microbiol.* **2016**, *7*, 1–11. [CrossRef]
46. Comitini, F.; Capece, A.; Ciani, M.; Romano, P. New insights on the use of wine yeasts. *Curr. Opin. Food Sci.* **2017**, *13*, 44–49. [CrossRef]
47. Knight, S.; Klaere, S.; Fedrizzi, B.; Goddard, M.R. Regional microbial signatures positively correlate with differential wine phenotypes: Evidence for a microbial aspect to terroir. *Sci. Rep.* **2015**, *5*, 1–10. [CrossRef] [PubMed]
48. Rollero, S.; Bloem, A.; Camarasa, C.; Sanchez, I.; Ortiz-Julien, A.; Sablayrolles, J.M.; Dequin, S.; Mouret, J.R. Combined effects of nutrients and temperature on the production of fermentative aromas by Saccharomyces cerevisiae during wine fermentation. *Appl. Microbiol. Biotechnol.* **2015**, *99*, 2291–2304. [CrossRef] [PubMed]
49. Gonzalez, R.; Morales, P. Wine secondary aroma: Understanding yeast production of higher alcohols. *Microb. Biotechnol.* **2017**, *10*, 1449–1450. [CrossRef] [PubMed]

50. Martin, V.; Valera, M.J.; Medina, K.; Boido, E.; Carrau, F. Oenological Impact of the Hanseniaspora/Kloeckera Yeast Genus on Wines—A Review. *Fermentation* **2018**, *4*, 76. [CrossRef]
51. Jolly, N.P.; Varela, C.; Pretorius, I.S. Not your ordinary yeast: Non-Saccharomyces yeasts in wine production uncovered. *FEMS Yeast Res.* **2014**, *14*, 215–237. [CrossRef] [PubMed]
52. Hu, L.; Wang, J.; Ji, X.; Liu, R.; Chen, F. Selection of non-Saccharomyces yeasts for orange wine fermentation based on their enological traits and volatile compounds formation. *J. Food Sci. Technol.* **2018**, *55*, 4001–4012. [CrossRef] [PubMed]
53. Rojas, V.; Manzanares, P.; Gil, V.; Pinaga, F. Studies on acetate ester production by non-Saccharomyces wine yeasts. *Int. J. Food Microbiol.* **2001**, *70*, 283–289. [CrossRef]
54. Carrau, F. Native yeasts for low input winemaking: Searching for wine diversity and increased complexity. In *Advances in Applied Microbiology V. 111*; Academic Press: Cambridge, MA, USA, 2006; pp. 33–39.
55. Benito, Á.; Calder, F.; Benito, S. The Influence of Non-Saccharomyces Species on Wine Fermentation Quality Parameters. *Fermentation* **2019**, *5*, 54. [CrossRef]
56. Martin, V.; Giorello, F.; Farin, L.; Minteguiaga, M.; Salzman, V.; Boido, E.; Aguilar, P.S.; Gaggero, C.; Dellacassa, E.; Mas, A.; et al. De Novo Synthesis of Benzenoid Compounds by the Yeast Hanseniaspora vineae Increases the Flavor Diversity of Wines. *J. Agric. Food Chem.* **2016**, *64*, 4574–4583. [CrossRef]
57. Martin, V.; Boido, E.; Giorello, F.; Mas, A.; Dellacassa, E.; Carrau, F. Effect of yeast assimilable nitrogen on the synthesis of phenolic aroma compounds by Hanseniaspora vineae strains. *Yeast* **2016**, *33*, 323–328. [CrossRef]
58. Moreira, N.; Mendes, F.; de Guedes Pinho, P.; Hogg, T.; Vasconcelos, I. Heavy sulphur compounds, higher alcohols and esters production profile of Hanseniaspora uvarum and Hanseniaspora guilliermondii grown as pure and mixed cultures in grape must. *Int. J. Food Microbiol.* **2008**, *124*, 231–238. [CrossRef] [PubMed]
59. Viana, F.; Gil, J.V.; Vallés, S.; Manzanares, P. Increasing the levels of 2-phenylethyl acetate in wine through the use of a mixed culture of Hanseniaspora osmophila and Saccharomyces cerevisiae. *Int. J. Food Microbiol.* **2009**, *135*, 68–74. [CrossRef] [PubMed]
60. Barbosa, C.; Mendes-Faia, A.; Lage, P.; Mira, N.P.; Mendes-Ferreira, A. Genomic expression program of Saccharomyces cerevisiae along a mixed-culture wine fermentation with Hanseniaspora guilliermondii. *Microb. Cell Fact.* **2015**, *14*, 1–17. [CrossRef] [PubMed]
61. Luan, Y.; Zhang, B.Q.; Duan, C.Q.; Yan, G.L. Effects of different pre-fermentation cold maceration time on aroma compounds of Saccharomyces cerevisiae co-fermentation with Hanseniaspora opuntiae or Pichia kudriavzevii. *LWT Food Sci. Technol.* **2018**, *92*, 177–186. [CrossRef]
62. Hu, K.; Jin, G.J.; Xu, Y.H.; Tao, Y.S. Wine aroma response to different participation of selected Hanseniaspora uvarum in mixed fermentation with Saccharomyces cerevisiae. *Food Res. Int.* **2018**, *108*, 119–127. [CrossRef]
63. Hu, K.; Jin, G.J.; Mei, W.C.; Li, T.; Tao, Y.S. Increase of medium-chain fatty acid ethyl ester content in mixed H. uvarum/S. cerevisiae fermentation leads to wine fruity aroma enhancement. *Food Chem.* **2018**, *239*, 495–501. [CrossRef]
64. De Benedictis, M.; Bleve, G.; Tristezza, M.; Tufariello, M.; Grieco, F. An optimized procedure for the enological selection of non-Saccharomyces starter cultures. Antonie van Leeuwenhoek. *Int. J. Gen. Mol. Microbiol.* **2011**, *99*, 189–200. [CrossRef]
65. Medina, K.; Boido, E.; Fariña, L.; Gioia, O.; Gomez, M.E.; Barquet, M.; Gaggero, C.; Dellacassa, E.; Carrau, F. Increased flavour diversity of Chardonnay wines by spontaneous fermentation and co-fermentation with Hanseniaspora vineae. *Food Chem.* **2013**, *141*, 2513–2521. [CrossRef]
66. Lleixà, J.; Martín, V.; del Portillo, M.C.; Carrau, F.; Beltran, G.; Mas, A. Comparison of fermentation and wines produced by inoculation of Hanseniaspora vineae and Saccharomyces cerevisiae. *Front. Microbiol.* **2016**, *7*, 1–12. [CrossRef]
67. Englezos, V.; Torchio, F.; Cravero, F.; Marengo, F.; Giacosa, S.; Gerbi, V.; Rantsiou, K.; Rolle, L.; Cocolin, L. Aroma profile and composition of Barbera wines obtained by mixed fermentations of Starmerella bacillaris (synonym Candida zemplinina) and Saccharomyces cerevisiae. *LWT Food Sci. Technol.* **2016**, *73*, 567–575. [CrossRef]
68. González-Royo, E.; Pascual, O.; Kontoudakis, N.; Esteruelas, M.; Esteve-Zarzoso, B.; Mas, A.; Canals, J.M.; Zamora, F. Oenological consequences of sequential inoculation with non-Saccharomyces yeasts (Torulaspora delbrueckii or Metschnikowia pulcherrima) and Saccharomyces cerevisiae in base wine for sparkling wine production. *Eur. Food Res. Technol.* **2015**, *240*, 999–1012. [CrossRef]
69. Calabretti, A.; La Cara, F.; Sorrentino, A.; Di Stasio, M.; Santomauro, F.; Rastrelli, L.; Gabrielli, L.; Limone, F.; Volpe, M.G. Characterization of volatile fraction of typical Irpinian wines fermented with a new starter yeast. *World J. Microbiol. Biotechnol.* **2012**, *28*, 1433–1442. [CrossRef] [PubMed]
70. Hu, K.; Zhu, X.L.; Mu, H.; Ma, Y.; Ullah, N.; Tao, Y.S. A novel extracellular glycosidase activity from Rhodotorula mucilaginosa: Its application potential in wine aroma enhancement. *Lett. Appl. Microbiol.* **2016**, *62*, 169–176. [CrossRef] [PubMed]
71. Wang, X.C.; Li, A.H.; Dizy, M.; Ullah, N.; Sun, W.X.; Tao, Y.S. Evaluation of aroma enhancement for "Ecolly" dry white wines by mixed inoculation of selected Rhodotorula mucilaginosa and Saccharomyces cerevisiae. *Food Chem.* **2017**, *228*, 550–559. [CrossRef] [PubMed]
72. Martínez-garcía, R.; García-martínez, T.; Puig-pujol, A.; Carlos, J.; Moreno, J. Changes in sparkling wine aroma during the second fermentation under CO2 pressure in sealed bottle. *Food Chem.* **2017**, *237*, 1030–1040. [CrossRef]
73. Loira, I.; Morata, A.; Comuzzo, P.; Callejo, M.J.; González, C.; Calderón, F.; Suárez-Lepe, J.A. Use of Schizosaccharomyces pombe and Torulaspora delbrueckii strains in mixed and sequential fermentations to improve red wine sensory quality. *Food Res. Int.* **2015**, *76*, 325–333. [CrossRef]

74. Renault, P.; Coulon, J.; Moine, V.; Thibon, C.; Bely, M. Enhanced 3-sulfanylhexan-1-ol production in sequential mixed fermentation with Torulaspora delbrueckii/Saccharomyces cerevisiae reveals a situation of synergistic interaction between two industrial strains. *Front. Microbiol.* **2016**, *7*, 293. [CrossRef]
75. Egue, L.A.N.; Bouatenin, J.K.M.; Florent, K.N.; Koussemon-camara, M. Microbial Pathogenesis Virulence factors and determination of antifungal susceptibilities of Candida species isolated from palm wine and sorghum beer. *Microb. Pthogenes* **2018**, *124*, 5–10. [CrossRef]
76. Morrison-whittle, P.; Lee, S.A.; Fedrizzi, B.; Goddard, M.R. Co-evolution as Tool for Diversifying Flavor and Aroma Profiles of Wines. *Front. Microbiol.* **2018**, *9*, 1–10. [CrossRef]
77. Soden, A.; Francis, I.L.; Oakey, H.; Henschke, P.A. Effects of co-fermentation with Candida stellata and Saccharomyces cerevisiae on the aroma and composition of Chardonnay wine. *Aust. J. Grape Wine Res.* **2000**, *6*, 21–30. [CrossRef]
78. Rocha, M.; Rodrigues, F.; Coutinho, P.; Delgadillo, I.; Coimbra, M.A. Volatile composition of Baga red wine Assessment of the identification of the would-be impact odourants. *Anal. Chim. Acta* **2004**, *513*, 257–262. [CrossRef]
79. Varela, C.; Barker, A.; Tran, T.; Borneman, A.; Curtin, C. Sensory profile and volatile aroma composition of reduced alcohol Merlot wines fermented with Metschnikowia pulcherrima and Saccharomyces uvarum. *Int. J. Food Microbiol.* **2017**, *252*, 1–9. [CrossRef] [PubMed]
80. Hranilovic, A.; Gambetta, J.M.; Jeffery, D.W.; Grbin, P.R.; Jiranek, V. Lower-alcohol wines produced by Metschnikowia pulcherrima and Saccharomyces cerevisiae co-fermentations: The effect of sequential inoculation timing. *Int. J. Food Microbiol.* **2020**, *329*, 108651. [CrossRef] [PubMed]
81. Escribano, R.; González-Arenzana, L.; Portu, J.; Garijo, P.; López-Alfaro, I.; López, R.; Santamaría, P.; Gutiérrez, A.R. Wine aromatic compound production and fermentative behaviour within different non-Saccharomyces species and clones. *J. Appl. Microbiol.* **2018**, *124*, 1521–1531. [CrossRef]
82. Manzanares, P. Past and Future of Non-Saccharomyces Yeasts: From Spoilage Microorganisms to Biotechnological Tools for Improving Wine Aroma Past and Future of Non-Saccharomyces Yeasts: From Spoilage Microorganisms to Biotechnological Tools for Improving Wine Aroma Co. *Front. Microbiol.* **2016**. [CrossRef]
83. Otero, R.R.C.; Pe, A.I.B. Optimization of a rapid method for studying the cellular location of b -glucosidase activity in wine yeasts. *J. Appl. Microbiol.* **2005**, 558–564. [CrossRef]
84. Prior, K.J.; Bauer, F.F.; Divol, B. The utilisation of nitrogenous compounds by commercial non-Saccharomyces yeasts associated with wine. *J. Food Microbiol.* **2019**, *79*, 75–84. [CrossRef]
85. Renault, P.; Miot-Sertier, C.; Marullo, P.; Hernández-Orte, P.; Lagarrigue, L.; Lonvaud-Funel, A.; Bely, M. Genetic characterization and phenotypic variability in Torulaspora delbrueckii species: Potential applications in the wine industry. *Int. J. Food Microbiol.* **2009**, *134*, 201–210. [CrossRef]
86. Ferrando, N.; Araque, I.; Ortís, A.; Thornes, G.; Bautista-Gallego, J.; Bordons, A.; Reguant, C. Evaluating the effect of using non-Saccharomyces on Oenococcus oeni and wine malolactic fermentation. *Food Res. Int.* **2020**, *138*, 1–9. [CrossRef]
87. Lonvaud-Funel, A. Lactic Acid Bacteria and Malolactic Fermentation in Wine. In *Biotechnology of Lactic Acid Bacteria: Novel Applications*, 2nd ed.; Wiley: Hoboken, NJ, USA, 2015; pp. 231–247.
88. Cappello, M.S.; Zapparoli, G.; Logrieco, A.; Bartowsky, E.J. Linking wine lactic acid bacteria diversity with wine aroma and flavour. *Int. J. Food Microbiol.* **2017**, *243*, 16–27. [CrossRef]
89. Dicks, L.M.T.; Dellaglio, F.; Collins, M.D. Proposal to reclassify Leuconostoc oenos as Oenococcus oeni [corrig.] gen. nov., comb. nov. *Int. J. Syst. Bacteriol.* **1995**, *45*, 395–397. [CrossRef] [PubMed]
90. Diez-Ozaeta, I.; Lavilla, M.; Amárita, F. Technological characterisation of potential malolactic starters from Rioja Alavesa winemaking region. *LWT* **2020**, *134*, 1–8. [CrossRef]
91. Badotti, F.; Moreira, A.P.B.; Tonon, L.A.C.; de Lucena, B.T.L.; de Gomes, F.C.O.; Kruger, R.; Thompson, C.C.; de Morais, M.A.; Rosa, C.A.; Thompson, F.L. *Oenococcus alcoholitolerans* sp. nov., a lactic acid bacteria isolated from cachaça and ethanol fermentation processes. Antonie van Leeuwenhoek. *Int. J. Gen. Mol. Microbiol.* **2014**, *106*, 1259–1267. [CrossRef]
92. Endo, A.; Okada, S. *Oenococcus kitaharae* sp. nov., a non-acidophilic and non-malolactic-fermenting oenococcus isolated from a composting distilled shochu residue. *Int. J. Syst. Evol. Microbiol.* **2006**, *56*, 2345–2348. [CrossRef] [PubMed]
93. Pripis-Nicolau, L.; De Revel, G.; Bertrand, A.; Lonvaud-Funel, A. Methionine catabolism and production of volatile sulphur compounds by Oenococcus œni. *J. Appl. Microbiol.* **2004**, *96*, 1176–1184. [CrossRef]
94. Lombardi, S.J.; Pannella, G.; Iorizzo, M.; Testa, B.; Succi, M.; Tremonte, P.; Sorrentino, E.; Di Renzo, M.; Strollo, D.; Coppola, R. Inoculum strategies and performances of malolactic starter lactobacillus plantarum M10: Impact on chemical and sensorial characteristics of fiano wine. *Microorganisms* **2020**, *8*, 516. [CrossRef]
95. Iorizzo, M.; Testa, B.; Lombardi, S.J.; García-Ruiz, A.; Muñoz-González, C.; Bartolomé, B.; Moreno-Arribas, M.V. Selection and technological potential of Lactobacillus plantarum bacteria suitable for wine malolactic fermentation and grape aroma release. *LWT Food Sci. Technol.* **2016**, *73*, 557–566. [CrossRef]
96. Hernandez-Orte, P.; Cersosimo, M.; Loscos, N.; Cacho, J.; Garcia-Moruno, E.; Ferreira, V. Aroma development from non-floral grape precursors by wine lactic acid bacteria. *Food Res. Int.* **2009**, *42*, 773–781. [CrossRef]
97. Devi, A.; Anu-Appaiah, K.A. Mixed malolactic co-culture (Lactobacillus plantarum and Oenococcus oeni) with compatible Saccharomyces influences the polyphenolic, volatile and sensory profile of Shiraz wine. *LWT* **2021**, *135*, 110246. [CrossRef]

98. Costello, P.J.; Siebert, T.E.; Solomon, M.R.; Bartowsky, E.J. Synthesis of fruity ethyl esters by acyl coenzyme A: Alcohol acyltransferase and reverse esterase activities in Oenococcus oeni and Lactobacillus plantarum. *J. Appl. Microbiol.* **2013**, *114*, 797–806. [CrossRef]
99. Knoll, C.; Fritsch, S.; Schnell, S.; Grossmann, M.; Rauhut, D.; Du Toit, M. Influence of pH and ethanol on malolactic fermentation and volatile aroma compound composition in white wines. *LWT Food Sci. Technol.* **2011**, *44*, 2077–2086. [CrossRef]
100. Lytra, G.; Miot-Sertier, C.; Moine, V.; Coulon, J.; Barbe, J.C. Influence of must yeast-assimilable nitrogen content on fruity aroma variation during malolactic fermentation in red wine. *Food Res. Int.* **2020**, *135*, 1–11. [CrossRef] [PubMed]
101. Liu, W.; Li, H.; Jiang, D.; Zhang, Y.; Zhang, S.; Sun, S. Effect of *Saccharomyces cerevisiae*, *Torulaspora delbrueckii* and malolactic fermentation on fermentation kinetics and sensory property of black raspberry wines. *Food Microbiol.* **2020**, *91*, 1–9. [CrossRef] [PubMed]
102. Zhao, C.; Su, W.; Mu, Y.; Jiang, L.; Mu, Y. Correlations between microbiota with physicochemical properties and volatile flavor components in black glutinous rice wine fermentation. *Food Res. Int.* **2020**, *138*, 109800. [CrossRef] [PubMed]
103. Antalick, G.; Perello, M.C.; De Revel, G. Characterization of fruity aroma modifications in red wines during malolactic fermentation. *J. Agric. Food Chem.* **2012**, *60*, 12371–12383. [CrossRef] [PubMed]
104. Guo, L.; Luo, Y.; Zhou, Y.; Bianba, C.; Guo, H.; Zhao, Y.; Fu, H. Exploring microbial dynamics associated with flavours production during highland barley wine fermentation. *Food Res. Int.* **2020**, *130*, 108971. [CrossRef]
105. Roda, A.; Lucini, L.; Torchio, F.; Dordoni, R.; De Faveri, D.M.; Lambri, M. Metabolite profiling and volatiles of pineapple wine and vinegar obtained from pineapple waste. *Food Chem.* **2017**, *229*, 734–742. [CrossRef]
106. Jiang, L.; Su, W.; Mu, Y.; Mu, Y. Major Metabolites and Microbial Community of Fermented Black Glutinous Rice Wine with Different Starters. *Front. Microbiol.* **2020**, *11*, 1–13. [CrossRef]
107. Gil-Sánchez, I.; Bartolomé Suáldea, B.; Victoria Moreno-Arribas, M. Malolactic Fermentation. In *Red Wine Technology*; Morata, A., Ed.; Academic Press: Cambridge, MA, USA, 2018; pp. 85–98, ISBN 9780128144008.
108. Brizuela, N.S.; Bravo-Ferrada, B.M.; La Hens, D.V.; Hollmann, A.; Delfederico, L.; Caballero, A.; Tymczyszyn, E.E.; Semorile, L. Comparative vinification assays with selected Patagonian strains of Oenococcus oeni and Lactobacillus plantarum. *LWT Food Sci. Technol.* **2017**, *77*, 348–355. [CrossRef]
109. Brizuela, N.; Tymczyszyn, E.E.; Semorile, L.C.; Valdes La Hens, D.; Delfederico, L.; Hollmann, A.; Bravo-Ferrada, B. Lactobacillus plantarum as a malolactic starter culture in winemaking: A new (old) player? *Electron. J. Biotechnol.* **2019**, *38*, 10–18. [CrossRef]
110. Balmaseda, A.; Rozès, N.; Leal, M.Á.; Bordons, A.; Reguant, C. Impact of changes in wine composition produced by non-Saccharomyces on malolactic fermentation. *Int. J. Food Microbiol.* **2021**, *337*, 108954. [CrossRef] [PubMed]
111. Strickland, M.T.; Schopp, L.M.; Edwards, C.G.; Osborne, J.P. Impact of Pediococcus spp. on pinot noir wine quality and growth of Brettanomyces. *Am. J. Enol. Vitic.* **2016**, *67*, 188–198. [CrossRef]
112. Kaur, B.; Kumar, B.; Kaur, G.; Chakraborty, D.; Kaur, K. Application of recombinant Pediococcus acidilactici BD16 (fcs+/ech+) in malolactic fermentation. *Appl. Microbiol. Biotechnol.* **2015**, *99*, 3015–3028. [CrossRef] [PubMed]
113. Matei, F.; Kosseva, M.R. Microbiology of Fruit Wine Production. In *Science and Technology of Fruit Wine Production*; Elsevier Inc.: Amsterdam, The Netherlands, 2017; pp. 73–103, ISBN 9780128010341.
114. Rollán, G.C.; Farías, M.E.; De Nadra, M.C.M. Characterization of two extracellular proteases from Leuconostoc oenos. *World J. Microbiol. Biotechnol.* **1995**, *11*, 153–155. [CrossRef]
115. Bekal, S.; Van Beeumen, J.; Samyn, B.; Garmyn, D.; Henini, S.; Diviès, C.; Prévost, H. Purification of Leuconostoc mesenteroides citrate lyase and cloning and characterization of the citCDEFG gene cluster. *J. Bacteriol.* **1998**, *180*, 647–654. [CrossRef]
116. Mateo, E.; Torija, M.J.; Mas, A.; Bartowsky, E.J. Acetic acid bacteria isolated from grapes of South Australian vineyards. *Int. J. Food Microbiol.* **2014**, *178*, 98–106. [CrossRef]
117. Bartowsky, E.J.; Xia, D.; Gibson, R.L.; Fleet, G.H.; Henschke, P.A. Spoilage of bottled red wine by acetic acid bacteria. *Lett. Appl. Microbiol.* **2003**. [CrossRef]
118. Nurgel, C.; Pickering, G.J.; Inglis, D.L. Sensory and chemical characteristics of Canadian ice wines. *J. Sci. Food Agric.* **2004**, *84*, 1675–1684. [CrossRef]
119. Bartowsky, E.J.; Henschke, P.A. Acetic acid bacteria spoilage of bottled red wine—A review. *Int. J. Food Microbiol.* **2008**, *125*, 60–70. [CrossRef]
120. Valera, M.J.; Torija, M.J.; Mas, A.; Mateo, E. Acetobacter malorum and Acetobacter cerevisiae identification and quantification by Real-Time PCR with TaqMan-MGB probes. *Food Microbiol.* **2013**, *36*, 30–39. [CrossRef]
121. Valera, M.J.; Laich, F.; González, S.S.; Torija, M.J.; Mateo, E.; Mas, A. Diversity of acetic acid bacteria present in healthy grapes from the Canary Islands. *Int. J. Food Microbiol.* **2011**, *151*, 105–112. [CrossRef] [PubMed]
122. Shamala, T.R.; Sreekantiah, K.R. Microbiological and biochemical studies on traditional Indian palm wine fermentation. *Food Microbiol.* **1988**, *5*, 157–162. [CrossRef]
123. Hommel, R.K. Gluconobacter. In *Encyclopedia of Food Microbiology*, 2nd ed.; Academic Press: Amsterdam, The Netherlands, 2014; Volume 2, pp. 99–105, ISBN 9780123847331.
124. Amaresan, N.; Annapurna, K.; Sankaranarayanan, A.M.; Senthil Kumar, K.K. (Eds.) Mitesh Dwivedi Gluconobacter. In *Beneficial Microbes in Agro-Ecology*; Academic Press: Cambridge, MA, USA, 2020; pp. 521–544, ISBN 978-0-12-823414-3.
125. Hyma, K.E.; Saerens, S.M.; Verstrepen, K.J.; Fay, J.C. Divergence in wine characteristics produced by wild and domesticated strains of Saccharomyces cerevisiae. *FEMS Yeast Res.* **2011**, *11*, 540–551. [CrossRef] [PubMed]

126. Belda, I.; Gobbi, A.; Ruiz, J.; de Celis, M.; Ortiz-Álvarez, R.; Acedo, A.; Santos, A. Microbiomics to Define Wine Terroir. In *Comprehensive Foodomics*; Cifuentes, A., Ed.; Elsevier: Amsterdam, The Netherlands, 2021; pp. 438–451, ISBN 978-0-12-816396-2.
127. Tilloy, V.; Cadière, A.; Ehsani, M.; Dequin, S. Reducing alcohol levels in wines through rational and evolutionary engineering of Saccharomyces cerevisiae. *Int. J. Food Microbiol.* **2015**, *213*, 49–58. [CrossRef] [PubMed]
128. Salinas, F.; Cubillos, F.A.; Soto, D.; Garcia, V.; Bergström, A.; Warringer, J.; Ganga, M.A.; Louis, E.J.; Liti, G.; Martinez, C. The Genetic Basis of Natural Variation in Oenological Traits in Saccharomyces cerevisiae. *PLoS ONE* **2012**, *7*, e49640. [CrossRef]
129. Pérez-Torrado, R.; Querol, A.; Guillamón, J.M. Genetic improvement of non-GMO wine yeasts: Strategies, advantages and safety. *Trends Food Sci. Technol.* **2015**, *45*, 1–11. [CrossRef]
130. Yang, X.; Guo, Y.; Zhu, J.; Niu, Z.; Shi, G.; Liu, Z.; Li, K.; Guo, X. Genetic diversity and association study of aromatics in grapevine. *J. Am. Soc. Hortic. Sci.* **2017**, *142*, 225–231. [CrossRef]
131. Mendes, I.; Sanchez, I.; Franco-Duarte, R.; Camarasa, C.; Schuller, D.; Dequin, S.; Sousa, M.J. Integrating transcriptomics and metabolomics for the analysis of the aroma profiles of Saccharomyces cerevisiae strains from diverse origins. *BMC Genom.* **2017**, *18*, 1–13. [CrossRef]
132. Barbosa, C.; García-Martínez, J.; Pérez-Ortín, J.E.; Mendes-Ferreira, A. Comparative transcriptomic analysis reveals similarities and dissimilarities in Saccharomyces cerevisiae wine strains response to nitrogen availability. *PLoS ONE* **2015**, *10*, e0122709. [CrossRef]
133. Liu, P.; Wang, Y.; Ye, D.; Duan, L.; Duan, C.; Yan, G. Effect of the addition of branched-chain amino acids to non-limited nitrogen synthetic grape must on volatile compounds and global gene expression during alcoholic fermentation. *Aust. J. Grape Wine Res.* **2018**, *24*, 197–205. [CrossRef]
134. Katarína, F.; Katarína, M.; Katarína, Ď.; Ivan, Š.; Fedor, M. Influence of yeast strain on aromatic profile of Gewürztraminer wine. *LWT Food Sci. Technol.* **2014**, *59*, 256–262. [CrossRef]
135. Tronchoni, J.; Curiel, J.A.; Morales, P.; Torres-Pérez, R.; Gonzalez, R. Early transcriptional response to biotic stress in mixed starter fermentations involving Saccharomyces cerevisiae and Torulaspora delbrueckii. *Int. J. Food Microbiol.* **2017**, *241*, 60–68. [CrossRef] [PubMed]
136. Tondini, F.; Lang, T.; Chen, L.; Herderich, M.; Jiranek, V. Linking gene expression and oenological traits: Comparison between Torulaspora delbrueckii and Saccharomyces cerevisiae strains. *Int. J. Food Microbiol.* **2019**, *294*, 42–49. [CrossRef] [PubMed]
137. Ruiz, J.; de Celis, M.; de Toro, M.; Mendes-Ferreira, A.; Rauhut, D.; Santos, A.; Belda, I. Phenotypic and transcriptional analysis of Saccharomyces cerevisiae during wine fermentation in response to nitrogen nutrition and co-inoculation with Torulaspora delbrueckii. *Food Res. Int.* **2020**, *137*, 1–12. [CrossRef] [PubMed]
138. Lin, J.; Massonnet, M.; Cantu, D. The genetic basis of grape and wine aroma. *Hortic. Res.* **2019**, *6*, 1–24. [CrossRef] [PubMed]
139. Moreno-Arribas, M.V.; Polo, M.C. Winemaking biochemistry and microbiology: Current knowledge and future trends. *Crit. Rev. Food Sci. Nutr.* **2005**, *45*, 265–286. [CrossRef]
140. Schmidtke, L.M.; Blackman, J.W.; Agboola, S.O. Production technologies for reduced alcoholic wines. *J. Food Sci.* **2012**, *77*. [CrossRef]
141. Ozturk, B.; Anli, E. Different techniques for reducing alcohol levels in wine: A review. *BIO Web Conf.* **2014**, *3*, 02012. [CrossRef]
142. Petruzzi, L.; Capozzi, V.; Berbegal, C.; Corbo, M.R.; Bevilacqua, A.; Spano, G.; Sinigaglia, M. Microbial resources and enological significance: Opportunities and benefits. *Front. Microbiol.* **2017**, *8*, 1–13. [CrossRef]
143. Bordiga, M.; Rinaldi, M.; Locatelli, M.; Piana, G.; Travaglia, F.; Coïsson, J.D.; Arlorio, M. Characterization of Muscat wines aroma evolution using comprehensive gas chromatography followed by a post-analytic approach to 2D contour plots comparison. *Food Chem.* **2013**, *140*, 57–67. [CrossRef]

*foods*

MDPI

*Article*

# Oral Release Behavior of Wine Aroma Compounds by Using In-Mouth Headspace Sorptive Extraction (HSSE) Method

**María Pérez-Jiménez, Carolina Muñoz-González and María Angeles Pozo-Bayón ***

Instituto de Investigación en Ciencias de la Alimentación (CIAL), CSIC-UAM, C/Nicolás Cabrera, 9, 28049 Madrid, Spain; maria.perez@csic.es (M.P.-J.); c.munoz@csic.es (C.M.-G.)
* Correspondence: m.delpozo@csic.es

**Abstract:** The oral release behavior of wine aroma compounds was determined by using an in-mouth headspace sorptive extraction (HSSE) procedure. For this, 32 volunteers rinsed their mouths with a red wine. Aroma release was monitored at three time points (immediately, 60 s, and 120 s) after wine expectoration. Twenty-two aroma compounds belonging to different chemical classes were identified in the mouth. Despite the large inter-individual differences, some interesting trends in oral release behavior were observed depending on the chemical family. In general, esters and linear alcohols showed rapid losses in the mouth over the three sampling times and therefore showed a low oral aroma persistence. On the contrary, terpenes, lactones, and $C_{13}$ norisoprenoids showed lower variations in oral aroma release over time, thus showing a higher oral aroma persistence. Additionally, and despite their low polarity, furanic acids and guaiacol showed the highest oral aroma persistence. This work represents the first large study regarding in-mouth aroma release behavior after wine tasting, using real wines, and it confirmed that oral release behavior does not only depend on the physicochemical properties of aroma compounds but also on other features, such as the molecular structure and probably, on the characteristics and composition of the oral environment.

**Keywords:** wine aroma; oral release; aroma persistence; in-mouth headspace sorptive extraction



**Citation:** Pérez-Jiménez, M.; Muñoz-González, C.; Pozo-Bayón, M.A. Oral Release Behavior of Wine Aroma Compounds by Using In-Mouth Headspace Sorptive Extraction (HSSE) Method. *Foods* **2021**, *10*, 415. https://doi.org/10.3390/foods10020415

Academic Editors: Enrique Durán-Guerrero and Remedios Castro-Mejías

Received: 30 December 2020
Accepted: 9 February 2021
Published: 13 February 2021



## 1. Introduction

Wine odorant molecules belong to very different chemical classes (esters, alcohols, aldehydes, terpenes, phenols, etc.) in which their physicochemical properties such as volatility, boiling point, polarity, hydrophobicity, and/or molecular structure all differ.

The different physicochemical properties of odorant molecules also determine their release from the wine matrix and therefore their transfer to the surrounding air that will carry these volatile chemical molecules to the olfactory receptors when breathing. This is the orthonasal route and it is the main reason for the perceived odor when we smell a wine. However, during wine tasting, odorant molecules are released in the oropharyngeal cavities, and the expiration flows carry them to the olfactory receptors by the so-called retronasal pathway. In this case, as well as the physicochemical characteristics of the odorant molecules and wine matrix composition, factors related to the individual oral physiology (breathing flow, dilution with saliva, interaction of odorants with salivary proteins, biochemical transformation by salivary enzymes, etc.) can also affect aroma release [1–5]. Moreover, the formation of wine residues or the adsorption of odorant molecules in the oral mucosa might be the origin of aroma reservoirs ready to be released in the successive swallowing–exhalation episodes of the remaining saliva in the mouth once the wine has been swallowed [6]. This is the origin of the long lasting aroma perception, also known as after-odor or aroma persistence [7].

The immediate and prolonged retronasal aroma perception is of great importance, since it is a key factor in determining wine quality and ultimately consumer preferences [8]. Because of this, in the last couple of decades, the study of the retronasal aroma of wine has gained popularity and the number of scientific works dealing with this topic has increased.

Most of these studies have been carried out using sensory analysis [9] and using in vitro headspace experiments simulating wine oral conditions [10–13]. More recently, the use of in vivo approaches to monitor retronasal aroma release in more realistic wine consumption situations have been used, although these works are still scarce [7,14–20].

Both types of approaches can provide us with information about the behavior of wine odorant compounds during real or simulated wine consumption conditions that might better correlate to wine aroma perception than when only using the volatile profile analysis of a wine. In this sense, previous works have attempted to study this behavior by using static or dynamic headspace analysis [10,11,21] to simulate oral conditions. From these studies, it was shown that saliva differently affected the rate of aroma release depending on the type of aroma compound and on the wine matrix composition. More recently, Piombino and co-workers (2019) [13] showed that the release of volatile compounds from wine was strongly related to hydrophobicity. While aroma compounds with $\log P < 0$ increased their retronasal release, odorants with $2 < \log P < 5$ showed an opposite trend, and aroma compounds with $0 < \log P < 2$ were the most affected compounds by the wine matrix components as residual sugars.

Although very valuable, in vitro oral conditions do not perfectly represent the physiological conditions of the oral cavity. Different works have found that saliva enzymes and oral microbiota can affect aroma compounds in the mouth in a different way depending on the individual, which in turn could be difficult to mimic by only using in vitro conditions. For instance, salivary enzymes are able to hydrolyze odorless glycosylated precursors [22,23] or metabolize odorants into their degradation products [4,24–29]—both cases give rise to different volatile odorant metabolites depending on individual differences in saliva composition.

In this sense, in vivo aroma analysis should be better suited to determine the retronasal release behavior of aroma compounds in the mouth during wine tasting. For instance, Esteban-Fernández et al. (2016) [15] used in vivo intraoral Solid Phase Microextraction (SPME) to compare the retronasal release behavior of six aroma compounds (isoamyl acetate, ethyl hexanoate, linalool, guaiacol, β-phenylethanol, and β-ionone) in the mouth of three individuals after tasting aromatised wines. They confirmed the impact of compound hydrophobicity on the degree of adsorption to oral mucosa, which provoked differences in in-mouth aroma release kinetics in the post-ingestion phase.

These works have a great scientific value since they represent the first analytical studies performed in real wine consumption conditions to monitor retronasal aroma release. However, because of the relatively low number of individuals employed to perform these assays, it would be interesting to set up new studies using a representative number of volunteers and using real wines, which will allow us to obtain more straightforward conclusions on the behavior of wine odorants in the mouth.

To do so, we utilized the in-mouth headspace sorptive extraction (HSSE) technique, which has been proven to be a reliable and feasible tool that allows for the monitoring of oral aroma release of a great number of different odorant chemical classes at real wine concentrations [18]. This methodology is based on the application of a polydimethyl siloxane (PDMS) twister in the mouth to perform the headspace intra-oral aroma extraction after wine intake. The twisters are further desorbed and analyzed by Gas Chromatography-Mass Spectrometry (GC-MS). One of the main advantages of this technique is that the twisters with the breathing aroma extract can be automatically desorbed in the thermal desorption unit (TDU) of the GC–MS, allowing for the analysis of a higher number of wine breath extracts and thus increasing the possibility of working with a greater number of individuals. Additionally, the lower in-mouth extraction times (30 s) compared to other in vivo methods [15] will allow us to perform more in-mouth aroma samplings in a shorter period of time (80 s) once the wine has been ingested.

In this frame, the aim of this work was to assess the oral release behavior of the naturally occurring wine aroma compounds, composed of different volatile chemical families at different concentrations, by using the in-mouth HSSE procedure. To do this, we instigated 32 volunteers to rinse their mouths with a red wine for 30 s. Subsequently, a glass device provided with a PDMS twister was placed in the mouth in order to monitor the aroma released at three different times (immediately, 60 s, and 120 s) after the wine was expectorated.

## 2. Materials and Methods

### 2.1. Wine Chemical Composition

A commercial red wine (Marqués de Murrieta, 2017) from the Tempranillo grape variety was employed in this study. The non-volatile composition of the wine: pH (4.0 ± 0.3), total polyphenol content (1917.3 ± 10.3 mg of gallic acid/L), free amino acids (561.2 ± 72.2 mg Leu/L), free amino acids plus peptides (328.9 ± 11.0 mg Leu/L), neutral polysaccharides (4014.1 ± 741.4 mg mannose/L), and procyanidins (1365.9 ± 42.0 mg catechin/L) was previously determined.

The volatile composition of the red wine was assessed using a headspace sorptive extraction method and gas chromatography analysis (HSSE-GC–MS) using a 20 mm length × 0.5 mm PDMS twister (Gerstel, Mülheim an der Ruhr Germany) as previously described [18]. For the quantification, a calibration curve for each aroma compound was carried out (Table S1) using the same conditions already described [18].

### 2.2. In-Mouth Aroma Analysis

#### 2.2.1. Subjects

Thirty-two subjects—15 females (47%) and 17 males (53%) between the ages of 18 and 72 years old—participated in this study. Fifty percent of the participants were younger than 35 years old and the other 50% were older than 50 years old. They did not have any known illnesses and they all had normal olfactory and gustatory functions. The participants were informed about the purpose of this study and they all gave their written consent to participate in this study. The project in which this study was enclosed was also approved by the Bioethical Committee of the Spanish Council of Research (CSIC).

#### 2.2.2. In-Mouth HSSE Procedure

To determine both the immediate and the prolonged oral aroma release, we applied the previously published in-mouth HSSE method [18]. Briefly, the volunteers performed gentle rinses with the wine (15 mL) and 30 s after, they spat out the wine. During these rinses, swallowing episodes were not allowed and the lips and the velum tongue had to stay closed. For the intra-oral aroma extraction, a customized glass holder device (Segainvex-UAM, Madrid, Spain) provided with a 20 mm × 0.5 mm (length × film thickness) PDMS twister inside was employed. The volunteers placed the glass device in their mouths 5 s after spitting out the wines. To avoid any contact between the twister and the oral surfaces, we ensured that aroma sampling was performed in the headspace of the mouth. This procedure was repeated 3 times in order to determine the intra-oral aroma release at 3 different times after the wine expectoration. These sampling times were performed immediately (T1), 60 (T2) seconds, and 120 (T3) seconds after the wine rinses. For this, 3 different in-mouth devices with 3 different twisters were employed. Between each aroma monitoring, the subjects kept their lips closed, breathed normally through their nose, and were told to only swallow twice: immediately before introducing the twister into the mouth and immediately after removing it. Figure 1 shows a schematic representation of the main steps of the sampling procedure. Volunteers were previously trained in the in-mouth HSSE procedure before the wine sampling evaluation.

The twisters were removed from the glass device with a magnet bar and desorbed in a thermal desorption unit (TDU) (Gerstel) after the oral aroma sampling. Twenty different PDMS twisters were used for the whole experiment. In order to check their variability, we previously tested them all using a synthetic aromatized wine. Differences among the PDMS twisters in their aroma extraction were always lower than 5%.

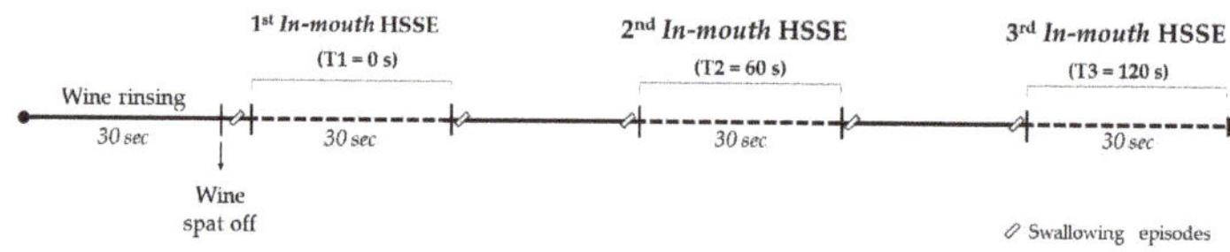


**Figure 1.** Sampling procedure followed for in-mouth HSSE aroma monitoring.

Each analysis was performed 3 times for each volunteer ($n$ = 32). Before the in-mouth aroma extraction and between wine replicates, volunteers washed their mouths by rinsing with a pectin-water solution (1 g/L). After rinses, they waited 15 min before starting a new in-mouth HSSE analysis.

#### 2.2.3. Thermal Desorption

The PDMS twisters were desorbed in the thermal desorption unit (TDU) coupled to a cooled injection system (CIS-4) (Gerstel). The TDU was configured in splitless mode. The TDU ramp temperature was first 40 °C, then increased at 60 °C/min to 240 °C and then held for 5 min. The CIS was employed for analyte cryofocusing by using liquid nitrogen. The CIS ramp temperature started at −100 °C, then heated to 240 °C at 12 °C/min, and held for 5 min. The injection was configured in solvent vent mode.

#### 2.2.4. GC–MS Analysis

Aroma analysis was carried out in a 6890 N GC coupled to a 5973 N mass spectrometer (Agilent). The stationary phase was a DB-WAX polar capillary column with dimensions of 30 m × 0.25 mm and film thickness of 0.50 µm (Agilent, J&W Scientific, Folsom, CA, USA), while the mobile phase was helium gas at a flow rate of 1 mL/min. The oven temperature conditions were the same as those previously described [18,30]. The oven started at 40 °C, then it raised to 130 °C at 4 °C/min, and finally it rose to 240 °C at 8 °C/min and kept for 5 min.

The temperatures of the MS system were configured in the following way: 270 °C for the transfer line, 150 °C for the quadrupole, and 230 °C for the ion source. The electron impact was fixed at 70 eV and the ionization current was fixed at 10 µA. Both selected ion mass monitoring (SIM) and full scan mode (mass range of 35–350 $m/z$) were used for data acquisition. The detected peaks were identified by comparing the mass spectra and the retention times with those of the same reference compounds analyzed with the same chromatographic conditions and by using the NIST 2.0 database.

As a result, the absolute peak areas (APAs) of the volatile compounds released in the mouth were obtained. APAs (3 repetitions) were used to compare the 3 in-mouth sampling points.

### 2.3. *Statistical Analysis*

For the statistical analysis, the XLSTAT 2020.4 software (Addinsoft, Paris, France) was employed. One-way ANOVA and Tukey test were applied to check for significant differences among APAs from the 3 in-mouth sampling points (T1, T2, and T3) and to check for mean comparison. A principal component analysis (PCA) was applied in order to examine how the physicochemical properties of the aroma compounds influenced their oral release in each sampling time. A significance level of $\alpha$ = 0.05 was fixed in all the statistical analyses.

## 3. Results and Discussion

### 3.1. *Oral Aroma Release Behavior after Wine Tasting*

To determine the oral aroma release behavior after wine consumption, we applied the in-mouth HSSE procedure [18]. For this, 32 volunteers rinsed their mouths with a red wine for 30 s and then spat it out. A PDMS twister was then placed inside the mouth to monitor aroma release for 30 s (T1). This procedure was also repeated for 60 s and 120 s (T2 and T3, respectively) after wine expectoration, as shown in Figure 1. Despite the large dilution of the aroma compounds in the mouth due to the exhalation flows following wine expectoration [31,32], all the aroma compounds detected in the headspace of the wine (Table S1) were also identified in the mouth of the volunteers (Table 1 and Figure S1). In order to check if there were significant differences ($p < 0.05$) in aroma release through the three monitoring times, we performed one-way ANOVA. Table 1 shows these results.

**Table 1.** Retention times, physicochemical properties, and oral release (average values of three repetitions of absolute peak areas) of the 22 aroma compounds detected in the mouth at three different sampling points: 0 (T1), 60 (T2), and 120 (T3) seconds after wine expectoration. The table also shows the results from one-way ANOVA to check for significant differences ($p < 0.05$) between the three extraction times.

| No. Compound | Aroma Compound | Retention Time (min) | Physicochemical Properties of Aroma Compounds | | | | | | Oral Aroma Release (APAs) Over the Three Sampling Points | | | |
|---|---|---|---|---|---|---|---|---|---|---|---|---|
| | | | MW [a] | logP [a] | BP [a] | VP [a] | WS [a] | OT [b] | T1 | T2 | T3 | Pr > F |
| | Esters | | | | | | | | | | | |
| 1 | Isoamyl acetate | 7.46 | 130.19 | 2.26 | 142.5 | 5.67 | 1100 | 30 [1] | 2,495,178 | 963,196 | 524,939 | <0.0001 |
| 2 | Ethyl butanoate | 5.53 | 116.16 | 1.85 | 121.5 | 14.6 | 2745 | 125 [1] | 317,302 | 301,709 | 258,753 | 0.681 |
| 3 | Ethyl pentanoate | 7.83 | 130.19 | 2.34 | 146.1 | 4.8 | 925.5 | >200 [1] | 96,840 | 152,040 | 141,734 | 0.369 |
| 4 | Ethyl hexanoate | 10.52 | 144.22 | 2.83 | 167 | 1.8 | 308.7 | 62 [1] | 599,957 | 260,843 | 166,670 | <0.0001 |
| 5 | Ethyl octanoate | 17.07 | 172.27 | 3.81 | 208.5 | 0.235 | 33.39 | 5 [1] | 492,678 | 226,935 | 151,675 | <0.0001 |
| 6 | Ethyl decanoate | 23.23 | 200.32 | 4.79 | 241.5 | 0.0428 | 3517 | 200 [1] | 115,138 | 60,170 | 49,934 | <0.0001 |
| 7 | Hexyl acetate | 11.97 | 144.22 | 2.83 | 171.5 | 1.45 | 308.7 | 1.5 [3] | 62,829 | 42,389 | 32,037 | 0.002 |
| 8 | Diethyl succinate | 24.09 | 174.2 | 1.39 | 217.7 | 0.147 | 5547 | 200,000 [4] | 257,106 | 112,972 | 66,785 | <0.0001 |
| 9 | Ethyl cinnamate | 31.05 | 176.22 | 2.99 | 271 | 0.00874 | 160.6 | 5.1 [1] | 509,575 | 30,132 | 16,649 | 0.012 |
| | *Alcohols* | | | | | | | - | | | | |
| 10 | 1-Pentanol | 10.01 | 88.15 | 1.33 | 137.9 | 2.65 | 20,890 | - | 7,468,430 | 5,012,622 | 3,629,045 | <0.0001 |
| 11 | 1-Hexanol | 14.58 | 102.17 | 2.03 | 157.6 | 0.881 | 6885 | 8000 [4] | 522,716 | 308,090 | 210,761 | <0.0001 |
| 12 | Phenylethyl ethanol | 28.54 | 122.17 | 1.57 | 218.2 | 0.0243 | 21,990 | 14,000 [1] | 2,289,276 | 1,332,593 | 994,164 | 0.000 |
| 13 | Z-3-Hexen-1-ol | 15.52 | 100.16 | 1.61 | 156.5 | 0.937 | 15,475 | 400 [1] | 59,232 | 117,719 | 116,501 | 0.163 |
| | *Furanic acids* | | | | | | | | | | | |
| 14 | Furfural | 17.78 | 96.09 | 0.83 | 161.7 | 2.32 | 53,580 | 14,100 [1] | 626,770 | 596,750 | 639,977 | 0.877 |
| 15 | 5-Methyl furfural | 20.42 | 110.11 | 0.67 | 187 | 0.644 | 29,110 | 20,000 [4] | 111,669 | 103,936 | 103,787 | 0.821 |
| 16 | Furfuryl alcohol | 23.68 | 98.1 | 0.45 | 171 | 0.409 | 221,000 | 2000 [4] | 596,677 | 568,836 | 744,321 | 0.524 |
| | *Terpenes* | | | | | | | | | | | |
| 17 | α-Pinene | 5.25 | 136.24 | 4.27 | 155.5 | 4.02 | 34,834 | - | 278,068 | 244,054 | 248,766 | 0.961 |
| 18 | Limonene | 9.49 | 136.24 | 4.83 | 178 | 1.45 | 4.581 | 15 [2] | 435,510 | 347,603 | 387,864 | 0.721 |
| | *C13 norisoprenoids* | | | | | | | | | | | |
| 19 | β-Ionone | 29.3 | 192.3 | 4.42 | 127 | 0.0227 | 25.16 | 0.09 [1] | 783,236 | 83,581 | 33,204 | 0.017 |
| | *Lactones* | | | | | | | | | | | |
| 20 | γ-Butyrolactone | 22.3 | 86.09 | -0.31 | 204 | 0.295 | 447,500 | 35,000 [4] | 128,301 | 104,228 | 101,701 | 0.304 |
| 21 | γ-Nonalactone | 30.1 | 156.23 | 2.08 | 136 | 0.0118 | 228.66 | 30 [1] | 98,120 | 65,455 | 56,939 | 0.039 |
| | *Volatile phenols* | | | | | | | | | | | |
| 22 | Guaiacol | 27.67 | 124.14 | 1.34 | 205 | 0.113 | 28,462 | 10 [1] | 91,729 | 74,640 | 78,168 | 0.106 |

[a] MW: molecular weight; logP: hydrophobic constant; BP: boiling point; VP: vapor pressure; WS: water solubility. Parameters estimated using the software EPI Suite 4.11-Estimation Program Interface (U.S. EPA 2012). [b] OT: odor threshold calculated in hydroalcoholic and/or synthetic wines. [1] [33]; [2] [34]; [3] [35]; [4] [36].

As it can be seen in Table 1, from the 22 compounds identified in the mouth, 12 of them showed significant variations ($p < 0.05$) depending on the sampling point (T1, T2, T3). Esters and alcohols were the most affected chemical groups by the sampling time. All of them, except ethyl butanoate, ethyl pentanoate, and Z-3-Hexen-1-ol, showed significant differences ($p < 0.05$) in the amount released depending on the sampling point. The fast decrease of esters and alcohols in the oral cavity as a function of time revealed that both groups of compounds showed a relatively low oral persistence. In addition to them, the compounds β-ionone and γ-nonalactone also showed significant differences ($p < 0.05$)

depending on the sampling point. Contrarily, other volatile compounds belonging to other chemical families, such as furanic acids, terpenes, or the volatile phenol guaiacol, did not show significant differences depending on the sampling point, suggesting that they remained in the oral cavity for longer times. Both the chemical structure of the aroma compounds and their physicochemical properties (also reported in Table 1) might be responsible for the differences in the aroma release behavior.

Therefore, in order to further investigate the in-mouth aroma release behavior of the different chemical groups, we calculated the percentage of oral aroma release (%OAR) at each sampling point. For this, the amount of aroma released in the first sampling point, immediately after wine expectoration (T1), was considered as 100% (%OARt1). From this, the percentage of aroma release after 60 s (%OARt2) and 120 s (%OARt3) was also calculated. Figure 2 shows these results together with the chemical structure of each aroma compound. It also depicts the average, minimum, and maximum %OAR values corresponding to three repetitions of the same wine tested for the 32 volunteers for each sampling point ($n$ = 96).

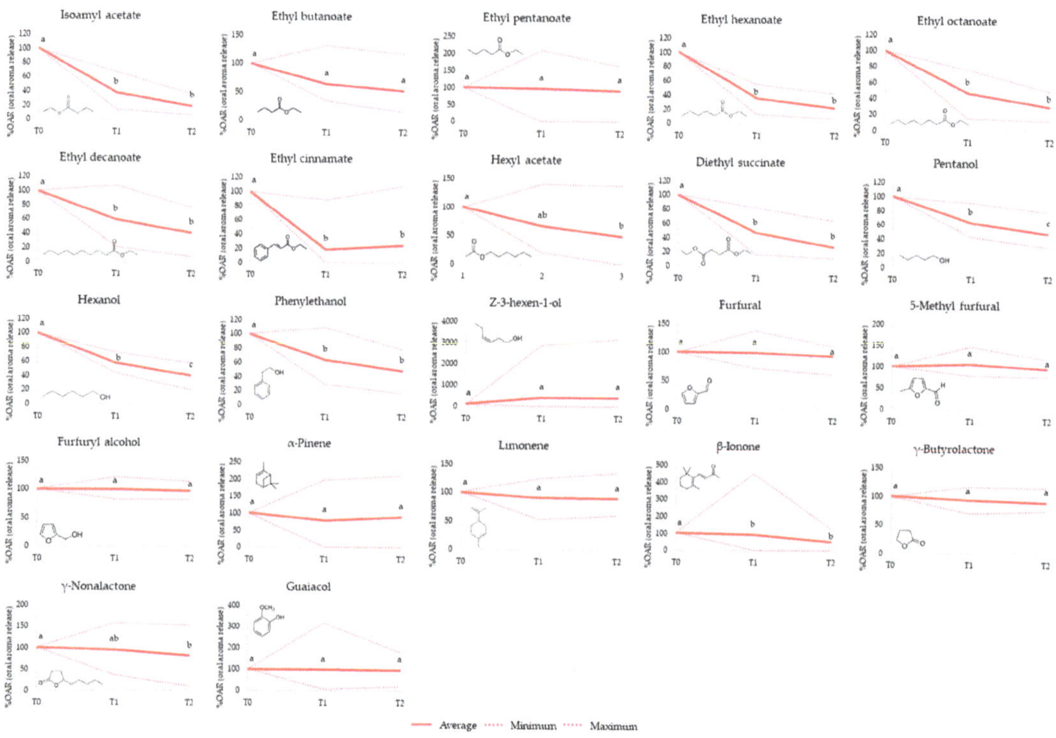


**Figure 2.** Oral aroma release profiles (%OAR) of 22 wine aroma compounds at 0 (T1), 60 (T2), and 120 (T3) seconds after of wine expectoration. The solid line shows the average value, while the dotted lines show the maximum and minimum values from three repetitions from the same wine and 32 volunteers ($n$ = 96). Different letters above the points denote different significance level ($p < 0.05$) from the Tukey test.

As it can be seen (Figure 2), very wide minimum and maximum %OAR intervals were observed for most wine volatile compounds. This was expected considering the large in-mouth aroma release variations that can be explained by differences in oral (breathing flows, saliva composition, oral cavity volumes) or other physiological features (age, gender), which have been related to significant differences in aroma release during food and beverage consumption [2,37,38]. Other factors related to variations from the instrumental methodology would be unlikely since all the participants were previously trained in the in-mouth HSSE procedure and the repeatability of this procedure was previously proven to be lower than 18% for a large number of different types of wine volatiles [18].

As can be seen in Figure 2, except for ethyl pentanoate, the rest of the esters showed a similar oral release behavior. This was characterized by a progressive reduction in oral aroma release over time, with the largest aroma losses (between 50 to 80%) in the third sampling time (T3) (Figure 2), which was 120 s after wine rinsing. Isoamyl acetate and ethyl hexanoate showed the lowest oral release in the third sampling point (T3), with %OAR of 19% and 22%, respectively (Figure 2). Thus, these esters exhibited the lowest oral persistence. These results are in agreement with the results from a previous work using intra-oral SPME approach and aromatized wines [15] and also confirmed the congruence between both in-mouth extraction procedures. Some reasons, such as the weaker interactions of these compounds to oral components (salivary proteins, oral mucosa cells) or the degradation of esters by salivary esterase enzymes might contribute to explaining the rapid loss of these compounds in the mouth [24,27,29,39]. However, as shown in Figure 2, some differences in the oral release behavior among the different esters were also found. For instance, in the case of ethyl decanoate (logP = 4.79), the most hydrophobic assayed linear ester, the %OARt3 was 41%, while in the case of shorter and less hydrophobic linear esters such as ethyl octanoate (logP = 3.81) and ethyl hexanoate (logP = 2.83), the percentages of %OARt3 were 29% and 22%, respectively. These results showed that 120 s after wine expectoration, the oral amount of ethyl decanoate was almost double that determined by the other two esters. These results are in agreement with those reported by Muñoz-González and co-authors (2019) [20], who also observed a higher persistence of ethyl decanoate compared to ethyl hexanoate and isoamyl acetate by using in-nose Proton Transfer Reaction-Mass spectrometry (PTR–MS). On the contrary, some esters with lower molecular weight, such as ethyl butanoate and ethyl pentanoate, did not significantly decrease over time. This could be mostly due to the large individual variations observed among volunteers. Thus, while some volunteers showed a decrease in the oral release of these esters over time, others showed an increase in their oral release. This dissimilar behavior among volunteers deserves further attention.

The linear alcohols (pentanol and hexanol) showed a similar behavior to that previously found for esters (Figure 2), which was characterized by a progressive decrease in their oral release over the three in-mouth aroma extractions (Figure 2). As shown in Figure 2, up to 40–50% of the initial aroma content remained 120 s after wine expectoration (%OARt3) (Figure 2). On the basis of these results, we can consider these alcohols as compounds with a relatively low-medium oral persistence. Although alcohols have not been reported to be metabolized by salivary enzymes [4,25], their low-medium oral persistence could be due to their weak interaction with oral physiology [15]. Interestingly, the linear alcohol cis-hexen-1-ol did not follow the same trend as the other alcohols. Not only did the release of this compound not decrease, but it increased in the second and third sampling points (Figure 2). This atypical behavior could be linked to the de novo formation of this metabolite from wine aroma precursors by oral microbiota or salivary enzymes. In fact, previous in vitro studies have proven the formation of cis-3-hexen-1-ol from glycosylated wine aroma precursors by isolated oral bacteria from saliva [23]. However, new experiments should be addressed in order to confirm these results. Regarding the cyclic alcohol phenylethanol, although a decrease in %OAR over time was observed, there were no differences between the percentage of aroma released 60 and 120 s after wine expectoration. This percentage was about 50% of the initial amount of this compound in the mouth, confirming its higher oral persistence compared to the linear alcohols, which was in agreement with results from a previous work using intra-oral SPME [15].

On the other hand, furanic acids (furfural logP = 0.83, 5-methyl furfural logP = 0.67, and furfuryl alcohol logP = 0.45) and the volatile phenol guaiacol (logP = 1.34) showed very little changes (<5%) in their release over the three sampling times (Figure 2). Thus, these compounds were still present in the mouth 120 s after wine expectoration, indicating a relatively high oral persistence. These results are in agreement with the high oral adsorption determined for guaiacol after the application of the spit off odorant measurement procedure [15]. The existence of interactions between the galloyl ring of the wine polyphenols that might be adsorbed onto the oral mucosa with the aromatic ring of the aroma molecules through π–π interactions could explain these results [12,17]. Results from the present study also confirm this explanation for other polar volatile molecules from the furanic chemical group (furfural, methyl furfural, and furfuryl alcohol), all of them characterized by the presence of an aromatic ring in their structure.

Regarding terpenes (limonene and α-pinene), which are characterized for being compounds with a high hydrophobicity (logP = 4.27 and 4.83, respectively) (Table 1), they showed very similar %OAR over the three sampling points, and therefore a high oral persistence. Only about 12% of the initial aroma amount was lost in the third in-mouth extraction (T3), and then 120 s after wine expectoration (Figure 2). As previously explained for the larger and more hydrophobic esters, stronger interactions between the most hydrophobic aroma compounds and the hydrophobic domains of the salivary mucins from the mucosal pellicle that covers the oral surface can be expected [28]. This might explain the higher oral aroma persistence observed for these wine aroma compounds.

On the other hand, the $C_{13}$ norisoprenoid β-ionone, which also presented a high hydrophobicity (logP = 4.42) (Table 1), showed minor changes in the %OARt2 (10% of aroma loss compared to the first sampling point), but its oral release decreased almost 50% in the third sampling point (Figure 2). In previous works and using different methodologies such as intra-oral SPME [15] and in-nose PTR–MS [20], this compound showed a high oral persistence exhibiting relatively low oral aroma losses (around 40%) even at 4 min after the oral exposure to the wine.

Finally, in the case of the wine lactones (γ-butyrolactone and γ-nonalactone), they also showed minor changes in their oral release over the three sampling points with discrete aroma losses lower than 20% in both T2 and T3 sampling points (Figure 2). Despite the very different polarities of both lactones, γ-butyrolactone logP = −0.31 and γ-nonalactone logP = 2.08, their behavior was quite similar in the mouth, characterized by a high in-mouth persistence (Figure 2). The formation of Schiff bases between ketones and oral proteins (mucins, α-amylase) [2] might favor the higher adsorption of these compounds to the oral mucosa, thus being responsible for their higher oral persistence. However, this behavior should be confirmed in new in-vivo studies using synthetic wines supplemented with different types of ketone compounds with linear and cyclic structures.

### *3.2. Relationship between Oral Aroma Release and the Physicochemical Properties of Aroma Compounds*

To further understand the relationship between the physicochemical properties of aroma compounds and the oral release behavior, we performed a principal component analysis (PCA). For this, some physicochemical properties such as hydrophobicity (logP) and boiling point, as well as other features such as the odor thresholds of the 22 wine aroma compounds identified in the mouth after wine rinsing (Table 1), were selected as independent variables. Additionally, the %OAR calculated after 60 (%OARt2) and 120 s (%OARt3) were used. The compound cis-3-hexen-1-ol was removed from this treatment, since, as previously shown, it showed atypical oral release behavior that deserves additional studies. Figure 3 shows the representation of this PCA.

As can be seen in Figure 3, the PCA biplot explained 70.43% of the total data variation. The first principal component (PC1) explained 43.49%, while the second one (PC2) explained 26.94% of the total variability. PC1 was positively correlated to the percentage of oral aroma release from both the second (%OARt2) and third (%OARt3) in-mouth sampling points. As can be seen, PC1 mainly separated the compounds according to their chemical family. The furanic group, guaiacol, the two lactones, and the two terpenes exhibited positive values for PC1, with the furanic group showing the highest factor loadings (0.8–0.9). On the contrary, the ester group (except ethyl pentanoate and hexyl acetate) and the alcohol group (βphenylethanol, hexanol, pentanol) were negatively related to this component.

This confirms that the functional group is one of the main determinants of the oral release behavior, as previously suggested [4].

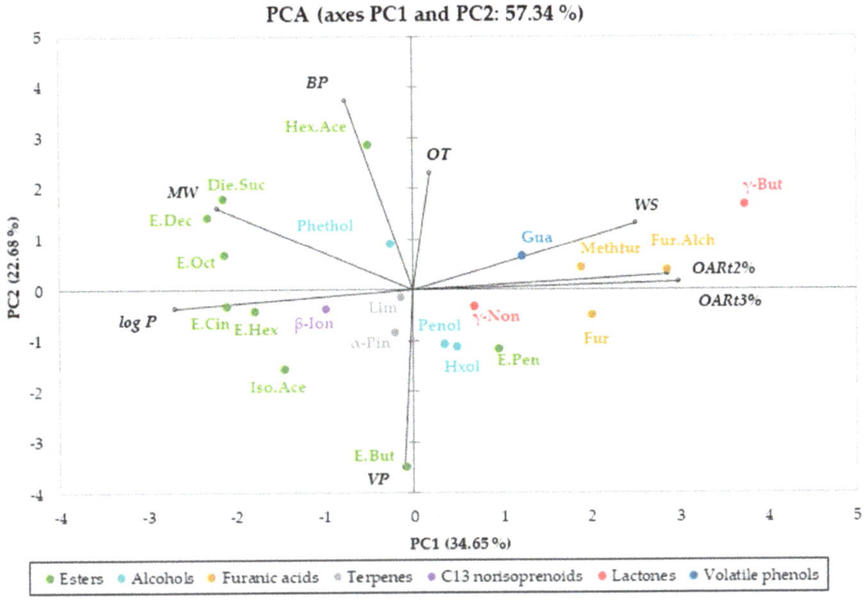


**Figure 3.** Principal component analysis (PCA) performed with some physicochemical features of the aroma compounds (boiling point (BP), hydrophobic constant (logP), odor threshold (OT)) and %OARt2 and %OARt3. logP: hydrophobic constant; BP: boiling point; OT: odor threshold; α-Pin: α-pinene; β-ion: β-ionone; γ-But: γ-butyrolactone; γ-Non: γ-nonalactone; Die.Suc: diethyl succinate; E.But: ethyl butanoate; E.Cin: ethyl cinnamate; E. Dec: ethyl decanoate; E.Hex: ethyl hexanoate; E. Oct: ethyl octanoate; E.Pen: ethyl pentanoate; Fur: furfural; Fur.Alc: furfuryl alcohol; Gua: guaiacol; Hex.Ace: hexyl acetate; Hxol: hexanol; Iso.Ace: isoamyl acetate; Lim: limonene; Methfur: 5-methyl furfural; Phethol: phenylethanol; Pntol: pentanol.

Additionally, PC1 was negatively correlated to the boiling point (BP) and hydrophobicity (logP) values. These results suggest that compounds with lower hydrophobicity (lower logP values), such as the furanic group, would show a higher oral aroma persistence compared to higher hydrophobic compounds. This could be because polar compounds are more easily dissolved in the water phase that covers the mucosal pellicle, which would produce an increase in their oral persistence, as recently suggested by Ployon and coworkers [28] using an ex vivo model. Thus, furanic acids (furfural, methyl furfural, and furfuryl alcohol), which were more polar in terms of logP (logP = 0.83, 0.67, and 0.45, respectively) (Table 1), were the compounds that showed the highest oral release in T2 and T3, exhibiting %OAR values higher than 95% in both sampling points (T2 and T3). However, this explanation does not seem valid for other polar compounds, such as the alcohols (with lower polarities compared to furanic compounds), in which the oral aroma persistence was relatively low (above 60% in both sampling points) compared to the furanic group. However, as previously shown, compound structure and the presence of aromatic rings in their molecule, as it is the case for some polar compounds (guaicol and furanic group), strongly determined their oral release behavior. In the same sense, the two compounds belonging to the terpene group, which have the highest logP values (4.27 and 4.83 for α-pinene and limonene, respectively), also showed higher oral aroma persistence compared to the alcohol group. This seems to corroborate the strong effect of the chemical structure on oral aroma persistence.

In the case of PC2, as shown in Figure 3, it practically did not show any correlation to the %OAR t2 and %OAR t3.

Nonetheless, it is also worth mentioning that the PCA results did not explain 100% of the data variation, which means that the physicochemical properties of the aroma compounds considered in this study were not the unique factor when explaining the oral aroma behavior of wine odorants after wine tasting. Other physicochemical features related to the molecular structure and the characteristics and composition of the oral environment might help explain the differences in oral aroma persistence among wine volatiles.

## 4. Conclusions

For the first time, using a large group of volunteers ($n$ = 32) and a real wine, we determined the oral release behavior of 22 wine aroma compounds in the mouth at three different points after wine tasting (immediate, 60 s, and 120 s) by using a previously validated in-mouth HSSE procedure. Even considering the large inter-individual variations as a consequence of physiological features, some interesting trends in oral release behavior were observed depending on the chemical compound. In general, esters and linear alcohols showed the highest variations (large decrease) in oral aroma release, and therefore a low oral aroma persistence. On the contrary, terpenes, lactones, and $C_{13}$ norisoprenoids showed lower variations in oral aroma release over the three sampling points, and therefore a higher oral aroma persistence. Additionally, and despite their low polarity, furanic acids and guaiacol showed the highest oral persistence. The oral release behavior was not only dependent on the physicochemical properties of the aroma compounds but also because of other features such as the molecular structure and probably on the characteristics and composition of the oral environment.

**Supplementary Materials:** The following are available online at https://www.mdpi.com/2304-8158/10/2/415/s1: Table S1: Volatile composition (average ± standard deviation in µg/L) of the red wine determined by HSSE-GC–MS and regression lines used for the quantification of aroma compounds in the wine. Figure S1. Chromatograms corresponding to the aroma profile of the wine (a) using HSSE and from the headspace of the mouth after spitting of the same wine using in-mouth HSSE (b). Numbers correspond to the compounds described in Table 1.

**Author Contributions:** Conceptualization and methodology, M.A.P.-B.; formal analysis, M.P.-J.; writing—original draft preparation, M.P.-J.; writing—review and editing, C.M.-G. and M.A.P.-B. All authors have read and agreed to the published version of the manuscript.

**Funding:** This research was funded by the Spanish MINECO through the AGL201678936-R and PID2019-111734RB-100 projects. M.P.-J. and C.M.-G. thank MINECO for their respective predoctoral and JdC-i contracts.

**Institutional Review Board Statement:** The study was conducted according to the guidelines of the Declaration of Helsinki, and approved by the Bioethical Committee of the Spanish National Council of Research (CSIC) (June, 2017).

**Informed Consent Statement:** Informed consent was obtained from all subjects involved in the study.

**Acknowledgments:** The authors greatly thank the volunteers who participated in this study and Celia Criado for her technical assistance.

**Conflicts of Interest:** The authors declare no conflict of interest. The funders had no role in the design of the study; in the collection, analyses, or interpretation of data; in the writing of the manuscript; or in the decision to publish the results.

## References

1. Pozo-Bayón, M.Á.; Muñoz-González, C.; Esteban-Fernández, A. Wine Preference and Wine Aroma Perception. In *Wine Safety, Consumer Preference, and Human Health*; Springer: Berlin/Heidelberg, Germany, 2016; pp. 139–162.
2. Ployon, S.; Morzel, M.; Canon, F. The role of saliva in aroma release and perception. *Food Chem.* **2017**, *226*, 212–220. [CrossRef]
3. Linforth, R.; Taylor, A.J. Persistence of volatile compounds in the breath after their consumption in aqueous solutions. *J. Agric. Food Chem.* **2000**, *48*, 5419–5423. [CrossRef] [PubMed]

4. Muñoz-González, C.; Feron, G.; Brulé, M.; Canon, F. Understanding the release and metabolism of aroma compounds using micro-volume saliva samples by ex vivo approaches. *Food Chem.* **2018**, *240*, 275–285. [CrossRef]
5. Muñoz-González, C.; Feron, G.; Canon, F. Physiological and oral parameters contribute prediction of retronasal aroma release in an elderly cohort. *Food Chem.* **2020**, *342*, 128355. [CrossRef]
6. Buettner, A.; Schieberle, P. Influence of mastication on the concentrations of aroma volatiles—some aspects of flavour release and flavour perception. *Food Chem.* **2000**, *71*, 347–354. [CrossRef]
7. Buettner, A. Investigation of potent odorants and afterodor development in two chardonnay wines using the buccal odor screening system (boss). *J. Agric. Food Chem.* **2004**, *52*, 2339–2346. [CrossRef]
8. King, E.S.; Kievit, R.L.; Curtin, C.; Swiegers, J.H.; Pretorius, I.S.; Bastian, S.E.; Francis, I.L. The Effect of multiple yeasts co-inoculations on sauvignon blanc wine aroma composition, sensory properties and consumer preference. *Food Chem* **2010**, *122*, 618–626. [CrossRef]
9. Baker, A.K.; Ross, C.F. Wine finish in red wine: The effect of ethanol and tannin concentration. *Food Qual. Prefer.* **2014**, *38*, 65–74. [CrossRef]
10. Genovese, A.; Piombino, P.; Gambuti, A.; Moio, L. Simulation of retronasal aroma of white and red wine in a model mouth system. Investigating the influence of saliva on volatile compound concentrations. *Food Chem.* **2009**, *114*, 100–107. [CrossRef]
11. Mitropoulou, A.; Hatzidimitriou, E.; Paraskevopoulou, A. Aroma release of a model wine solution as influenced by the presence of non-volatile components. Effect of Commercial tannin extracts, polysaccharides and artificial saliva. *Food Res. Int.* **2011**, *44*, 1561–1570. [CrossRef]
12. Perez-Jiménez, M.; Esteban-Fernández, A.; Muñoz-González, C.; Pozo-Bayón, M.A. Interactions among odorants, phenolic compounds, and oral components and their effects on wine aroma volatility. *Molecules* **2020**, *25*, 1701. [CrossRef] [PubMed]
13. Piombino, P.; Moio, L.; Genovese, A. Orthonasal vs. retronasal: Studying how volatiles' hydrophobicity and matrix composition modulate the release of wine odorants in simulated conditions. *Food Res. Int.* **2019**, *116*, 548–558. [CrossRef]
14. Buettner, A.; Mestres, M. Investigation of the retronasal perception of strawberry aroma aftersmell depending on matrix composition. *J. Agric. Food Chem.* **2005**, *53*, 1661–1669. [CrossRef] [PubMed]
15. Esteban-Fernández, A.; Rocha-Alcubilla, N.; Muñoz-González, C.; Moreno-Arribas, M.V.; Pozo-Bayón, M.Á. Intra-oral adsorption and release of aroma compounds following in-mouth wine exposure. *Food Chem.* **2016**, *205*, 280–288. [CrossRef] [PubMed]
16. Muñoz-González, C.; Pérez-Jiménez, M.; Pozo-Bayón, M.Á. Oral persistence of esters is affected by wine matrix composition. *Food Res. Int.* **2020**, *135*, 109286. [CrossRef] [PubMed]
17. Perez-Jiménez, M.; Chaya, C.; Pozo-Bayón, M.Á. Individual differences and effect of phenolic compounds in the immediate and prolonged in-mouth aroma release and retronasal aroma intensity during wine tasting. *Food Chem.* **2019**, *285*, 147–155. [CrossRef]
18. Perez-Jimenez, M.; Pozo-Bayon, M.A. Development of an in-Mouth headspace sorptive extraction method (hsse) for oral aroma monitoring and application to wines of different chemical composition. *Food Res. Int.* **2019**, *121*, 97–107. [CrossRef] [PubMed]
19. Muñoz-González, C.; Martín-Álvarez, P.J.; Moreno-Arribas, M.V.; Pozo-Bayón, M.Á. Impact of the nonvolatile wine matrix composition on the in vivo aroma release from wines. *J. Agric. Food Chem.* **2014**, *62*, 66–73. [CrossRef] [PubMed]
20. Munoz-Gonzalez, C.; Canon, F.; Feron, G.; Guichard, E.; Pozo-Bayon, M.A. Assessment wine aroma persistence by using an in vivo Ptr-Tof-Ms approach and its relationship with salivary parameters. *Molecules* **2019**, *24*, 1277. [CrossRef]
21. Muñoz-González, C.; Feron, G.; Guichard, E.; Rodríguez-Bencomo, J.J.; Martín-Álvarez, P.J.; Moreno-Arribas, M.V.; Pozo-Bayón, M.Á. Understanding the Role of saliva in aroma release from wine by using static and dynamic headspace conditions. *J. Agric. Food Chem.* **2014**, *62*, 8274–8288. [CrossRef] [PubMed]
22. Mayr, C.M.; Parker, M.; Baldock, G.A.; Black, C.A.; Pardon, K.H.; Williamson, P.O.; Herderich, M.J.; Francis, I.L. Determination of the importance of in-mouth release of volatile phenol glycoconjugates to the flavor of smoke-tainted wines. *J. Agric. Food Chem.* **2014**, *62*, 2327–2336. [CrossRef] [PubMed]
23. Muñoz-González, C.; Cueva, C.; Pozo-Bayon, M.A.; Moreno-Arribas, M.V. Ability of Human oral microbiota to produce wine odorant aglycones from odourless grape glycosidic aroma precursors. *Food Chem.* **2015**, *187*, 112–119. [CrossRef]
24. Buettner, A. Influence of human salivary enzymes on odorant concentration changes occurring in vivo. 1. Esters and thiols. *J. Agric. Food Chem.* **2002**, *50*, 3283–3289. [CrossRef] [PubMed]
25. Buettner, A. Influence of human saliva on odorant concentrations. 2. Aldehydes, alcohols, 3-alkyl-2-methoxypyrazines, methoxyphenols, and 3-hydroxy-4, 5-dimethyl-2 (5 h)-furanone. *J. Agric. Food Chem.* **2002**, *50*, 7105–7110. [CrossRef] [PubMed]
26. Lasekan, O. A comparative analysis of the influence of human salivary enzymes on odorant concentration in three palm wines. *Molecules* **2013**, *18*, 11809–11823. [CrossRef] [PubMed]
27. Pérez-Jiménez, M.; Rocha-Alcubilla, N.; Pozo-Bayón, M.Á. Effect of saliva esterase activity on ester solutions and possible consequences for the in-mouth ester release during wine intake. *J. Agric. Food Chem.* **2018**, *50*, 62–70.
28. Ployon, S.; Brule, M.; Andriot, I.; Morzel, M.; Canon, F. Understanding retention and metabolization of aroma compounds using an in vitro model of oral mucosa. *Food Chem.* **2020**, *318*, 126468. [CrossRef] [PubMed]
29. Muñoz-González, C.; Brulé, M.; Feron, G.; Canon, F. Does Interindividual variability of saliva affect the release and metabolization of aroma compounds ex vivo? The particular case of elderly suffering or not from hyposalivation. *J. Texture Stud.* **2019**, *50*, 36–44. [CrossRef] [PubMed]
30. Weldegergis, B.T.; Tredoux, A.G.J.; Crouch, A.M. Application of a headspace sorptive extraction method for the analysis of volatile components in south African wines. *J. Agric. Food Chem.* **2007**, *55*, 8696–8702. [CrossRef] [PubMed]

31. Buettner, A.; Schieberle, P. Exhaled odorant measurement (Exom)—a new approach to quantify the degree of in-mouth release of food aroma compounds. *LWT* **2000**, *33*, 553–559. [CrossRef]
32. Linforth, R.; Martin, F.; Carey, M.; Davidson, J.; Taylor, A.J. Retronasal transport of aroma compounds. *J. Agric. Food Chem.* **2002**, *50*, 1111–1117. [CrossRef]
33. San Juan, F. Estrategias Óptimas Para La Caracterización De Las Moléculas Odorantes Del Vino: Modelización De La Calidad Y Atributos Sensoriales. Ph.D. Thesis, Universidad de Zaragoza, Zaragoza, Spain, 2012.
34. Peng, C.-T.; Wen, Y.; Tao, Y.-S.; Lan, Y.-Y. Modulating the formation of meili wine aroma by prefermentative freezing process. *J. Agric. Food Chem.* **2013**, *61*, 1542–1553. [CrossRef] [PubMed]
35. Li, H.; Tao, Y.-S.; Wang, H.; Zhang, L. Impact odorants of chardonnay dry white wine from Changli county (China). *Eur. Food Res. Technol.* **2008**, *227*, 287–292. [CrossRef]
36. Juan, F.S.; Cacho, J.; Ferreira, V.; Escudero, A. Aroma Chemical composition of red wines from different price categories and its relationship to quality. *J. Agric. Food Chem.* **2012**, *60*, 5045–5056. [CrossRef] [PubMed]
37. Piombino, P.; Genovese, A.; Esposito, S.; Moio, L.; Cutolo, P.P.; Chambery, A.; Severino, V.; Moneta, E.; Smith, D.P.; Owens, S.M. Saliva from obese individuals suppresses the release of aroma compounds from wine. *PLoS ONE* **2014**, *9*, e85611. [CrossRef] [PubMed]
38. Mishellany-Dutour, A.; Woda, A.; Laboure, H.; Bourdiol, P.; Lachaze, P.; Guichard, E.; Feron, G. Retro-nasal aroma release is correlated with variations in the in-mouth air cavity volume after empty deglutition. *PLoS ONE* **2012**, *7*, e41276. [CrossRef]
39. Pérez-Jiménez, M.; Muñoz-González, C.; Pozo-Bayón, M.Á. Understanding human salivary esterase activity and its variation under wine consumption conditions. *RSC Adv.* **2020**, *10*, 24352–24361.

foods

MDPI

*Article*

# Evaluation of the Effect of a Grape Seed Tannin Extract on Wine Ester Release and Perception Using In Vitro and In Vivo Instrumental and Sensory Approaches

**Carolina Muñoz-González *, Celia Criado, María Pérez-Jiménez and María Ángeles Pozo-Bayón**

Instituto de Investigación en Ciencias de la Alimentación (CIAL), CSIC-UAM, C/Nicolás Cabrera, 9, 28049 Madrid, Spain; celia.criado@csic.es (C.C.); maria.perez@csic.es (M.P.-J.); m.delpozo@csic.es (M.Á.P.-B.)
* Correspondence: c.munoz@csic.es

**Abstract:** This study aimed to systematically evaluate the effect of a commercial grape seed tannin extract (GSE) fully characterized (53% monomers, 47% procyanidins) on wine ester release and perception using a global approach. The behavior of two esters (ethyl hexanoate, ethyl decanoate) was studied in a control wine or in the same wine supplemented with the GSE in preconsumption (in vitro headspace-stir bar sorptive extraction-gas chromatography mass spectrometry (HS-SBSE-GC/MS) and orthonasal perception) and consumption (intraoral-HS-SBSE-GC/MS and dynamic retronasal perception) conditions. For the compound ethyl hexanoate, no significant differences ($p > 0.05$) among wines were observed in the in vitro analyses while they were observed in the three in vivo experiments ($p < 0.05$). Thus, the wine supplemented with the GSE showed lower (35%) in vivo release and ortho (36%) and retronasal (16%) perception scores than the control wine. Overall, this suggests that components of the GSE could interact with this compound, directly and/or through complexes with oral components, affecting its release and conditioning its perception. However, perceptual interactions and effects of polyphenols on oral esterases cannot be discarded. On the contrary, the compound ethyl decanoate was not significantly affected by the addition of GSE. In conclusion, the addition of tannin extracts to wines can modulate aroma perception in a compound-dependent manner.

**Keywords:** wine; tannins; volatile compounds; polyphenol-aroma interactions; saliva; in vitro release; in vivo release; retronasal aroma; time-intensity; HS-GC/MS



**Citation:** Muñoz-González, C.; Criado, C.; Pérez-Jiménez, M.; Pozo-Bayón, M.Á. Evaluation of the Effect of a Grape Seed Tannin Extract on Wine Ester Release and Perception Using In Vitro and In Vivo Instrumental and Sensory Approaches. *Foods* **2021**, *10*, 93. https://doi.org/10.3390/foods10010093

Received: 6 December 2020
Accepted: 2 January 2021
Published: 5 January 2021



## 1. Introduction

Grape polyphenols can be naturally present in wines, or they can be added as oenological extracts. Grape tannin extracts usually contain condensed tannins (or proanthocyanidins) and other components, depending on their purity [1]. Their addition to wines was related to providing color stabilization, contributing to wine structure and mouthfeel [1,2] or to wine flavor [3], among others. Apart from their effects on astringency and bitterness [3], recent studies have suggested that grape tannin extracts could modulate wine aroma [4,5], which is one of the main drivers for wine preferences [6].

The mechanisms behind the modulator effect of grape tannin extracts on wine aroma are unknown, but different hypotheses were proposed in the literature. On the one hand, the capacity of components present on the tannin extracts to interact with some volatile compounds was proven using in vitro conditions in model wines [7–9]. In 1999, Dufour and Bayonove [7] observed that the addition of a condensed tannin fraction (0–14 g/L) to a hydroalcoholic solution provoked a small reduction of benzaldehyde volatility and an increase of limonene volatility while had no effect on two esters (isoamyl acetate, ethyl hexanoate). Accordingly, Mitropoulou and coworkers [8] observed that the addition of a condensed tannin extract (0–12 g/L) to a model wine had almost no influence on the volatility of some esters (isoamyl acetate and ethyl hexanoate) while others (ethyl octanoate, ethyl decanoate, ethyl dodecanoate) suffer a salting-out or retention effect depending on the tannin concentration assayed. In 2013, Villamor and coworkers [9] found that increasing tannin addition (500 mg/L–1500 mg/L) to a model wine increased the release of high molecular weight compounds to the headspace (1-octen-3-one, 2-methoxyphenyl, eugenol) but decreased the release of alcohols (3-methyl-1-butanol, 1-hexanol). From these studies, it could be presumed that these interactions could affect the availability of the aroma compounds to reach the olfactory receptors through the orthonasal pathway, affecting most likely wine odor [9]. However, olfaction is a complex phenomenon that can also be affected by several biases, such as the way sniffing is performed ([10]), and therefore, studies are needed to verify this point.

During consumption, wine is introduced in the oral cavity and components of the tannin extracts can interact with salivary proteins in the mouth (free or bound to the mucosal pellicle) [11–14] or directly to the epithelial mucosa cells [15]. As side effects, these interactions may influence the oral retention of volatiles and their transfer to the olfactory receptors through the retronasal pathway. Thus, it was suggested that some polyphenols already adsorbed onto the mucosal pellicle might bind certain aroma molecules with aromatic rings, such as guaiacol [16], increasing their retention in the mouth. Apart from physicochemical interactions, polyphenols present in the tannin extracts could affect the activity of salivary enzymes [17]. This would be of special relevance in the case of enzymes involved in the metabolization of aroma compounds [18]. In this regard, inhibition of esterases by grape seed tannin extracts was recently suggested [4,17]. Thus, the influence of tannin extracts on salivary enzymes deserves more research. Finally, aroma perception during wine consumption is a complex phenomenon produced as a result of the integration in the brain of multiple sensory signals [19,20]. Therefore, to understand the meaning of the polyphenol-volatile interactions in a real consumption context, wine retronasal aroma perception must be assessed.

Additionally, it is important to notice that the technical effects provided by the tannin extracts will be dependent upon their composition. In this regard, there is a wide range of oenological extracts in the market, although, in general, there is a lack of information about their composition [1]. This makes it a challenge to attribute the effects of these extracts to specific components, making it difficult the comparison across studies. The aim of this work was to systematically evaluate the effect of a grape seed tannin extract (GSE) previously characterized [21] on the release and perception of two target volatile compounds using in vitro and in vivo instrumental and sensory approaches. To do this, a wine with a low-aroma profile was enriched with two esters (20 mg/L), which are typical ubiquitous volatile compounds from alcoholic fermentation and relevant compounds contributing to the wine aroma profile. The aromatized rosé wine (control wine) was supplemented with the GSE (at 500 mg/L). Ester release was studied in vitro, through headspace-stir bar sorptive extraction-gas chromatography mass spectrometry (HS-SBSE-GC/MS), and in vivo, through intra-oral-HS-SBSE-GC/MS ($n$ = 10). Additionally, orthonasal and the dynamics of retronasal aroma perception during wine tasting were evaluated through sensory evaluation ($n$ = 10). The same panelists participated in all the experiments. To our knowledge, this is the first time that such a global approach was applied to understand the effect of oenological tannins on wine aroma.

## 2. Materials and Methods

### 2.1. Wine Samples

A commercial rosé wine (D.O. Navarra) (pH of 3.03, 13% $v/v$ ethanol, total polyphenol content of 355.88 ± 1.74 mg Eq Gallic acid/L and procyanidin content of 66.37 ± 0.01 mg/L) was selected for this study by its low polyphenol content. A commercial tannin extract from grape seeds (GSE) (Vitaflavan®) (D.R.T. Les Dèrives Resiniques and Terpéniques, Vielle-Saint-Girons, France) mainly composed of 53% monomers and 47% procyanidins was added to this control wine (CW) at a concentration of 500 mg/L, to obtain another wine with a higher polyphenol content (PW). The individual phenolic composition of this extract was previously published [21]. The GSE was deodorized under an $N_2$ current for 30 min in order to remove endogenous aromas.

Two food-grade linear esters (Sigma-Aldrich, Steinheim, Germany) with different physicochemical properties (Table 1) were added to the wine glasses to obtain a final concentration of 20 mg/L. The technique HS-HSSE-GC/MS [22] was used to evaluate the presence of endogenous aroma compounds in the original wine that was insignificant compared to the concentration of the aroma added (ethyl hexanoate: 0.37 ± 0.02 mg/L; ethyl decanoate: not detected).

**Table 1.** Characteristics of the esters added to the wines.

| Compounds | CAS Number | Physicochemical Characteristics | | | | Sensory Descriptor [d] |
|---|---|---|---|---|---|---|
| | | Chemical Formula | Mw [a] | *log P* [b] | BP [c] | |
| Ethyl hexanoate | 123-66-0 | $C_8H_{16}O_2$ | 144 | 2.83 | 167 | Fruity |
| Ethyl decanoate | 110-38-3 | $C_{12}H_{24}O_2$ | 200 | 4.79 | 248 | Grape |

[a] Molecular weight (g/mol). [b] Hydrophobic constant estimated with EPI suite (US EPA 2000–2007). [c] Boiling point (°C) estimated with EPI Suite (US EPA 2000–2007). [d] Flavornet (http://www.flavornet.org) database.

### 2.2. Panel

For this study, ten young individuals (8 females, 2 males) were recruited. A triangular test with aromatized hydroalcoholic solutions was performed to evaluate their absence of anosmia. All the individuals received information on the nature of the study and provided written consent form before their participation. They were instructed not to eat or drink one hour before the instrumental or sensory experiments. The CSIC Bioethics Committee for Research approved the experimental protocol.

### 2.3. Instrumental Evaluation

#### 2.3.1. In Vitro Ester Release

A previous method [22] was followed with slight modifications. Briefly, wine samples (1-mL) were placed in headspace vials (20-mL) containing a glass insert with a preconditioned PDMS (polydimethylsiloxane) stir bar (Twister) (Gerstel, Mülheim, Germany). Extraction was performed for 15 min, and then, the twisters were removed for thermal desorption. The experiments were done in triplicate.

#### 2.3.2. In Vivo Ester Release

An instrumental procedure developed by Pérez-Jimenez and Pozo-Bayón [22] was used. Briefly, each individual introduced the wine samples (15 mL) into their mouths, kept them for 30 s and then expectorated them. Two aroma extractions were performed. The first one, five seconds after expectoration, and the second one, sixty seconds later. The extractions were performed for 30 s with a PDMS twister (Gerstel, Mülheim, Germany) introduced in the oral cavity using glass made twister holders. Once the extraction was finished, the twisters were removed for thermal desorption. Each wine sample was analyzed in triplicate by each individual. Controls of the oral cavity before wine rinsing were also performed to ensure the absence of the esters under study.

#### 2.3.3. Thermal Desorption and Gas Chromatography Mass Spectrometry (GC–MS) Analyses

The twisters were desorbed using a TDU, and a CIS-4 injector (Gerstel, Mülheim, Germany) coupled to a 6890 N GC—5973 N mass spectrometers (Agilent). Compounds were separated in a DB-WAX column (30 m × 0.25 mm i.d. × 0.50 μm film thickness) (J&W Scientific, Folsom, CA, USA) using the following oven temperatures: 40 °C, held 1 min, 10 °C/min to 240 °C and 240 °C held 1 min (run time: 18 min). The acquisitions were performed in the scan (from $m/z$ 35 to 350) and SIM modes. For specific details on the GC–MS conditions, see Pérez-Jimenez and Pozo-Bayón [22].

Aroma identification was done by using retention times and mass spectra provided by the NIST 2.0 database. The aroma release between wine samples was done on the basis of absolute peak areas (APAs).

### 2.4. Sensory Evaluation

#### 2.4.1. Training Sessions

First, individuals were trained to recognize orthonasally and retronasally the two aroma descriptors of the study (fruity and grape) in a wine matrix containing the respective chemical references (ethyl hexanoate and ethyl decanoate). Second, they received specific training on the use of the intensity scale (15 cm unstructured scale delimited at the ends) and on the discrimination between different aroma intensities (corresponding to 1, 3, 9 and 27 mg/L of each aroma compound). Third, individuals were trained in the protocol for wine consumption to evaluate retronasal aroma perception and in the use of tablet devices using the Compusense software for time-intensity (TI) measurements. Six training sessions were performed by all the individuals, while three of them received 2 additional sessions to be fully trained. The serving temperature of the samples was 18 °C. Samples were served in wine glasses covered with plastic Petri dishes to prevent volatile loss. Water (Nestle Aquarel, Barcelona, Spain) and unflavored unsalted crackers (ARO, Madrid, Spain) were provided to the individuals among samples.

#### 2.4.2. Orthonasal Evaluation

Wine samples were equilibrated for 15 min. Then, individuals were instructed to remove the plastic Petri dishes over the wine glasses, to smell them and to evaluate the intensity of the two descriptors (fruity and grape) using an unstructured scale. Samples were evaluated in triplicate and in randomized order.

#### 2.4.3. Retronasal Evaluation by Time-Intensity (TI)

Individuals were trained to follow the instructions: introduce the wine samples (15 mL) into their mouths and expectorated them after 5 s. After this time, individuals started the TI evaluation of the two aroma descriptors (fruity and grape) using the Compusense software (version 19.0.7236.30304, Guelph, Canada) that lasted 60 s. To do so, they had to move the cursor along the unstructured scale (15 cm) to note the intensity of aroma descriptors perceived. Data were recorded at a frequency of 1 s. Samples were evaluated in triplicate and in randomized order.

Mean curves of dynamic intensity scores for each wine were determined by averaging the data of all the individuals at each time point.

### 2.5. Statistical Analyses

One-way analysis of variance (ANOVA) and Tukey's test for mean comparison were used to check the influence of the GSE on the in vitro and in vivo ester release and on the orthonasal and retronasal ester perception. The significance level was $p < 0.05$, $\alpha < 0.05$ throughout the study. The XLSTAT program (v.19.01) (Addinsoft, Paris, France) was used for data processing.

## 3. Results

### 3.1. *Effects of a Grape Seed Tannin Extract (GSE) on the* In Vitro *Ester Release from Wine*

To determine the effect of the GSE in preconsumption (in vitro) conditions, the APAs of esters recovered in the headspace above the CW and PW were submitted to a one-way ANOVA. The results are shown in Figure 1. As can be seen, no significant differences were observed between wines for any of the two esters assayed ($p > 0.05$). Although the release of both esters was slightly higher in the CW compared to the PW, these differences were not statistically significant ($p > 0.05$).

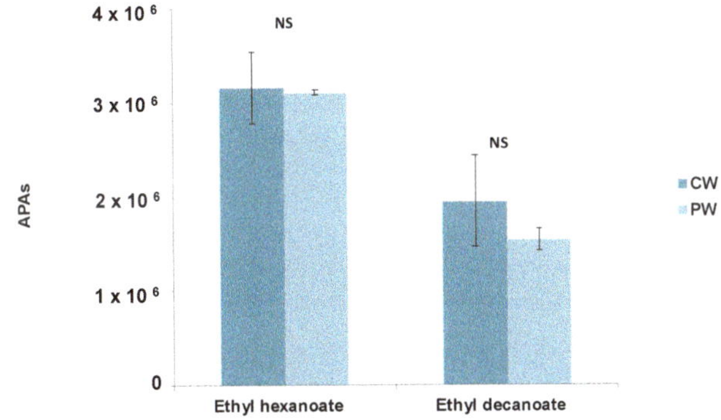


**Figure 1.** Comparison of the absolute peak areas (APAs) of esters released to the headspace in the control wine (CW) and in the wine with grape seed tannin extract (GSE) added (PW) using in vitro conditions (NS: no significant differences ($p > 0.05$) between wine samples).

### 3.2. *Effects of a Grape Seed Tannin Extract (GSE) on the* In Vivo *Oral Ester Release from Wine*

To check the influence of the GSE in consumption conditions, the esters released into the oral cavity of ten individuals after they rinsed their mouths with the two wines (CW and PW) was followed at two sampling points (0 and 60 s). The APAs of esters recovered intra-orally were submitted to a one-way ANOVA. The results can be observed in Figure 2.

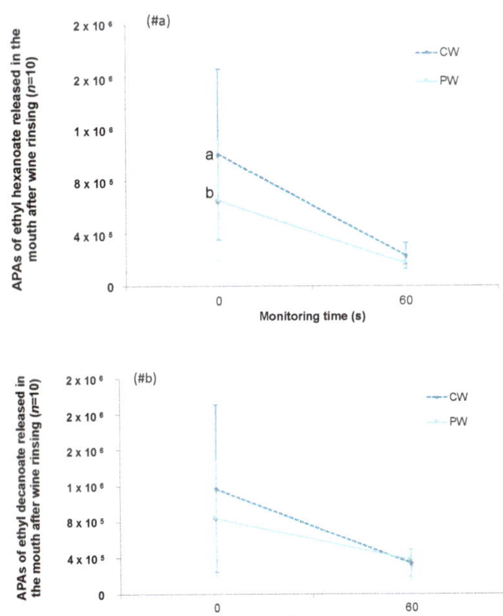


**Figure 2.** Comparison of the absolute peak areas (APAs) of ethyl hexanoate (**#a**) and ethyl decanoate (**#b**) released in the mouth after oral rinsing with the control wine (CW) and with the wine with the GSE added (PW) using in vivo conditions. Different letters across the different time points indicate significant differences between wines ($p < 0.05$, $\alpha < 0.05$).

As expected, a decrease in ester release was observed over time in the oral cavity, although the rate of the decrease seemed faster for the CW than for the PW. Moreover, large interindividual differences on the intra-oral ester release were observed for both compounds, mostly on the first monitoring time. The presence of the GSE seemed to decrease the intra-oral release of both esters. Although a similar trend was observed for both esters (Figure 2), results of the ANOVA test only showed a significant effect ($p < 0.05$) of the GSE for the compound ethyl hexanoate on the first monitoring time (0 s) (being a 35% less released in PW compared to CW). In the second monitoring time (60 s), no significant differences between wines were observed for any of the esters assayed.

### 3.3. Effects of a Grape Seed Tannin Extract (GSE) on the Orthonasal Perception of Wine Esters

A one-way ANOVA was performed to evaluate the effect of the GSE addition on the orthonasal perception of the two aroma descriptors (fruity and grape) by 10 individuals (Figure 3). The evaluation of the two aroma descriptors was done independently of one another.

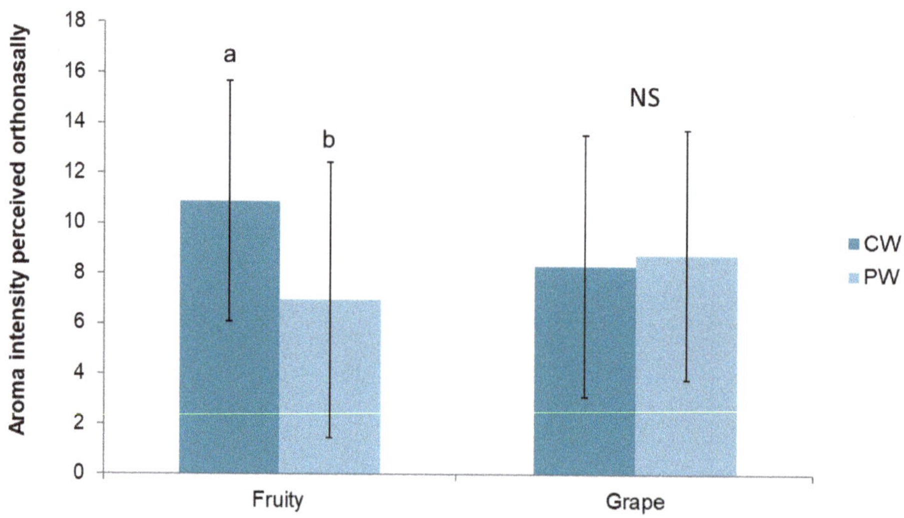


**Figure 3.** Intensity scores of the aroma descriptors determined orthonasally by the trained panel ($n$ = 10) in the control wine (CW) and in the wine supplemented with the GSE (PW). Different letters across the different time points indicate significant differences between wines ($p < 0.05$, $\alpha < 0.05$); NS: no significant differences between wines ($p > 0.05$).

As it can be seen, significant differences were observed in the perception of fruity note (from ethyl hexanoate) between wines, with PW showing a significantly lower score (36%) for fruity intensity than CW. In contrast, no significant differences were observed for the perception of grape note (ethyl decanoate) between wines.

### 3.4. Effects of a Grape Seed Tannin Extract (GSE) on the Retronasal Perception of Wine Esters

A one-way ANOVA was carried out to check the effect of the GSE addition on the perception of the aroma descriptors elicited by the two esters through a dynamic sensory TI evaluation. The evaluation of the two aroma descriptors was done independently of one another. As it can be seen (Figure 4), a decrease in the intensity notation of both attributes was observed over time. The GSE seemed to suppress the perception of the fruity attribute (ethyl hexanoate), while no effect was observed for the grape note (ethyl decanoate). However, the differences observed for the fruity note were only significant in the monitoring time of 10 s (Figure 4), with a difference in the intensity score between PW and CW of 16%.

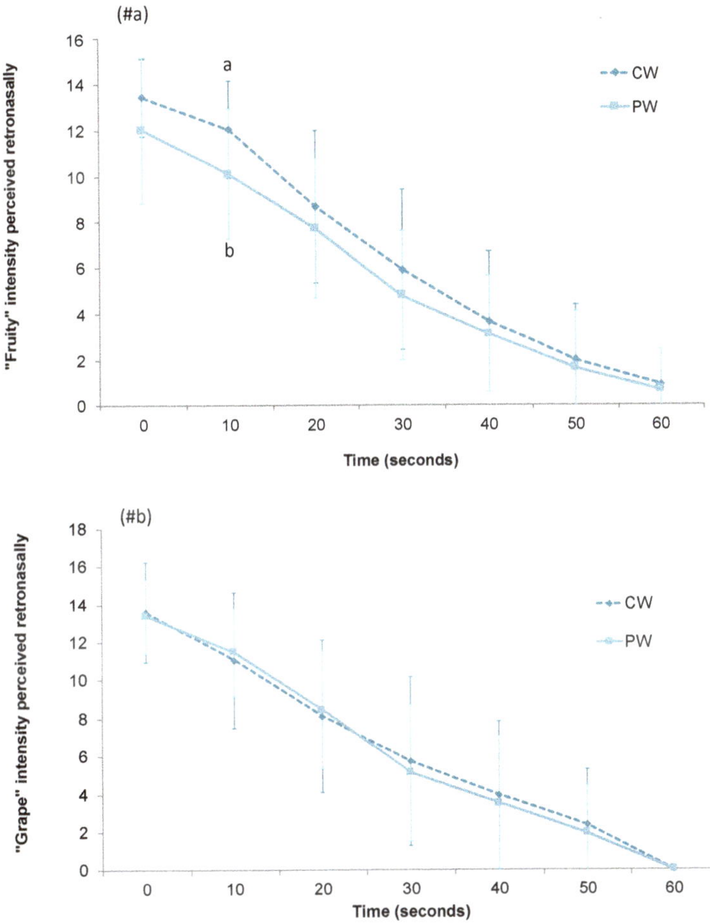


**Figure 4.** Intensity scores of the fruity (**#a**) and grape (**#b**) descriptors determined retronasally by the trained panel ($n$ = 10) in the control wine (CW) and in the wine supplemented with the GSE (PW). Different letters across the different time points indicate significant differences between wines ($p < 0.05$, $\alpha < 0.05$).

## 4. Discussion

The effects of tannin extracts on wine aroma are hardly understood, among other reasons, because the comparison across studies is difficult. Usually, studies on this topic have employed commercial extracts of unknown/not specified composition, which generates complications to compare the results and to extract conclusions. Apart from that, the fact that the studies on this topic are made with different experimental protocols and/or techniques or with wines with a different chemical composition (volatile and matrix) adds a degree of uncertainty to this topic. For these reasons, a global and systematic approach was followed in this study. A grape seed tannin extract (GSE) (53% monomers, 47% procyanidins), fully characterized previously [21], was used to evaluate its influence on wine aroma release and perception. A rosé wine with (PW) or without (CW) the GSE was systematically assayed using a wide array of instrumental and sensory techniques. The wine was aromatized with two esters, ethyl hexanoate (fruity note) and ethyl decanoate (grape note), that were selected for being compounds from the same chemical family (esters) presenting different physicochemical properties (Table 1) and because they are associated with pleasant aroma notes easily recognizable (ortho and retronasally) by consumers. The same individuals ($n = 10$) participated in all in vivo (instrumental and sensory) experiments, and they were previously trained to follow specific tasting protocols. Before the experimentation, the absence of anosmia was checked for all the individuals in preliminary triangular tests.

The effects of the GSE on wine esters were firstly evaluated in pre-consumption conditions, that is, before the wine is introduced in the mouth. One of the first sensory signals perceived by wine consumers is related to its odor, produced when volatiles are released from the wine matrix and reach the olfactory receptors by the orthonasal pathway. This phenomenon was evaluated through two approaches: one instrumental by measuring the esters released to the headspace above the wines (CW, PW) in in vitro conditions, and one sensory approach by measuring the orthonasal perception of the individuals ($n = 10$) when smelling the wines. Results from the in vitro static headspace analyses did not show a significant effect ($p > 0.05$) of the GSE on the release of esters to the headspace (Figure 1). These results are in agreement with those from Dufour and Bayonove and Mitropoulou et al. [7,8], who found that the addition of a tannin extract to model wines did not affect the release of ethyl hexanoate at any of the assayed tannin extract concentrations. On the contrary, Mitroupoulu and collaborators [8] did observe a salting-out effect on ethyl decanoate upon tannin addition, which was not observed in the present study. Divergences among studies could be due to compositional differences among the extracts assayed and/or the concentrations employed (0.5 g/L vs. 0–12 g/L), although no direct comparison can be made since in work from Mitropoulou et al. [8] the composition of the tannin extract was not specified. Regarding the results of the orthonasal experiment (Figure 3), significant differences were observed between wines (CW and PW) for the fruity note provided by the compound ethyl hexanoate, while no effects were observed for the grape descriptor associated with the compound ethyl decanoate. Thus, the fruity note was less (36%) intensely rated by the individuals in the PW compared to the CW. Although to the authors' knowledge, the effects of procyanidins on the orthonasal perception of wine esters has not been elucidated to date, the effects of monomers and, in particular, of catechin and gallic acid (also constituents of the grape seed extract) on the suppression of perception of esters was confirmed in several works [23–25]. Thus, this suppressed perception could be due to the monomers present in the extract, although the role of procyanidins cannot be discarded.

These apparently contradictory results between the instrumental and sensory experiment performed in preconsumption conditions were also observed by Lorrain and colleagues [23] when studying the effects of phenolic compounds on the volatility and sensory perception of red wine esters in model solutions. In that article, the authors found a significant effect of the compound catechin on the orthonasal perception of specific esters, although the same effects were much less clear on the headspace experiments. They related this contradiction to a conflict on sensitivity between mass spectrometry and the human nose, being the latter more sensitive than the former. Moreover, they stated that the weakness of the polyphenol-volatile hydrophobic interactions (also suggested by Dufour and Bayonove [7]) could have produced that small changes between instrumental and sensory approaches (volumes, temperatures, extraction times) would lead to a disruption of the interactions between polyphenols and volatiles. Overall, this could also have occurred in the present work.

During consumption, wine is introduced into the oral cavity. From here, aroma compounds interact with oral components before traveling to the olfactory receptors located in the nose to be perceived. In an attempt to decipher the effects of the GSE on wine aroma in a consumption situation, two complementary instrumental and sensory approaches were used. First, the esters released in the oral cavity of the individuals after wine rinsing were monitored through a previously developed technique called intraoral-HS-SBSE-GC/MS [22]. Figure 2 shows that the intra-oral ester release decreased over time, which means that esters disappeared from the oral cavity after wine rinsing, as was expected. Interestingly, the decrease rate seemed faster in CW than in PW, which could indicate that the GSE addition could have an effect on aroma persistence. This could be related to an inhibitory effect of polyphenols (present in the GSE) on salivary enzymes as it was recently suggested [17], which will deserve more research. Another interesting finding was related to the fact that the intra-oral release was significantly different among wines during the first monitoring time, with PW releasing less ethyl hexanoate to the oral headspace than CW (35% difference). This could be the consequence of the polyphenol-volatile interaction observed in the orthonasal analyses and would indicate that a lower amount of ethyl hexanoate molecules could be available to reach the olfactory receptors immediately after wine rinsing in PW than in CW. In addition, it could not be discarded that interactions between components of the GSE and oral components (salivary proteins, buccal cells) [8,9,16] could have affected, as side effects, ethyl hexanoate volatility. In this regard, a negative relationship between flavan-3-ol content in wines and intra-oral aroma release of esters, and especially of ethyl hexanoate, was previously observed [16]. In that study, it was suggested that this phenomenon could be due to the formation of salivary proteins-polyphenol-carbohydrate complexes able to encapsulate aroma compounds in the oral cavity. On the contrary, for the compound ethyl decanoate, no differences between wines were found for any of the time points assayed. This result differed from that found by Muñoz-Gonzalez and coworkers [4], who observed an increase in the oral volatility of this compound after drinking wine with a similar extract added. Nevertheless, it should be noticed that the monitoring times and the concentration of the extract assayed were very different in both experiments, which confirms that the experimental conditions employed are determinant to compare the effects of GSE on wine aroma.

Second, the retronasal perception of the individuals after rinsing their mouths with the two wines was evaluated using the dynamic TI approach. In this case (Figure 4), the individuals rated with lower scores the fruity descriptor (ethyl hexanoate) in the PW compared to the CW, although significant differences were only observed at one evaluation time (10 s) (16% difference), maybe due to the high inter-individual differences observed. As in the case of the in vivo release experiments, no significant differences between wines were observed for the compound ethyl decanoate. These results are in agreement with those of Cliff et al. [26] that found suppression of fruity aroma in wine in the presence of a GSE (0–5 g/L). Similarly, Pérez-Jimenez and colleagues [5] observed a lower score of fruity notes (derived from esters) in wines with a GSE added (150 mg/L). The lower intensity of fruity perception observed for PW in comparison to CW could be related to the lower intra-oral release observed for this wine, derived from polyphenol-volatile or polyphenol-oral components-volatile interactions. However, it is important to highlight that during wine drinking, many senses are activated at a time which would inevitably modulate the perceived intensities of each other through cross-modal sensory interactions. Taking into account the taste-aroma interactions in the process of drinking wine [27], the bitterness and astringency brought by polyphenols contained in the GSE may have influenced the perception of aroma. Thus, it could not be discarded that this phenomenon could have had an effect on the suppression of the fruity descriptor when it was retronasally perceived.

## 5. Conclusions

This research work has demonstrated that the addition of a GSE to a rosé wine influenced the release and perception of esters in a compound- dependent manner. For the compound ethyl hexanoate, the addition of the GSE did not significantly affect its release from wine measured in in vitro (headspace) conditions. However, the addition of the GSE decreased the in vivo release of ethyl hexanoate after wine rinsing measured intra-orally. Moreover, the addition of the GSE affected the orthonasal and retronasal fruity perception associated with this compound. Thus, participants scored with a lower fruity intensity the wine with the GSE added compared to the control wine. This suppression observed by using instrumental and sensory approaches suggests that some components of the extract (procyanidins and/or monomers of catechin, epicatechin, ... ) can interact directly and/or through complexes with oral components with the compound ethyl hexanoate. Apart from the physicochemical nature of these interactions, the existence of other phenomena such as perceptual interactions or some effects of polyphenols on esterase activity cannot be discarded. For the compound ethyl decanoate, no effects were observed due to the GSE addition.

**Author Contributions:** Conceptualization, C.M.-G. and M.Á.P.-B.; formal analysis, C.C. and M.P.-J.; data curation, C.M.-G.; writing—original draft preparation, C.M.-G.; writing—review and editing, C.M.-G., C.C. and M.Á.P.-B. All authors have read and agreed to the published version of the manuscript.

**Funding:** This work was supported by the Spanish MINECO through the AGL201678936-R and PID2019-111734RB-100 projects. C.M.-G. and M.P.-J. thank MINECO for their respective JdC and predoctoral contracts.

**Institutional Review Board Statement:** The study was conducted according to the guidelines of the Declaration of Helsinki, and approved by the Institutional Review Board (or Ethics Committee) of CSIC.

**Informed Consent Statement:** Informed consent was obtained from all subjects involved in the study.

**Data Availability Statement:** The data presented in this study are available on request from the corresponding author. The data are not publicly available due to ethical reasons.

**Acknowledgments:** Authors greatly thank the volunteers that participated in this study and María Martín Arias for her technical assistance.

**Conflicts of Interest:** The authors declare no conflict of interest. The funders had no role in the design of the study; in the collection, analyses, or interpretation of data; in the writing of the manuscript, or in the decision to publish the results.

## References

1. Obreque-Slíer, E.; Peña-Neira, A.; López-Solís, R.; Ramírez-Escudero, C.; Zamora-Marín, F. Phenolic characterization of commercial enological tannins. *Eur. Food Res. Tech.* **2009**, *229*, 859–866. [CrossRef]
2. Vignault, A.; Pascual, O.; Gombau, J.; Jourdes, M.; Moine, V.; Fermaud, M.; Roudet, J.; Canals, J.; Teissedre, P.-L.; Zamora, F. New insight about the functionality of oenological tannins; Main results of the working group on oenological tannins. *BIO Web Conf.* **2019**, *12*, 02005. [CrossRef]
3. Landon, J.L.; Weller, K.; Harbertson, J.F.; Ross, C.F. Chemical and sensory evaluation of astringency in Washington State red wines. *Am. J. Enol. Vitic.* **2008**, *59*, 153–158.
4. Muñoz-González, C.; Canon, F.; Feron, G.; Guichard, E.; Pozo-Bayón, M.A. Assessment Wine Aroma Persistence by Using an in Vivo PTR-ToF-MS Approach and Its Relationship with Salivary Parameters. *Molecules* **2019**, *24*, 1277. [CrossRef]
5. Perez-Jiménez, M.; Chaya, C.; Pozo-Bayón, M.Á. Individual differences and effect of phenolic compounds in the immediate and prolonged in-mouth aroma release and retronasal aroma intensity during wine tasting. *Food Chem.* **2019**, *285*, 147–155. [CrossRef]
6. Charters, S.; Pettigrew, S. The dimensions of wine quality. *Food Qual. Pref.* **2007**, *18*, 997–1007. [CrossRef]
7. Dufour, C.; Bayonove, C.L. Interactions between wine polyphenols and aroma substances. An insight at the molecular level. *J. Agric. Food Chem.* **1999**, *47*, 678–684. [CrossRef]
8. Mitropoulou, A.; Hatzidimitriou, E.; Paraskevopoulou, A. Aroma release of a model wine solution as influenced by the presence of non-volatile components. Effect of commercial tannin extracts, polysaccharides and artificial saliva. *Food Res. Int.* **2011**, *44*, 1561–1570. [CrossRef]
9. Villamor, R.R.; Evans, M.A.; Mattinson, D.S.; Ross, C.F. Effects of ethanol, tannin and fructose on the headspace concentration and potential sensory significance of odorants in a model wine. *Food Res. Int.* **2013**, *50*, 38–45. [CrossRef]
10. Genva, M.; Kenne Kemene, T.; Deleu, M.; Lins, L.; Fauconnier, M.-L. Is It Possible to Predict the Odor of a Molecule on the Basis of its Structure? *Int. J. Mol. Sci.* **2019**, *20*, 3018. [CrossRef]
11. Bate-Smith, E. Haemanalysis of tannins: The concept of relative astringency. *Phytochemistry* **1973**, *12*, 907–912. [CrossRef]
12. Soares, S.; Vitorino, R.; Osorio, H.; Fernandes, A.; Venâncio, A.; Mateus, N.; Amado, F.; de Freitas, V. Reactivity of human salivary proteins families toward food polyphenols. *J. Agric. Food Chem.* **2011**, *59*, 5535–5547. [CrossRef]
13. Gibbins, H.; Proctor, G.; Yakubov, G.; Wilson, S.; Carpenter, G. Concentration of salivary protective proteins within the bound oral mucosal pellicle. *Oral Dis.* **2014**, *20*, 707–713. [CrossRef] [PubMed]
14. Silva, M.S.; Garcia-Estevez, I.; Brandão, E.; Mateus, N.; de Freitas, V.; Soares, S. Molecular interaction between salivary proteins and food tannins. *J. Agric. Food Chem.* **2017**, *65*, 6415–6424. [CrossRef] [PubMed]

15. Payne, C.; Bowyer, P.K.; Herderich, M.; Bastian, S.E. Interaction of astringent grape seed procyanidins with oral epithelial cells. *Food Chem.* **2009**, *115*, 551–557. [CrossRef]
16. Esteban-Fernández, A.; Muñoz-González, C.; Jiménez-Girón, A.; Pérez-Jiménez, M.; Pozo-Bayón, M.Á. Aroma release in the oral cavity after wine intake is influenced by wine matrix composition. *Food Chem.* **2018**, *243*, 125–133. [CrossRef]
17. Perez-Jiménez, M.; Esteban-Fernández, A.; Muñoz-González, C.; Pozo-Bayón, M.A. Interactions among Odorants, Phenolic Compounds, and Oral Components and Their Effects on Wine Aroma Volatility. *Molecules* **2020**, *25*, 1701. [CrossRef]
18. Muñoz-González, C.; Feron, G.; Brulé, M.; Canon, F. Understanding the release and metabolism of aroma compounds using micro-volume saliva samples by ex vivo approaches. *Food Chem.* **2018**, *240*, 275–285. [CrossRef]
19. Niimi, J.; Liland, K.H.; Tomic, O.; Jeffery, D.W.; Bastian, S.E.; Boss, P.K. Prediction of wine sensory properties using mid-infrared spectra of Cabernet Sauvignon and Chardonnay grape berries and wines. *Food Chem.* **2020**, 128634. [CrossRef]
20. Sáenz-Navajas, M.-P.; Campo, E.; Valentin, D.; Fernández-Zurbano, P.; Ferreira, V. Sensory Interactions in Wine: Effect Of Nonvolatile Molecules on Wine Aroma and Volatiles on Taste/Astringency Perception. In *The Sense of Taste*; Nova Science Publishers, Inc.: New York, NY, USA, 2015; p. 97.
21. Sánchez-Patán, F.; Cueva, C.; Monagas, M.; Walton, G.E.; Gibson, G.R.; Martín-Álvarez, P.J.; Moreno-Arribas, M.V.; Bartolomé, B. Gut microbial catabolism of grape seed flavan-3-ols by human faecal microbiota. Targetted analysis of precursor compounds, intermediate metabolites and end-products. *Food Chem.* **2012**, *131*, 337–347. [CrossRef]
22. Pérez-Jiménez, M.; Pozo-Bayón, M.Á. Development of an in-mouth headspace sorptive extraction method (HSSE) for oral aroma monitoring and application to wines of different chemical composition. *Food Res. Int.* **2019**, *121*, 97–107. [CrossRef] [PubMed]
23. Lorrain, B.; Tempere, S.; Iturmendi, N.; Moine, V.; de Revel, G.; Teissedre, P.-L. Influence of phenolic compounds on the sensorial perception and volatility of red wine esters in model solution: An insight at the molecular level. *Food Chem.* **2013**, *140*, 76–82. [CrossRef] [PubMed]
24. Aronson, J.; Ebeler, S.E. Effect of polyphenol compounds on the headspace volatility of flavors. *Am. J. Enol. Vitic.* **2004**, *55*, 13–21.
25. Lund, C.; Nicolau, L.; Gardner, R.; Kilmartin, P.A. Effect of polyphenols on the perception of key aroma compounds from Sauvignon Blanc wine. *Aust. J. Grape Wine Res.* **2009**, *15*, 18–26. [CrossRef]
26. Cliff, M.A.; Stanich, K.; Edwards, J.E.; Saucier, C.T. Adding grape seed extract to wine affects astringency and other sensory attributes. *Food Qual.* **2012**, *35*, 263–271. [CrossRef]
27. Arvisenet, G.; Guichard, E.; Ballester, J. Taste-aroma interaction in model wines: Effect of training and expertise. *Food Qual. Pref.* **2016**, *52*, 211–221. [CrossRef]

foods

MDPI

Article

# 1,1,6-Trimethyl-1,2-dihydronaphthalene (TDN) Sensory Thresholds in Riesling Wine

**Andrii Tarasov [1,*], Nicoló Giuliani [1], Alexey Dobrydnev [2], Christoph Schuessler [1], Yulian Volovenko [2], Doris Rauhut [3] and Rainer Jung [1]**

[1] Department of Enology, Hochschule Geisenheim University (HGU), Von-Lade-Straße 1, 65366 Geisenheim, Germany; nicolo.giuliani@live.com (N.G.); Christoph.Schuessler@hs-gm.de (C.S.); Rainer.Jung@hs-gm.de (R.J.)

[2] Department of Organic Chemistry, Faculty of Chemistry, Taras Shevchenko National University of Kyiv, Volodymyrska str. 60, 48000 Kyiv, Ukraine; alexey.pierrot@gmail.com (A.D.); yulian_volovenko@ukr.net (Y.V.)

[3] Department of Microbiology and Biochemistry, Hochschule Geisenheim University (HGU), Von-Lade-Straße 1, 65366 Geisenheim, Germany; Doris.Rauhut@hs-gm.de

* Correspondence: Andrii.Tarasov@hs-gm.de; Tel.: +49-6722-502-187

Received: 6 April 2020; Accepted: 1 May 2020; Published: 9 May 2020

**Abstract:** 1,1,6-Trimethyl-1,2-dihydronaphthalene (TDN) is an aroma compound responsible for the kerosene/petrol notes in Riesling wines. In the current article, three sensory thresholds for TDN were determined in young Riesling wine: *detection threshold* (about 4 μg/L), *recognition threshold* (10–12 μg/L), and *rejection threshold* (71–82 μg/L). It was demonstrated that an elevated content of free $SO_2$ in wine may have a certain masking effect on the TDN aroma perception. In addition, the influence of wine serving temperature on the recognition of kerosene/petrol notes was studied. It was found, that a lower wine serving temperature (about 11 °C) facilitated identification of the TDN aroma compared to the same wine samples at room temperature.

**Keywords:** 1,1,6-Trimethyl-1,2-dihydronaphthalene (TDN); wine; sensory threshold; serving temperature

## 1. Introduction

1,1,6-Trimethyl-1,2-dihydronaphthalene (TDN) is one of the key wine aroma components in Riesling wines, and it belongs to the $C_{13}$-norisoprenoids. With the kerosene/petrol aroma, TDN is considered controversial from the consumers' preference perspective. Low and medium TDN concentrations contribute to the complexity of the wine bouquet, while high TDN content often evokes negative impressions caused by the strong kerosene/petrol odor dominance.

The level of TDN in wine increases during bottle aging due to the transformations of carotenoid-derived precursors originating from grapes [1–3]. The quantity of TDN precursors depends on the viticulture practices such as grape clusters defoliation [4,5], soil fertilization [6], water irrigation [7,8], and the selection of vine clones [9]. Global climate change, warmer temperatures, and higher sun exposure of the grapes may intensify formation of TDN in the succeeding Riesling wines [10,11]. The option of yeast strains can also affect the formation of TDN in wine, probably due to the pathways of the precursors' conversion [9,12]. The TDN level in wine can also be managed by the selection of bottle closures, which are able to absorb a significant amount of TDN from the wine [13–16]. Finally, it was demonstrated that wine storage conditions, e.g., elevated temperature, can accelerate the formation of TDN [17,18].

The typical content of TDN in European Riesling wines is usually between 1 and 50 μg/L, while in Australian wines it can reach up to 250 μg/L and more [19–22]. The sensory threshold of TDN in wine was defined in several studies as being in a range of values between 2 and 20.6 μg/L (Table 1).

In the first publication, Simpson (1978) reported the *flavor threshold* in Riesling at 20 μg/L, however, no details regarding the panel were described. Several decades later, the TDN *odor detection threshold* (*ODT*) was determined at a significantly lower level, 2 μg/L [20]. Trained panelists evaluated model and Chardonnay spiked wines in a series of 3-AFC tests. The succeeding work also utilized 3-AFC tests but with an untrained panel of consumers using spiked Riesling wines. As a result, the defined *ODT* values were close to the initial one, about 20 μg/L [23]. In addition, the *consumer rejection threshold* (*CRejT*) in this research was also determined to be 157 μg/L and 82.4 μg/L depending on the wine vintage (2010 and 2011) and the country in which the tests were conducted (New Zealand and the USA), respectively. A recent study revealed the following values of TDN perception thresholds in Riesling wine: 3.1 μg/L for *ODT* by trained panellists and 14.7 μg/L for *consumer detection threshold* (*CDT*) [24]. In the same work the *CRejT* was found to be 60 μg/L and 91 μg/L for young and aged Riesling wines, respectively.

**Table 1.** Summary of the studies devoted to the 1,1,6-trimethyl-1,2-dihydronaphthalene (TDN) sensory thresholds determination.

| References | Panel | Base Wine | Wine Temperature | Sensory Method | Sensory Thresholds, μg/L |
|---|---|---|---|---|---|
| [18] | n/a | Riesling | n/a | triangle tests | *Flavor threshold:* 20 |
| [20] | Trained | Model wine; Chardonnay | n/a | 3-AFC tests | *ODT*: 2 |
| [23] | Untrained consumers | Riesling, 1-year-old wine | 23 °C | 3-AFC tests | *ODT*: 20.6 (2010 vintage, NZ), 18.2 (2011 vintage, USA) *CRejT* [1]: 157 (2010 vintage, NZ), 82.4 (2011 vintage, USA) |
| [24] | Trained consumers | Riesling, 1-year-old wine (2015 vintage) | 15 °C | 3-AFC tests | *ODT*, Trained panel: 3.1 *CDT*, Consumers: 14.7 *CRejT* [1] 1-year-old wine: 60 *CRejT* [1] 8-year-old wine: 91 |

[1] *CRejT* was determined by the preference tests.

The variability of the reported TDN sensory thresholds values can be related to both diverse concepts of sensory thresholds and variations in sensory evaluation methods. In the current work, we aimed to define and study *detection* and *recognition thresholds* of TDN:

- *detection threshold* (*DT*) implies the lowest level at which a stimulus can be detected, but not necessarily recognized;
- *recognition threshold* (*RT*) corresponds to the level when a stimulus can be recognized and identified; it is usually higher than *DT* [25].

In addition our goal was to determine a TDN *rejection threshold* (*RejT*) and to compare it with the previously reported *CRejT* values. This issue was of interest since various approaches to the evaluation of *rejection thresholds* are still under discussion [26].

Finally, this study investigated the influence of free $SO_2$ levels and wine serving temperature on the perception of TDN aroma. These factors can vary significantly in reality and, to our knowledge, have not been previously studied. Earlier, it was demonstrated that ethanol levels and carbonation can enhance the *odor detection threshold* of TDN in certain matrices [24].

## 2. Materials and Methods

### 2.1. Chemicals and Materials

The following chemicals were used for the experiment and analyses: ethanol absolute AnalaR NORMAPUR® ACS, ≥99.5% (VWR Chemicals); sodium chloride (Carl Roth GmbH, Germany); $\beta$-ionone-$d_3$, ≥95% (aromaLAB GmbH, Germany); and 1,1,6-trimethyl-1,2-dihydronaphthalene ≥ 95% (own synthesis [27]). Parafilm "M"® was purchased from Carl Roth GmbH, Germany.

Transparent green glass bottles (1 L volume) with MCA finish type were supplied by Richard Wagner GmbH + Co. KG, Alzey (Germany). Screw caps of MCA type were supplied by Rheingauer Winzerbedarf GmbH. The base wine, Riesling Villa Monrepos from the Rheingau region (Germany) of 2016 vintage, was bottled in April 2017 in the winery of the Hochschule Geisenheim University. Analysis of the wine after the bottling revealed the following: alcohol content 12.3% (*v*/*v*), titratable acidity 7.3 g/L, sugar content 7.5 g/L, and pH 3.2. The TDN concentration in the wine was 2.2 µg/L (analysis before the *sensory sessions*). The adjustment of the $SO_2$ content in the wine resulted in free/total $SO_2$ concentrations of 40/120 mg/L for *Sensory Session 1* (high free $SO_2$ content) and 10/65 mg/L for *Sensory Session 2* (low free $SO_2$ content). The high free $SO_2$ content of 40 mg/L is relatively elevated compared to many wines globally, but is typical for many Riesling wine producers in the Rheingau region. This particularity is explained by the local practices and expectations of a longer aging time of these wines.

Young Riesling wine, 2016, was selected as the base wine for all the sensory tests in order to avoid an elevated initial level of TDN. Wines produced from other international grape varieties were not used since they typically do not possess noticeable amounts of TDN [20] and kerosene/petrol aroma. Furthermore, the composition of other wine matrices may affect the TDN perception thresholds values, which makes them inapplicable for Riesling wines.

The free $SO_2$ content in wine is one of the parameters that can have an impact on the wine aroma perception. The influence of *low* and *high* free $SO_2$ content on TDN recognition in Riesling wine was studied in the current research, since no information on this issue was found in previous publications. The same Riesling wine with *high* (40 mg/L) and *low* (10 mg/L) content of free $SO_2$ was used in the tests of *Sensory Sessions 1* and 2, respectively.

### 2.2. Preparation of TDN Stock Solutions

The TDN stock solution was prepared by the addition of 9.1 mg of TDN into a 50 mL volumetric flask, which was then filled with ethanol to the 50 mL mark, resulting in a TDN concentration of about 0.182 mg/mL. The stock solution was stored in a refrigerator at 4 °C with a ground glass stopper, additionally sealed with Parafilm®. Before the wine spiking procedures, the TDN stock solution was kept for about 15–30 min outside the refrigerator at room temperature.

### 2.3. Panels for Sensory Sessions

Two panels participated in the *sensory sessions*. Panel 1 (*Sensory Session 1*) consisted of 20 tasters: 11 male and 9 female. Panel 2 (*Sensory Session 2*) comprised 22 tasters: 13 male and 9 female. The age of the tasters was in the range between 21 and 45 years. All the panelists, employees or students of the Hochschule Geisenheim University (Germany), were regular wine consumers. Both panels were international (more than 15 nationalities from European, Asian, and the American countries). All the tasters were familiar with Riesling wines and their typical aromas, therefore, no special training or panelists selection was done. The sensory evaluations were conducted in June 2017 in the specialized well-lit (white light) and odor-free sensory analysis room in the Department of Enology of the Hochschule Geisenheim University (Germany). There were 30 separated booths, specialized individual places for panelists. The room temperature was about 22 °C.

Each *sensory session* consisted of two parts: *thresholds determination test* (for *DT*, *RT*, and *RejT*) and a *series of 3-AFC tests* (Figure 1). The structure and content of both *sensory sessions* were identical except for the level of free $SO_2$ in the wine samples. The evaluation of the wine samples in all the tests was orthonasal.

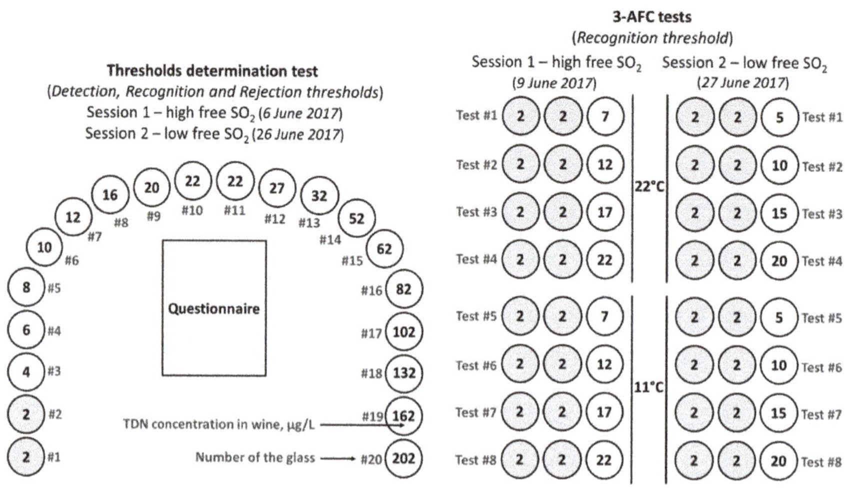


**Figure 1.** Design of the experiment. Scheme of the glasses with TDN spiked Riesling wines presented for the *sensory sessions*. The gray background of the circle indicates the base wine without TDN spiking.

## 2.4. Thresholds Determination Tests

The *thresholds determination test* [28] was preferred to the 3-AFC test methods (according to ISO 13301:2018 [29] and ASTM (American Society for Testing and Materials) E679-19 [30]) for two reasons. First, we aimed to compare our sensory thresholds outcomes with the previously reported values defined by the 3-AFC test methods (Table 1). Second, the offered approach allowed a convenient determination of the three sensory thresholds in a single test.

### 2.4.1. Preparation of Wine Samples

Twenty bottles (1 L) of Riesling wine samples with various concentrations of TDN were prepared in the morning before each *sensory session*. The wine was preliminarily homogenized in a stainless steel container and transferred back to the bottles. Wine samples were spiked with the TDN stock solution and mixed in order to reach the target TDN concentrations of 4–202 µg/L according to the design of the experiment (Figure 1). The prepared wine samples were kept in 1 L bottles not more than 2–3 h at room temperature before pouring into the glasses.

The highest TDN concentration was limited to 202 µg/L. This value was chosen on the basis of the maximal reported *CRejT* being 157 µg/L [23] and included a necessary margin. The TDN content was increased in small steps between 2 and 22 µg/L in the first ten glasses since *DT* and *RT* were expected to be around these values and required a precise determination (Figure 1). For the last ten glasses, the difference of TDN concentration between the wine samples was larger. This was a compromise in order to cover a bigger range of values (22–202 µg/L) for the determination of *RejT* and to keep a reasonable number of glasses in the test.

### 2.4.2. Performance of Thresholds Determination Test

The method of *DT* and *RT* thresholds determination described by Busch-Stockfisch (2002) [28] was reviewed and approved in the Department of Enology, Hochschule Geisenheim University. As a modification to this method, the determination of *RejT* was added and the questionnaire was redesigned (Figure 2). Twenty wine tasting glasses (ISO 3591) containing the corresponding wine samples were placed on the table in front of each panelist. Each glass contained about 35 mL of one of the wine samples poured 30–45 min before the start of the test and immediately covered with a plastic lid. The wine temperature during the sensory evaluation was 22 ± 1 °C. The glasses with serial numbers on the plastic lids were presented in the order of increasing TDN concentration. The panelists were informed that the wine sample in glass #1 was a control sample and each following glass contained the same base wine with the content of TDN equal or higher compared to the previous one ([TDN] "glass n+1" ≥ [TDN] "glass n"). The concepts of *DT*, *RT*, and *RejT* were clarified to the panelists. The task of the test was to evaluate the wine samples orthonasaly one-by-one, starting from the 2nd glass using the paper questionnaire. If the wine in the glass was perceived to be the same as the control (glass #1), an indication should be made in the column "Cont.". If the following wine sample was different from the control, but no TDN related aromas were recognized, the column "Det." should be chosen (*detection threshold*). The "Recognition threshold" column was provided for the wine samples in which TDN aroma could be identified (Descriptor) and evaluated by intensity. The last column "Rejection" was introduced for the *rejection threshold*. It was explained as a concentration of TDN, at which aroma intensity was not acceptable (too high and unpleasant) in the bouquet of the wine. The first markings in the "Det." and "Recognition Threshold" columns were considered as panelist's personal *DT* and *RT*, respectively. Once judge reached the last column "Rejection", the test was finished.

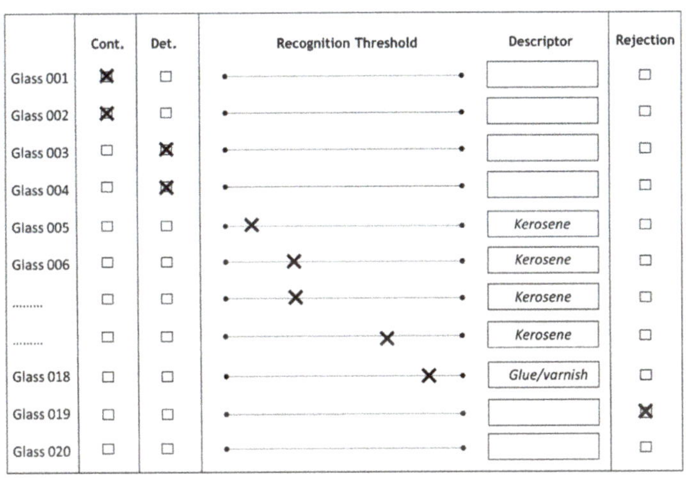


**Figure 2.** An example of a completed *threshold determination test* questionnaire.

During the test, panelists were not allowed to return to the previous glasses. This measure was applied due to the possibility of a panelists' adaptation to a higher TDN content and changes in the assessment of preceding samples. In addition, it was recommended to do not more than 3–4 sniffs per glass and to agitate the wine sample only after the first sniff. The panelists evaluated the wine samples at an individually convenient pace with pauses between samples, if necessary. No specific training was carried out prior to the *sensory sessions*. However, after the explanation of the test rules, the panelists were asked to do a trial test run, which was followed by a 15 min break. Later, the main test was performed. According to the method [28], the threshold values were determined in two ways: by the lowest value found by 50% of tasters and by the geometric mean value based on the answers of all panelists (Table 2).

**Table 2.** Results of the thresholds determination tests.

| Sensory Sessions (Accepted Questionnaires) | Thresholds | Calculation Approaches | | |
|---|---|---|---|---|
| | | Geometric Mean | 50% Panelists | Median |
| Session 1<br>high free $SO_2$,<br>($n$ = 16) | *Detection (DT)*<br>*Recognition (RT)*<br>*Rejection (RejT)* | 4 µg/L<br>12 µg/L<br>79 µg/L | 4 µg/L<br>12 µg/L<br>82 µg/L | 5 µg/L<br>12 µg/L<br>82 µg/L |
| Session 2<br>low free $SO_2$,<br>($n$ = 20) | *Detection (DT)*<br>*Recognition (RT)*<br>*Rejection (RejT)* | 3 µg/L<br>11 µg/L<br>71 µg/L | 4 µg/L<br>10 µg/L<br>82 µg/L | 4 µg/L<br>10 µg/L<br>82 µg/L |

## *2.5. 3-AFC Tests*

### 2.5.1. Preparation of Wine Samples

The wine samples preparation for the *3-AFC tests* was based on the *RT* values defined in the *thresholds determination tests*. Each *3-AFC test* comprised two control samples (base wine) and one spiked wine (Figure 1). Four levels of TDN spiking were applied for each *sensory session* in order to reach the following concentrations: "*RT*-5 µg/L", "*RT*", "*RT*+5 µg/L", and "*RT*+10 µg/L". Two sets of 1 L bottles with the wine samples were prepared in the morning 2–3 h before each *sensory session*. One set of the bottles was kept at room temperature, while the other set was cooled in the refrigerator in order to reach the wine serving temperature of 11 ± 1 °C.

### 2.5.2. Performance of 3-AFC Tests

Eight *3-AFC tests* were performed with the same panel within each *sensory session*. Wine samples for the first four tests were served at room temperature (22 ± 1 °C) in order to re-check the *RT* found in the *thresholds determination test*. The last four tests contained the same wine samples as the first ones but at a serving temperature of 11 ± 1 °C. Each glass contained about 35 mL of the wine, which was served immediately before each *3-AFC test*. The panelists were asked the following question: "Which sample has a more intense kerosene/TDN aroma?".

## *2.6. Processing of the Data*

Only the completely filled out questionnaires were accepted and subsequently statistically analyzed. Therefore, their number in the sensory tests can differ from the total number of panelists (Table 2, Figures 3 and 4). Questionnaires were prepared and processed using Fizz software 2.51a 86 (2016, Biosystemes, Couternon, France). The same software was used for the statistical analysis of data of the *3-AFC tests*. Text, calculations and figures for the *thresholds determination tests* and *3-AFC tests* were done using Microsoft Office Standard 2013 programs (Version 15.0.5153.1000, Microsoft Corporation, Redmond, Washington, DC, USA).

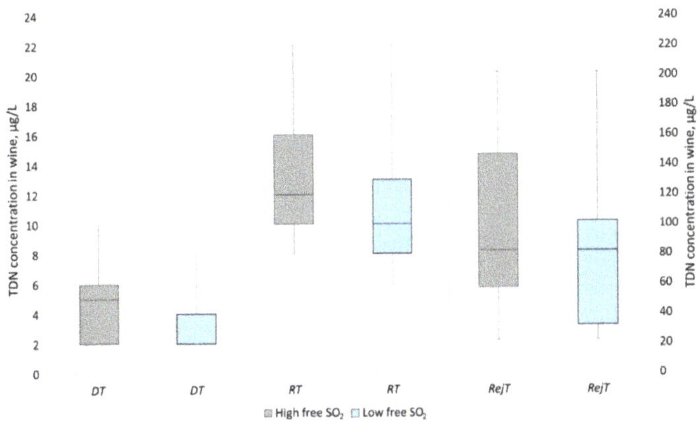


**Figure 3.** Distribution of the panelists' answers for the determination of the *detection threshold* (*DT*), *recognition threshold* (*RT*), and *rejection threshold* (*RejT*) of TDN in Riesling wine. Median is represented by the horizontal line. The bottom and top of the box represent the 1st and the 3rd quartiles, respectively. The whiskers represent the minimum and maximum values.

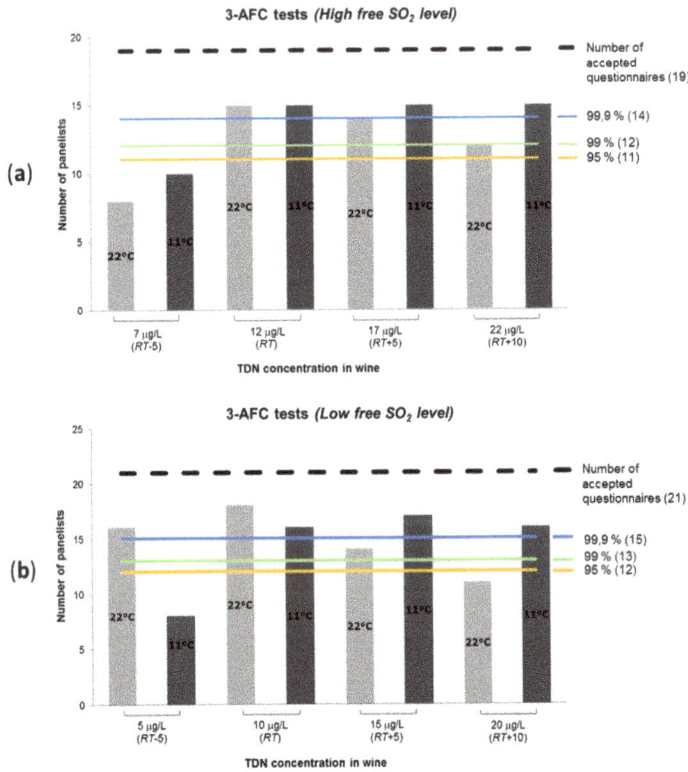


**Figure 4.** Results of the *3-AFC tests*. The dashed lines represent the total number of panelists whose questionnaires were accepted for the statistical analysis: (**a**) $n = 19$; (**b**) $n = 21$. The colored lines represent the three levels of significance of 95%, 99%, and 99.9% with the number of required panelists (accepted questionnaires).

### 2.7. *Analysis of TDN Content in Wine*

The level of TDN was analyzed in the base wine and validated in the selected spiked wine samples by GC-MS (SBSE) analysis according to the standard operation procedure at the Hochschule Geisenheim University [16]. In particular, the selected samples with TDN content close to the determined *recognition threshold* had the following TDN concentrations (expected/validated): *thresholds determination test* Sample #6 (10 μg/L/9.9 μg/L), Sample #7 (12 μg/L/12.0 μg/L).

## 3. Results and Discussion

### 3.1. *Thresholds Determination Tests (DT, RT, and RejT)*

According to the methodology, the TDN sensory thresholds values were determined as the lowest ones reported by 50% of the panelists and by the geometric means. Both methods in the current study revealed close results (Table 2). The box plots are presented for the demonstration of the panelists' answers distributions (Figure 3). The calculated medians were equal to the 50% panelists values, except for the *DT* (*high* free $SO_2$): 5 μg/L and 4 μg/L, respectively.

In general, the sensory thresholds for Riesling wine with *high* and *low* free $SO_2$ content were similar. At the same time, a trend towards smaller values was observed in the distribution of panelists' responses for the wine samples with the *low* free $SO_2$ content (Figure 3).

The determined *detection thresholds* at about 4 μg/L were close to the *ODT* found by the trained panelists at 3.1 μg/L in Riesling wine [24] and somewhat higher than the reported *ODT* in neutral Chardonnay or model wines at 2 μg/L [20]. The latter difference was predictable since Riesling was the base wine in the current study and it initially possessed 2.2 μg/L of TDN. In general, many Riesling wines contain about 4 μg/L of TDN, however, according to the concept of *DT*, it does not mean that these wines have a noticeable kerosene/petrol aroma. In addition, the comparison of TDN sensory thresholds in Chardonnay and Riesling wines is not relevant, since the latter wines are usually much more aromatic. Therefore, the TDN thresholds in Riesling wines are expected to be higher due to the wine matrix effects.

The *recognition thresholds* for the wines with *high* and *low* free $SO_2$ content were identified at 12 μg/L and 10–11 μg/L, respectively. These values are almost two times lower compared to the initially reported *flavor threshold* of 20 μg/L [18] or *ODT* at about 18–21 μg/L [23] and slightly lower than the 14.7 μg/L *CDT* defined by consumers in Riesling wines [24]. The minimal and maximal *RT* values indicated by the panelists were similar for both wines with the *high* and *low* free $SO_2$ content: 8–22 μg/L and 6–22 μg/L, respectively. At the same time, the distribution of the panelists' answers within the 1st and 3rd quartiles comprised somewhat smaller values for the *low* free $SO_2$ wines compared to the *high* free $SO_2$ counterparts: 8–13 μg/L vs. 10–16 μg/L, respectively. An excessive free $SO_2$ content in wine may partially mask TDN aroma, especially when the typical smell of sulfur dioxide is perceivable, as it was on the *Sensory Session 1*.

The *rejection thresholds* of the individual panelists varied significantly between 22 and 202 μg/L in both sessions. Both calculation approaches revealed similar results, about 80 μg/L for the *high* free $SO_2$ wine and around 70–80 μg/L for the *low* free $SO_2$ samples. These results were comparable with the *CRejT* values found in two 1-year-old Riesling wines at 60 μg/L [24] and 82.4 μg/L, but about two times lower than the other reported *CRejT* value of 157 μg/L [23] (Table 1). The variation of the presented values is not surprising, since the optimal approaches of *rejection threshold* determination remain under discussion. For example, in the recent comment regarding the preference tests for *CRejT* evaluation, it was remarked that even if one sample is not preferred sensorially over another, this does not always mean that a non-preferred sample is rejected [26]. Instead of preference tests, the panelists of the current research were asked to identify when the Riesling wine starts to possess a not acceptable (too high and unpleasant) level of TDN aroma. It is noteworthy that at high TDN concentrations, close to the individual *RejT*, some of the panelists switched from the kerosene/petrol aroma descriptors to those of solvent, glue, varnish, rubber, and pharmacy (Figure 2).

The influence of TDN content on the evaluation of Riesling wines remains to be demonstrated. Among the factors that can affect the acceptance/rejection of elevated TDN content in Riesling wine are different TDN aroma tolerances in various groups of people, regional consumers' habits, and variability of wine matrices (vintages, young, and aged wines). The vivid demonstration of these effects is the almost twofold difference of *CRejT* values for 1-year-old Riesling wines (157 μg/L and 82.4 μg/L) depending on the vintage (2010 and 2011) and the country in which the sensory tests took place (New Zealand and the USA), respectively, [23]. Another example is the Australian Riesling wine, which despite the great TDN content of 246 μg/L received a high sensory quality score [19]. In addition, it is not excluded that there are other volatile compounds in wine apart from TDN, which can be associated with the kerosene/petrol aroma.

### 3.2. 3-AFC Tests, Confirmation of the Recognition Threshold, and Influence of Wine Serving Temperature on the TDN Aroma Recognition

Since the *recognition threshold* is essential in terms of the wine aroma composition, it was decided to confirm the determined *RT* values with an alternative sensory method. The utilized *3-AFC tests* implied comparison of the spiked wines with the control samples according to the intensity of the kerosene/TDN aroma perception. The wine samples spiking was designed to reach the TDN concentrations below, above, and equal to the defined *RT* (Figures 1 and 4). The serving temperature of the wine was the same as for the *thresholds determination tests*, i.e., 22 ± 1 °C. Additionally, the identical *3-AFC tests* were conducted at a lower serving temperature, 11 ± 1 °C, in order to check whether the TDN recognition is temperature-dependent.

The results of the *3-AFC tests* demonstrated that all the spiked samples at the level of *RT* (12 μg/L with *high* and 10 μg/L with *low* free $SO_2$ content) were statistically significantly different from the control samples (Figure 4). At the same time, the panelists were not able to distinguish the "*RT*-5" spiked samples from the control wines at *high* free $SO_2$ level, which confirms that the *RT* was higher than 7 μg/L. In the case of *low* free $SO_2$ wine, the tasters identified the spiked wine "*RT*-5" at room temperature (99.9% significance). However, it does not mean that the *RT* for the wine with *low* free $SO_2$ was 5 μg/L. The panelists were able to identify the spiked sample, but not necessarily recognized that it had the kerosene/petrol aroma, i.e., the effect of *detection threshold*. This suggestion is supported by the results of the *3-AFC test* at low temperature, whereby the "*RT*-5" spiked wine was not distinguished from the control samples at both levels of free $SO_2$ contents.

The panelists' ability to distinguish TDN aroma at concentrations ≥ *RT* at lower temperature was always highly significant, 99.9%. At the same time, the recognition of the spiked wine samples at room temperature dropped down in the sequence of tests with the TDN concentrations "*RT*", "*RT*+5", and "*RT*+10". This phenomenon can be related to olfactory fatigue (sensory adaptation), but not exclusively, since the same effect was not observed at the lower wine serving temperature. The other possible reason is the particularity of wine aromas' volatility at different temperatures. Thus, the air in the glass at a lower temperature should be enriched with hydrophobic molecules such as TDN, which are more volatile compared to hydrophilic compounds of similar molecular weight. Hence, this should facilitate the sensorial recognition of the TDN aroma. In the case of a higher temperature, the volatility of all wine aroma components rises, but not proportionally. Therefore, the fraction of TDN with regard to other volatile compounds in the air inside the glass can decrease, which complicates the sensorial identification of the TDN aroma. The results of this phenomena were also observed during the scalping process of TDN by wine stoppers, whereby TDN was absorbed noticeably faster at lower temperatures in the vertical bottle position [16]. In addition, some aroma compounds that become sensorially noticeable only at higher temperatures can also cause certain masking effects.

## 4. Conclusions

The modified *thresholds determination method* demonstrated a convenient approach to define three sensory thresholds of TDN in the Riesling wine in one run: *detection threshold* (*DT*), about 4 μg/L, *recognition threshold* (*RT*), 10–12 µg/L, and *rejection threshold* (*RejT*), 71–82 µg/L. The *RT* values were additionally confirmed by the series of *3-AFC tests*. In comparison with the earlier defined TDN sensory thresholds, the current *RT* was somewhat lower than the previously reported *consumer detection threshold*, while the determined *RejT* values were close to some of the recently published *CRejT* values (Table 1). Nevertheless, no direct comparison of the latter values can be done since the wine acceptance/rejection concepts were diverse in different studies.

Variation of free $SO_2$ content in the wine did not affect substantially the TDN sensory thresholds, however, the noticeable smell of sulfur dioxide at *high* level of free $SO_2$ tended to mask the perception of the kerosene/petrol aroma. Finally, it was shown that the TDN aroma recognition was easier in cooled wine, about 11 °C, which is likely related to the particularities of odorants' volatility depending on temperature.

**Author Contributions:** Conceptualization, Funding acquisition: R.J., D.R., A.T.; Methodology, Project administration: R.J., A.T.; Investigation: A.T., N.G., C.S.; Formal analysis: C.S., A.T.; Resources: A.D., Y.V., D.R.; Writing—original draft: A.T., N.G.; Writing—review and editing: R.J., D.R., A.D. All authors have read and agreed to the published version of the manuscript.

**Funding:** This work was supported by the research grant provided by OIV: International Organization of Vine and Wine (January 2017–March 2018).

**Acknowledgments:** We thank Antonio Giovanni Tirelli (Università degli studi di Milano) for his contribution to this work, Steffen Stiehl (Hochschule Geisenheim University) for assisting with the GC-MS analyses, Melanie Brandt (Hochschule Geisenheim University) for analytical support, and Anja Abd Elrehim (Hochschule Geisenheim University) for proofreading. We thank Anna Yanchenko and Eugeniy Kamenev for the preparation of the graphical abstract.

**Conflicts of Interest:** The authors declare no conflict of interest.

## References

1. Daniel, M.A.; Capone, D.L.; Sefton, M.A.; Elsey, G.M. Riesling acetal is a precursor to 1,1,6-trimethyl-1,2-dihydronaphthalene (TDN) in wine. *Aust. J. Grape Wine Res.* **2009**, *15*, 93–96. [CrossRef]
2. Knapp, H.; Straubinger, M.; Stingl, C.; Winterhalter, P. Analysis of Norisoprenoid Aroma Precursors. In *Carotenoid-Derived Aroma Compounds*; ACS Symposium Series; American Chemical Society: Washington, DC, USA, 2001; Volume 802, pp. 20–35. ISBN 0-8412-3729-8.
3. Mendes-Pinto, M.M. Carotenoid breakdown products the—Norisoprenoids—In wine aroma. *Arch. Biochem. Biophys.* **2009**, *483*, 236–245. [CrossRef] [PubMed]
4. Kwasniewski, M.T.; Vanden Heuvel, J.E.; Pan, B.S.; Sacks, G.L. Timing of Cluster Light Environment Manipulation during Grape Development Affects C13 Norisoprenoid and Carotenoid Concentrations in Riesling. *J. Agric. Food Chem.* **2010**, *58*, 6841–6849. [CrossRef] [PubMed]
5. Schüttler, A.; Guthier, C.; Stoll, M.; Darriet, P.; Rauhut, D. Impact of grape cluster defoliation on TDN potential in cool climate Riesling wines. *BIO Web Conf.* **2015**, *5*, 01006. [CrossRef]
6. Linsenmeier, A.W.; Löhnertz, O. Changes in Norisoprenoid Levels with Long-term Nitrogen Fertilisation in Different Vintages of Vitis vinifera var. Riesling Wines. *S. Afr. J. Enol. Vitic.* **2007**, *28*, 17–24. [CrossRef]
7. Bindon, K.A.; Dry, P.R.; Loveys, B.R. Influence of Plant Water Status on the Production of C13-Norisoprenoid Precursors in Vitis vinifera L. Cv. Cabernet Sauvignon Grape Berries. *J. Agric. Food Chem.* **2007**, *55*, 4493–4500. [CrossRef] [PubMed]
8. Oliveira, C.; Silva Ferreira, A.C.; Mendes Pinto, M.; Hogg, T.; Alves, F.; Guedes de Pinho, P. Carotenoid Compounds in Grapes and Their Relationship to Plant Water Status. *J. Agric. Food Chem.* **2003**, *51*, 5967–5971. [CrossRef] [PubMed]
9. Sponholz, W.R.; Hühn, T. Factors influencing the aging of the Riesling wine. Clonal material and used yeast strain. *Vitic. Enol. Sci.* **1997**, *52*, 103–108.
10. Gerdes, S.M.; Winterhalter, P.; Ebeler, S.E. Effect of Sunlight Exposure on Norisoprenoid Formation in White Riesling Grapes. In *Carotenoid-Derived Aroma Compounds*; ACS Symposium Series; American Chemical Society: Washington, DC, USA, 2001; Volume 802, pp. 262–272. ISBN 0-8412-3729-8.

11. Marais, J.; van Wyk, C.J.; Rapp, A. Effect of Sunlight and Shade on N orisoprenoid Levels in Maturing Weisser Riesling and Chenin blanc Grapes and Weisser Riesling Wines. *S. Afr. J. Enol. Vitic.* **1992**, *13*, 23–32. [CrossRef]
12. Fischer, U.; Ziegler, M.; Gök, R.; Bechtloff, P.; Wacker, M.; Schmarr, H.G.; Winterhalter, P. Petrolfehlnote in Rieslingweinen—Pflanzenphysiologische, chemische und sensorische Ansätze zur Minimierung; 2017. FEI-Jahrestagung. Institut für Weinbau und Oenologie, DLR Rheinpfalz, 2017. Available online: https://www.fei-bonn.de/veranstaltungen-termine/jahrestagungen/jahrestagung-2017 (accessed on 2 April 2020).
13. Capone, D.; Sefton, M.; Pretorius, I.; Hoj, P. Flavour "scalping" by wine bottle closures—The "winemaking" continues post vineyard and winery. *Aust. N. Z. Wine Ind. J.* **2003**, *18*, 16–20.
14. Scrimgeour, N. *VINOLOK (VINOSEAL) Closure Evaluation Stage 1: Fundamental Performance Assessment*. 2014. Available online: http://www.columbit.com.au/wp-content/uploads/2016/09/awri-otr-taint-scalp-testing-final-report.pdf (accessed on 2 April 2020).
15. Skouroumounis, G.K.; Kwiatkowski, M.J.; Francis, I.L.; Oakey, H.; Capone, D.L.; Duncan, B.; Sefton, M.A.; Waters, E.J. The impact of closure type and storage conditions on the composition, colour and flavour properties of a Riesling and a wooded Chardonnay wine during five years' storage. *Aust. J. Grape Wine Res.* **2005**, *11*, 369–377. [CrossRef]
16. Tarasov, A.; Giuliani, N.; Dobrydnev, A.; Müller, N.; Volovenko, Y.; Rauhut, D.; Jung, R. Absorption of 1,1,6-trimethyl-1,2-dihydronaphthalene (TDN) from wine by bottle closures. *Eur. Food Res. Technol.* **2019**, *245*, 2343–2351. [CrossRef]
17. Marais, J.; van Wyk, C.J.; Rapp, A. Effect of Storage Time, Temperature and Region on the Levels of 1, 1,6-Trimethyl-1,2-dihydronaphthalene and Other Volatiles, and on Quality of Weisser Riesling Wines. *S. Afr. J. Enol. Vitic.* **1992**, *13*, 33–44. [CrossRef]
18. Simpson, R.F. 1,1,6-Trimethyl-1,2-dihydronaphthalene: An important contributor to the bottle aged bouquet of wine. *Chem. Ind.* **1978**, *1*, 37.
19. Black, C.; Leigh, F.; Prue, H.; Capone, D.; Anderson, S.; Day, M.; Holt, H.; Pearson, W.; Herderich, M.; Johnson, D. Aged Riesling and the development of TDN. *Wine Vitic. J.* **2012**, *27*, 20–26.
20. Sacks, G.L.; Gates, M.J.; Ferry, F.X.; Lavin, E.H.; Kurtz, A.J.; Acree, T.E. Sensory Threshold of 1,1,6-Trimethyl-1,2-dihydronaphthalene (TDN) and Concentrations in Young Riesling and Non-Riesling Wines. *J. Agric. Food Chem.* **2012**, *60*, 2998–3004. [CrossRef]
21. Simpson, R.F.; Miller, G.C. Aroma composition of aged Riesling wine. *Vitis* **1983**, *22*, 51–63.
22. Winterhalter, P. Die Rolle von Glykosidischen Vorstufen für das Aroma von Sekt und Wein und deren technologische Beeinflussung. 2015. Available online: https://www.fei-bonn.de/veranstaltungen-termine/jahrestagungen/jahrestagung-2015 (accessed on 2 April 2020).
23. Ross, C.F.; Zwink, A.C.; Castro, L.; Harrison, R. Odour detection threshold and consumer rejection of 1,1,6-trimethyl-1,2-dihydronaphthalene in 1-year-old Riesling wines. *Aust. J. Grape Wine Res.* **2014**, *20*, 335–339. [CrossRef]
24. Ziegler, M.; Gök, R.; Bechtloff, P.; Winterhalter, P.; Schmarr, H.G.; Fischer, U. Impact of matrix variables and expertise of panelists on sensory thresholds of 1,1,6-trimethyl-1,2-dihydronaphthalene known as petrol off-flavor compound in Riesling wines. *Food Qual. Prefer.* **2019**, *78*, 103735. [CrossRef]
25. Delwiche, J. Psychological Considerations in Sensory Analysis. In *The Sensory Evaluation of Dairy Products*; Clark, S., Costello, M., Drake, M., Bodyfelt, F., Eds.; Springer: New York, NY, USA, 2009; pp. 7–15. ISBN 978-0-387-77408-4.
26. Filho, T.L.; Minim, V.P.R.; dos Santos Navarro, R.D.C.; Della Lucia, S.M.; Minim, L.A. Methodology for determination of two new sensory thresholds: Compromised acceptance threshold and rejection threshold. *Food Res. Int.* **2015**, *76*, 561–566. [CrossRef]
27. Dobrydnev, A.; Tarasov, A.; Müller, N.; Volovenko, Y.; Rauhut, D.; Jung, R. An optimized method for synthesis and purification of 1,1,6-trimethyl-1,2-dihydronaphthalene (TDN). *MethodsX* **2020**, *7*, 100768. [CrossRef] [PubMed]
28. Busch-Stockfisch, M. Sensorik Praxishandbuch. In *der Produktentwicklung und Qualitätssicherung*; Behr's Verlag: Hamburg, Germany, 2002; ISBN 978-3-86022-958-3.

29. ISO. *ISO 13301:2018 Sensory Analysis—Methodology—General Guidance for Measuring Odour, Flavour and Taste Detection Thresholds by a Three-Alternative Forced-Choice (3-AFC) Procedure*; International Organization for Standardization: Geneva, Switzerland, 2018.
30. ASTM International. *ASTM E679-19, Standard Practice for Determination of Odor and Taste Thresholds By a Forced-Choice Ascending Concentration Series Method of Limits*; ASTM International: West Conshohocken, PA, USA, 2019.

*foods*

MDPI

*Article*

# Use of Sensory Analysis to Investigate the Influence of Climate Chambers and Other Process Variables in the Production of Sweet Wines

**M. Jesús Ruiz-Bejarano, Enrique Durán-Guerrero *, Remedios Castro, Carmelo G. Barroso and M. Carmen Rodríguez-Dodero**

Analytical Chemistry Department, Faculty of Sciences-IVAGRO, University of Cadiz, Agrifood Campus of International Excellence (ceiA3), Pol. Río San Pedro, s/n, Puerto Real, 11510 Cadiz, Spain; mariajesus.ruiz@uca.es (M.J.R.-B.); remedios.castro@uca.es (R.C.); carmelo.garcia@uca.es (C.G.B.); maricarmen.dodero@uca.es (M.C.R.-D.)

* Correspondence: enrique.duranguerrero@uca.es; Tel.: +34-956-016456

Received: 17 March 2020; Accepted: 1 April 2020; Published: 3 April 2020

**Abstract:** In this study, a climate chamber, as an alternative method, has been used to dry raisins and the sensory profiles of the sweet sherry wines obtained have been evaluated. Other important factors, namely grape variety, vintage, vinification conditions, as well as the ageing method and its length of time, have also been considered. When heavy rainfall had been registered, the musts extracted from grapes dried under controlled conditions in a climate chamber showed a lower intensity of the musty off-odor compared to those elaborated with sun-dried grapes. The wine fermented at low temperature with *Saccharomyces bayanus* scored the highest in citric and floral notes, and this was preferred over all the other wines that were evaluated. The wines aged in oak barrels were preferred to both, wines aged in the presence of oak chips as well as those aged without any wood contact. The use of climate chambers to dry the grapes that are going to be used for the elaboration of sweet wines appears to be an advantageous alternative to the traditional method, since it allows a more precise control of the process and highly valued sweet wines from a sensory point of view are obtained thereby.

**Keywords:** sensory analysis; sweet wine; raisining; climate chamber

## 1. Introduction

The production of sweet wines is commonly carried out with dehydrated grapes and there are three different processes to perform the water loss of the grapes: dehydration at a controlled temperature, humidity and ventilation; drying with non-controlled conditions and withering with eventual control of temperature and humidity and natural ventilation [1]. The sweet wines elaborated in Andalusia, Spain, are made from different white grape varieties, mainly Muscat and Pedro Ximénez, which are subjected to an ancient traditional method with non-controlled conditions: harvested grapes are dried in the open air, exposed to direct sunlight. This process is known as "asoleo" or sunning, where bunches of grapes are spread out on "redores"—esparto grass mats. During this sun-drying phase, which may take between seven and twenty days depending on the specific weather conditions [2,3], the bunches are turned over on a regular basis and covered at night. When the grapes are exposed to hours of intense sunshine, they gradually lose water and attain sugar concentration levels over 300 g/L; this process has an impact on their flavor profiles [4].

When this traditional procedure is applied, grapes may suffer undesired alterations that would have an impact on the quality of the final product; rainfalls particularly may contribute to the growth of certain fungi, which would lead to the loss of a considerable proportion of the treated grapes.

In a worse scenario, highly toxic metabolites might appear in the wines that have been elaborated using such grapes [5–8]. For that reason, novel and more advanced techniques to be employed in the sweet wine industry should be sought.

There are several alternatives for a controlled dehydration of the grape, which would bring about a number of considerable advantages. On the one hand, withering methods with natural ventilation have been widely employed for the production of sweet wines in Italy, where the use of artificially heated air is not allowed [9–14]. On the other hand, climate chambers with forced convection of hot air have also been used for the raisining of grapes and other fruits [6,7,15–20]. Regardless of external weather conditions, both alternatives (with or without forced ventilation) could allow the adjustment of temperature and humidity over the drying process, thus reducing the length of time required for the desired raisining stage.

Research studies on Pedro Ximénez and Muscat sweet musts and wines made from grapes dried both under sunlight and within climate chambers with forced ventilation can be found in the literature [21–26], however, no studies have been found to comprise the subsequent optimization of the whole process, namely, fermentation conditions, ageing method and evolution of the product under wood contact.

With a view to proposing the use of climate chambers as an alternative to the traditional sun-drying method, we have evaluated the sensory profiles of sweet sherry wines elaborated from grapes dried in climatic chambers. Some important factors, such as grape variety, vintage, fermentation conditions, as well as the ageing method and time, have been considered in this study.

## 2. Materials and Methods

### 2.1. Production of the Wines

#### 2.1.1. Raisining

Two grape varieties (Muscat, M, and Pedro Ximénez, PX) from three consecutive vintages (V01, V02 and V03) were used for this research. The grapes were subjected to two different drying procedures: the traditional natural sun-drying method (T, either on a terrace in our research centre or at the vineyard) and controlled drying in a climate chamber (C).

For chamber drying, about 2000 kg of ripe grape bunches of each grape variety and vintage were collected from a local winery in the Jerez-Xérès-Sherry D.O. (Denomination of Origin) region. The grapes were dried in a climate chamber (Ibercex A.S.L., S.A., La Poveda, Spain) for about 5 days at 40 °C and 10% relative humidity. The bunches were uniformly distributed in a single layer inside the chamber. For the sun drying method, the grapes were spread out also in a single layer on "esparto" grass mats to dry under sunlight for about 10–15 days. They were turned over every day and covered at night. Grape weight loss was periodically monitored for both methods and the drying process was considered as completed when such weight loss reached about 35% of the original weight, at 20–21° Baume.

After this, grapes were separately destemmed, crushed and pressed by means of a vertical press (300 bars maximum). The initial pH of the must ranged between 3.6 and 3.8 and it was different for each experience, so it was adjusted to 3.5 by adding tartaric acid (Agrovin, Alcázar de San Juan, Spain). The concentration of total sulphur dioxide was also set at 120 mg/L by adding potassium metabisulfite (Agrovin).

#### 2.1.2. Fermentation

Five different conditions were tested for the partial fermentation of the musts obtained from vintage 01 Muscat grapes dried in the climate chamber (experiments E1 to E5, Table 1). In experiment E1, a *Saccharomyces cerevisiae* yeast inoculum (40 g/hL, Lalvin D254, Lallemand, Montreal, Canada) was employed. The fermentation was carried out at room temperature (less than 30 °C). In order to evaluate the effect of the addition of nitrogen, the conditions for experiment E2 were the same as for experiment E1 with the addition of some yeast nutrients (diammonium phosphate, 10 g/hL; Actimax Plus, Agrovin). For the evaluation of the employment of pellicular maceration, in experiment E3 the conditions were the same as for experiment E1 but the grapes were crushed without previous destemming and submitted to prefermentative pellicular maceration for 24 h at 4 °C using 3 g/hL of pectolytic enzymes (Enozym Arome, Agrovin), and then pressed. In order to check the influence of the type of yeast in the fermentation, the must from experiment E5 was fermented under the same conditions as for experiment E3 but employing *Saccharomyces bayanus* (40 g/hL, Uvaferm 43, Lallemand). Finally, to investigate the effect of the temperature in the fermentation, in experiment E4, the must was fermented at low temperature (less than 10 °C) with *S. bayanus*. Taking into account the recommendations from the supplier, this yeast was more suitable than *S. cerevisiae* to carry out the fermentation at lower temperatures.

**Table 1.** Experimental conditions of the 5 fermentation assays carried out with Muscat must extracted from grapes dried in a climatic chamber in vintage 01.

| Assay | Yeast | Nitrogen | Pellicular Maceration with Pectolytic Enzymes | Fermentation Temperature |
|---|---|---|---|---|
| **E1** | *S. cerevisiae* | No | No | Room |
| **E2** | *S. cerevisiae* | Yes | No | Room |
| **E3** | *S. cerevisiae* | No | Yes | Room |
| **E4** | *S. bayanus* | No | No | Low |
| **E5** | *S. bayanus* | No | Yes | Room |

The fermentations processes were carried out in duplicate. For vintages 02 and 03 the fermentation conditions were the same as those previously selected for vintage 01, that is, low temperature (about 10 °C) with *S. Bayanus* yeast. When the wine sugar content was around 90–100 g/L, alcohol up to 17–18° alc. was added to prevent any further fermentation. Alcohol content was determined by a distillation procedure and subsequent measurement of the density of the distillate.

2.1.3. Ageing

After that, the selected Muscat wine from vintage 01 was subjected to ageing for 1 year using the following 3 parallel methods: For the first method, the wine was poured into 30 L medium toast American oak barrels (samples B); for the second method, the wine was put in contact with oak chips (Roblemor T, medium toast, Agrovin) at a dose of 4 g/L (samples identified by Ch) and for the third method, ageing took place without any contact with oak, in stainless steel tanks (samples S).

The sweet wines (Muscat and Pedro Ximénez) produced from vintages 02 and 03 were all aged in 30 L medium toasted American oak barrels (B). During this phase, all the wines were maintained in the same room at about 20 °C.

During each ageing phase, the samples were taken once every month over a whole year (S0 to S12). All of the samples were stored at 4 °C until their analysis.

*2.2. Sensory Evaluation Methodology*

The exclusively orthonasal sensory evaluation of the musts and wines was carried out in duplicate by a panel of between 13 and 17 members depending on the test. The samples were not tasted with the mouth, so the retronasal sensory evaluation was not performed. All the panel members had a medium level of experience in sensory analysis of sweet wines [27]. In addition, in order to validate the reproducibility of the judges' assessment, the descriptive profiles of 2 reference samples were obtained; for each descriptor, a two-factor ANOVA (judges × samples) of the descriptive data was performed to determine the consistency of the assessments. The tasting sessions were held in a standard tasting room equipped with separate booths [28] at 22 °C.

Quantitative descriptive analysis [29], and in some cases triangle discrimination [30] or ranking tests [31], were performed.

Twenty-milliliter samples were presented in blue glasses (typically used for olive oil sensory analysis [32]), so that color would not influence the panel members' assessments. The sample containers were topped with a watch-glass to minimize any possible aroma losses. The samples were identified by numerical codes composed by three random figures, and different for each judge. For triangle discrimination, samples were presented in order to let appear each possible disposition of the samples an equal number of times, but for descriptive analysis and ranking tests a randomized presentation was employed.

After performing the triangle test, the judges were also asked to provide a quantitative assessment of the detected differences. The following 5-point scale was used to rate such differences: not present (0), low (1), medium (2), strong (3) and very strong (4).

In respect to descriptive profile tests, as a preliminary stage to determine the appropriate descriptors that would better define the samples, a representative number of must and wine samples from the whole set were presented to the judges for them to provide qualitative descriptions for each one. The descriptors with a mention frequency over 5 were selected; i.e., fruity, citric, ripe fruit, raisin, floral, honey, herbaceous, vinous, lactic, musty, chemical character and oak. The final worksheet also included aromatic intensity and olfactory quality as descriptors. Each descriptor was scored according to a nine-point scale (0: absent; 2: light; 4: medium; 6: intense; 8: very intense), and also the olfactory quality of each sample was evaluated according to a structured nine-point scale (0: bad; 2: mediocre; 4: acceptable; 6: good; 8: very good) [33]. Table 2 presents the definitions agreed by all the judges for each descriptor as well as the standard used to recognize and quantify intensity (8) for each one of them. Both citric and floral were connected to a Muscat grape distillate as a standard because these were the two main descriptors of this cultivar. Therefore, the same standard could be perfectly used for the recognition of both descriptors.

**Table 2.** Aroma descriptors used for sample assessment.

| Descriptor | Definition | Standard |
|---|---|---|
| Aromatic intensity | Intensity of overall olfactory perception by the orthonasal route | Commercial Muscat |
| Fruity | Raw material (grapes) | Muscat and PX grapes |
| Citric | Reminiscence of the common note in citrus fruits | Muscat grape distillate, 30% alc. (*v*/*v*) |
| Ripe fruit | Reminiscence of white stone fruit in an advanced state of ripeness | Hydroalcoholic solution 30% alc. (*v*/*v*) with pieces of ripe apple and pear |
| Raisin | Reminiscence of the dried raw material | Raisins Muscat and PX |
| Floral | Common note in flowers, whatever the species | Muscat grape distillate, 30% alc. (*v*/*v*) |
| Honey | Reminiscence of honey | Flower honey |
| Herbaceous | Sharp green note | Commercial herbaceous pomace |
| Vinous | Reminiscence of recently fermented sherry wine | Freshly fermented sherry wine (and frozen until tasting) |
| Lactic | Characteristic note of wines that undergo malolactic fermentation | White wine that has recently undergone lactic fermentation (and frozen until tasting) |
| Musty | Olfactory defect with earthy or mushroomy notes | Hydroalcoholic mixture 15% alc. (*v*/*v*) with addition of mushroomy extract from Le nez du Vin * |
| Chemical character | Olfactory defect with notes of solvent, glue or medicinal | Hydroalcoholic mixture 15% alc. (*v*/*v*) with addition of aromatic glue and medicinal extracts from Le nez du Vin * |
| Oak | Reminiscence of American oak wood | Hydroalcoholic solution 15% alc. (*v*/*v*) with American oak chips |
| Olfactory quality | Overall measurement by the orthonasal route of aromatic complexity and intensity and the absence of defects | Commercial Muscat and PX wines |

* Le nez du Vin, Jean Lenoir Ed., 2006.

### 2.3. Statistical Analysis

The statistical analyses were carried out using Statistica 8.0 (StatSoft GmbH, Hamburg, Germany), and Microsoft Office Excel 2010 applications.

The treatment of the data from the triangle test was based on the tabulated statistic data [30] with respect to difference tests, by setting the $\alpha$-error at 0.05. Friedman test was applied for the treatment of the data from the olfactory quality ranking, as specified by the corresponding ISO standard [31].

An analysis of variance (ANOVA) was performed on the quantitative sensory data from the descriptive assessments. However, an acceptable dispersion of sensory data could complicate the discrimination between groups of samples for a particular descriptor (especially when professional or very trained panels are not employed). Therefore, it is proposed the subsequent application of multivariate statistical techniques that take into account the whole group of differences and similarities for all variables.

A Principal Component Analysis (PCA) was performed in order to highlighting the similarity of the samples, and to determine the main contributors to any of the differences found between them. Missing data were replaced by the average value of that variable in the group. The minimum eigenvalue to select a principal component was set at 1.0, and a factor rotation according to varimax normalized method was applied to confirm a correlation with the PCA. Loadings greater than 0.7 identified those variables well correlated to PCs.

In order to identify those descriptors that better differentiate between clusters of samples, a Linear Discriminant Analysis (LDA) was performed according to Wilks' lambda statistic, and the so-called forward stepwise method was employed. According to this methodology, the discrimination model is built step-by-step by reviewing all the variables at each step and evaluating which one contributes the most to the discrimination between clusters. The F-to-enter and F-to-remove were set at 1 and 0.

## 3. Results and Discussion

### 3.1. Effect of the Raisining Method on the Aroma of Raisin Musts

The musts obtained after traditional drying of the grapes from the first year (V01) were compared with those dried in a climatic chamber by means of a triangle test ($N = 13$ judges, $\alpha = 0.05$). In the case of Muscat, 10 out of the 13 judges identified the different sample, while in the discrimination of Pedro Ximénez musts, 8 judges detected the different sample. Given that according to the standard, 8 is the minimum number of coincidental judgments required to confirm a significant difference between the samples, we could conclude that the Muscat musts and also those of Pedro Ximénez presented a significantly different aromatic profile if the grapes had been raisined in the sun or in a climate chamber. The differences were valued by the judges as being of a moderate intensity ($1.7 \pm 0.4$ on a 0–4 scale).

An ANOVA was then applied to the scores obtained from the descriptive test on these V01 musts with the objective of identifying the aromatic notes that were responsible for the differences that had previously been confirmed by the triangle test. The resulting data are shown in Table 3. Of all the positive notes considered, only fruity and raisins could statistically differentiate ($p < 0.05$) the musts from Muscat grapes treated by the 2 drying methods tested, being higher the scores for the musts obtained under controlled conditions. From Pedro Ximénez musts, the results were similar, although the differences were not significant. Having said that, the most interesting result is that those musts obtained from grapes dried in chambers were perceived as having a lower intensity in the olfactory fungal defect (musty). The detection of such a defect in the musts from either a drying method, seems to indicate that the grapes had been previously contaminated, probably due to the rainfalls during the days before that year's harvesting period. As a result, the musts of grapes dried in climate chambers were better evaluated (aromatic quality).

When the data from the second vintage (V02) musts were analyzed, it was found that non-significant differences between the average scores for some of the defects in the Muscat musts could justify a better olfactory rating when their grapes had been dried in a climate chamber. However, this preference did not reach statistical significance. On the other hand, no differences could be confirmed for V03 musts that had been dried either in a climate chamber or by means of the traditional sun-drying method.

It should be taken into account that the grapes from these vintages (V02, and particularly in V03) arrived at the pilot plant in a better sanitary condition. This could explain their similar evolution regardless of the raisining method employed, as opposed to what happened with V01 grapes, which was characterized by intense rainfalls on the days prior to their harvest. Other authors neither found significant sensory differences between musts obtained by grapes dried traditionally or in a controlled way [34].

Due to the intrinsic variance of sensory data, it is often difficult to confirm differences between samples based on the results obtained from an analysis of variance. Therefore, in order to evaluate the minor differences between each of these descriptors that may come into sight when comparing "asoleo" and climate chamber musts, a PCA was applied to the sensory scores of all the must samples from V01, V02 and V03. The PCA allowed determining four principal components that account for 86.7% of the total variance. The loads (data not shown) confirmed a relationship between PC1 (40.1% of the variance explained) and the drying method, given its correlation with olfactory descriptors such as ripe fruit and raisins; and PC2 (22.6% of the variance explained) that represents a greater aromatic complexity, as it correlates with aromatic intensity and citric and floral notes, which contribute to better aromatic quality. On the other hand, PC3 and PC4 (with 16.1% and 7.9% of the explained variance, respectively) are related to aromatic defects (lactic, musty, herbaceous to PC3 and chemical to PC4).

When the sensory profiles were transferred to this new representation, it can be seen that for PC1 (Figure 1a), the raisin musts from the climatic chamber (C, in bold type) were located to the right of their corresponding raisins obtained by means of the sun-drying method (T). Muscat musts were positioned at the highest PC2 values, which means that their aroma was perceived as having a greater complexity, since they exhibited a greater intensity in highly appreciated citric and floral notes. It is well known that the sensory profile of musts depends on their variety [35,36]. Interestingly, the only Pedro Ximénez must with high PC2 values was the one made from V01 grapes dried in a climate chamber (V01.PX.C).

On the other hand, PC3 and PC4 allowed us to differentiate the musts according to their harvest (Figure 1b). As can be seen, the musts from harvest 03 (V03) were located in the most distant area from the defects (lactic, musty, herbaceous and chemical character), with no influence from the variety or the raisining method. This would confirm the important influence of the vintage on the quality of raisin musts, as some authors had already pointed out in previous studies [37,38].

**Table 3.** ANOVA of sensory scores of musts, vintages 01–03. Values are expressed as average ± standard deviation. Differences are significant at $p < 0.05$.

| Descriptor | Muscat V01 | | | Pedro Ximénez V01 | | | Muscat V02 | | | Pedro Ximénez V02 | | | Muscat V03 | | | Pedro Ximénez V03 | | |
|---|---|---|---|---|---|---|---|---|---|---|---|---|---|---|---|---|---|---|
| | Sun-Drying | Climate Chamber | *p* | Sun-Drying | Climate Chamber | *p* | Sun-Drying | Climate Chamber | *p* | Sun-Drying | Climate Chamber | *p* | Sun-Drying | Climate Chamber | *p* | Sun-Drying | Climate Chamber | *p* |
| Aromatic intensity | 4.5 ± 1.4 | 4.2 ± 1.3 | 0.152 | 4.5 ± 1.4 | 4.3 ± 1.0 | 0.817 | 3.9 ± 1.2 | 4.3 ± 1.4 | 0.223 | 4.2 ± 0.9 | 4.3 ± 1.2 | 0.463 | 6.0 ± 0.9 | 4.5 ± 1.2 | 0.276 | 3.0 ± 0.6 | 4.5 ± 0.9 | 0.465 |
| Fruity | 3.9 ± 1.2 | 4.7 ± 0.8 | 0.045 | 4.0 ± 1.5 | 5.5 ± 2.1 | 0.079 | 3.9 ± 0.7 | 4.2 ± 1.1 | 0.251 | 2.9 ± 0.8 | 3.5 ± 1.0 | 0.407 | 4.1 ± 1.1 | 4.2 ± 0.8 | 0.533 | 1.5 ± 0.8 | 2.0 ± 0.7 | 0.937 |
| Citric | 1.1 ± 1.4 | 0.9 ± 1.3 | 0.215 | 1.2 ± 1.6 | 0.7 ± 1.0 | 0.535 | 2.2 ± 0.9 | 2.3 ± 0.7 | 0.655 | 0.7 ± 0.4 | 0.6 ± 0.2 | 0.845 | 3.0 ± 0.8 | 2.5 ± 0.6 | 0.753 | 0.5 ± 0.4 | 0.4 ± 0.4 | 0.207 |
| Ripe fruit | 3.4 ± 2.2 | 2.6 ± 2.0 | 0.276 | 3.3 ± 1.8 | 3.0 ± 1.8 | 0.876 | 2.1 ± 1.2 | 2.4 ± 0.9 | 0.811 | 3.0 ± 1.1 | 3.4 ± 0.8 | 0.310 | 2.5 ± 0.7 | 2.5 ± 0.3 | 0.194 | 2.5 ± 0.8 | 3.0 ± 0.7 | 0.359 |
| Raisin | 3.3 ± 1.2 | 4.5 ± 1.5 | 0.049 | 2.0 ± 1.6 | 2.3 ± 1.7 | 0.096 | 1.1 ± 0.3 | 1.9 ± 0.7 | 0.339 | 3.4 ± 1.0 | 2.7 ± 1.2 | 0.976 | 2.5 ± 1.0 | 1.0 ± 0.9 | 0.893 | 3.0 ± 0.6 | 2.9 ± 1.1 | 0.788 |
| Floral | 2.0 ± 1.9 | 1.7 ± 1.6 | 0.557 | 1.3 ± 1.6 | 2.2 ± 2.1 | 0.516 | 2.1 ± 0.4 | 2.7 ± 0.2 | 0.158 | 0.9 ± 0.5 | 0.9 ± 0.6 | 0.617 | 8.1 ± 1.7 | 5.2 ± 2.1 | 0.750 | 0.6 ± 0.6 | 1.1 ± 0.5 | 0.263 |
| Honey | 1.2 ± 1.6 | 2.3 ± 2.5 | 0.345 | 1.8 ± 2.4 | 2.3 ± 1.9 | 0.701 | 1.1 ± 0.3 | 1.6 ± 1.3 | 0.729 | 1.9 ± 0.6 | 1.3 ± 1.1 | 0.363 | 2.1 ± 0.7 | 2.4 ± 0.8 | 0.537 | 0.2 ± 0.4 | 0.4 ± 0.7 | 0.417 |
| Herbaceous | 1.9 ± 1.4 | 1.2 ± 1.9 | 0.219 | 1.6 ± 0.6 | 1.7 ± 0.7 | 0.305 | 1.0 ± 0.7 | 0.7 ± 0.4 | 0.922 | 0.7 ± 0.4 | 1.0 ± 0.4 | 0.669 | 0.5 ± 0.6 | 0.4 ± 0.5 | 0.722 | 0.5 ± 0.7 | 1.0 ± 0.8 | 0.675 |
| Vinous | 0.9 ± 1.7 | 1.2 ± 1.9 | 0.337 | 1.2 ± 2.0 | 0.3 ± 0.8 | 0.375 | 0.3 ± 0.6 | 0.2 ± 0.5 | 0.976 | 0.9 ± 0.7 | 0.7 ± 1.1 | 0.721 | 0.3 ± 0.5 | 0.2 ± 0.6 | 0.690 | 1.2 ± 1.1 | 1.5 ± 0.5 | 0.287 |
| Lactic | 0.4 ± 0.9 | 0.5 ± 1.1 | 0.216 | 0.5 ± 0.8 | 0.0 ± 0.0 | 0.174 | 0.1 ± 0.5 | 0.1 ± 0.6 | 0.873 | 0.3 ± 0.6 | 0.2 ± 0.8 | 0.683 | 0.0 ± 0.0 | 0.0 ± 0.0 | 0.972 | 0.2 ± 0.5 | 0.5 ± 0.6 | 0.866 |
| Musty | 1.9 ± 2.2 | 1.3 ± 1.8 | 0.071 | 3.2 ± 1.8 | 1.2 ± 0.9 | 0.061 | 0.6 ± 0.4 | 0.1 ± 0.3 | 0.234 | 0.2 ± 0.3 | 0.2 ± 0.3 | 0.886 | 0.0 ± 0.0 | 0.0 ± 0.0 | 0.859 | 0.0 ± 0.0 | 0.0 ± 0.0 | 0.512 |
| Chemical character | 0.1 ± 0.5 | 0.2 ± 0.5 | 0.125 | 0.3 ± 0.5 | 0.0 ± 0.0 | 0.144 | 0.4 ± 0.5 | 0.2 ± 0.5 | 0.287 | 1.2 ± 0.5 | 1.0 ± 0.5 | 0.997 | 0.0 ± 0.0 | 0.0 ± 0.0 | 0.989 | 0.0 ± 0.0 | 0.0 ± 0.0 | 0.644 |
| Olfactory quality | 2.8 ± 1.9 | 3.6 ± 1.7 | 0.115 | 2.5 ± 2.3 | 4.7 ± 0.8 | 0.051 | 2.8 ± 1.1 | 5.1 ± 1.2 | 0.088 | 3.7 ± 0.4 | 3.9 ± 0.2 | 0.134 | 7.1 ± 1.1 | 6.4 ± 1.3 | 0.402 | 4.1 ± 0.9 | 4.5 ± 0.6 | 0.230 |

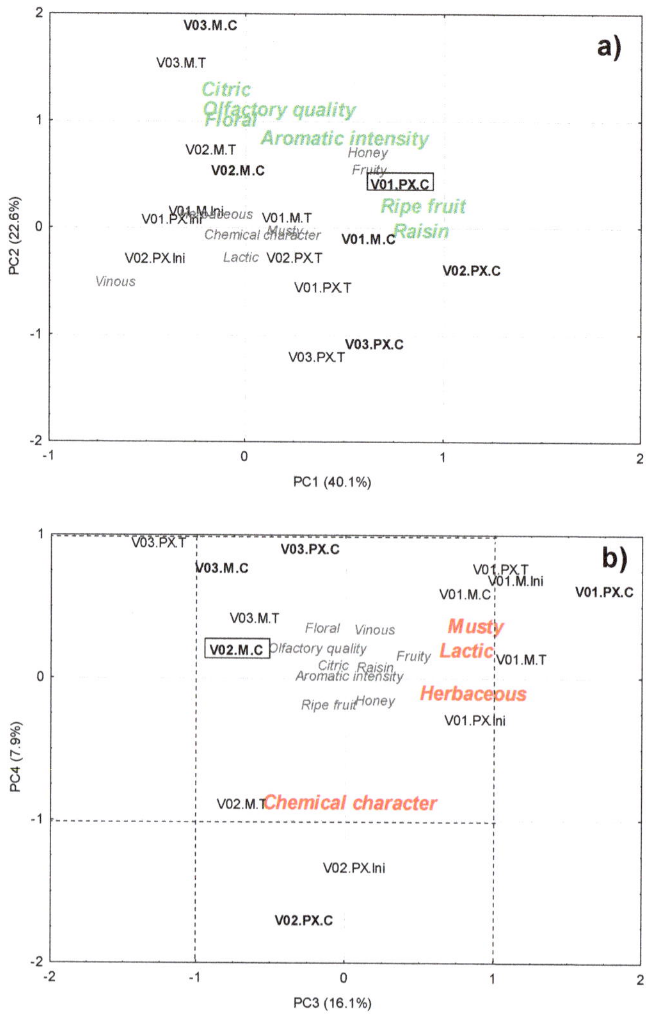


**Figure 1.** Principal components analysis of the sensory scores of musts from raisins obtained by the two different drying methods tested. Representation of both, the samples scores and the principal component loadings, onto the new space defined by: (**a**) PC1 and PC2 and (**b**) PC3 and PC4. The names of the samples indicate the vintage.variety.raisining method.

In any case, the most interesting result from our research is that the Muscat grape musts from the 02 vintage (V02.M.C), whose grapes were treated in a climate chamber, were also located in this defect-free area. It is therefore confirmed that the use of a climate chamber for the raisining phase of the grapes avoided some of the risks associated to adverse weather conditions, by allowing the hygienic drying of the grapes and preventing the appearance of fungi and its associated aromatic defects, while food safety was also preserved. In addition, in some cases the musts would exhibit a greater aromatic complexity. It should also be added that the drying time was reduced by 50–66%. At this point, it could be concluded that the capacity of the sensory analysis as a methodology to control the hygienic and sanitary state of musts was demonstrated.

### 3.2. Selecting the Fermenting Conditions for Sweet Wines Made from Grapes Dried in a Climate Chamber

The following step was the evaluation of some fermenting conditions in the production of sweet wine such as the type of yeast, the addition of nitrogen, the employment of pellicular maceration or the fermentation temperature (Table 1). The wines obtained were compared with each other by means of a ranking test according to olfactory quality. Thus, since the threshold of statistical significance ($\alpha = 0.05$) for the 17-member panel that participated in the test was 9.5, and that the experimental value reached 9.8, it could then be confirmed that some differences between the wines in terms of olfactory quality were perceived. The outcome of the ranking test was $E4^1$ (63) > $E1^1$ (56) > $E3^{1,2}$ (55) > $E5^2$ (43) > $E2^2$ (36), where the different superscripts indicate that these wines were perceived differently with regards to their olfactory qualities. The wine that had been fermented at low temperature with *S. bayanus* (E4) was the best rated, although the difference with the wine fermented with *S. cerevisiae* at room temperature (E1) was not significant. The nitrogen-added wine was granted the lowest score (E2), which agreed with the conclusions by other authors [39]. Finally, the wines subjected to pellicular maceration (E3 and E5) presented intermediate ratings, although the one fermented employing *S. cerevisiae* (E3) was slightly better scored.

The wines were also evaluated by means of descriptive tests. Table 4 shows the results from the ANOVA, according to which, the floral and citric notes have the greatest power of discrimination (lowest $p$-values) between the wines produced. These are attributed the highest score in the ranking test for wine E4 (fermented with *S. bayanus* at low temperature), which was characterized by high scores in Muscat typical *citric* and *floral* notes (in agreement with the observations of other authors regarding this yeast [40]), while the wine with the lowest score (E2, with added nitrogen) was characterized by very low intensities in these positive notes, as well as a clear olfactory defect of a chemical nature. Likewise, the fruity and floral notes of wines E3 and E5 (both subjected to pellicular maceration with pectolytic enzymes), were high. This was in agreement with the results observed by other authors regarding the volatile composition and sensory profile of Muscat wines employing this type of enzymes [41,42]. Comparing the obtained evaluation for these two wines (E3 and E5) obtained under the same conditions, but with different yeast strains, slight differences can be observed, so the effect of the employed yeast seems to be not so relevant, unless the temperature of fermentation is low (E4).

**Table 4.** ANOVA of sensory scores of wines for vintage 01 made under different fermentation conditions, as described in the text. Mean values and standard deviations are shown.

| Descriptor | E1 | E2 | E3 | E4 | E5 | $p$ |
|---|---|---|---|---|---|---|
| **Aromatic intensity** | 4.1 ± 0.5 | 4.2 ± 0.6 | 4.4 ± 0.6 | 4.1 ± 0.5 | 4.2 ± 0.5 | 0.97 |
| **Fruity** | 4.2 ± 0.3 | 2.7 ± 0.4 | 3.8 ± 0.3 | 4.1 ± 0.4 | 3.8 ± 0.5 | 0.11 |
| **Citric** * | $2.2 \pm 0.7^{2}$ | $1.4 \pm 0.5^{1}$ | $2.8 \pm 0.3^{2,3}$ | $3.1 \pm 0.4^{3}$ | $2.7 \pm 0.3^{2,3}$ | 0.04 |
| **Ripe fruits** | 1.9 ± 0.4 | 1.5 ± 0.6 | 2.1 ± 0.4 | 1.7 ± 0.5 | 1.4 ± 0.4 | 0.87 |
| **Raisins** | 1.9 ± 0.5 | 1.6 ± 0.5 | 2.2 ± 0.4 | 2.3 ± 0.4 | 1.1 ± 0.3 | 0.84 |
| **Floral** * | $3.6 \pm 0.8^{1,2}$ | $3.0 \pm 0.7^{1}$ | $3.9 \pm 0.5^{1,2}$ | $4.1 \pm 0.6^{2}$ | $4.0 \pm 0.8^{2}$ | 0.07 |
| **Honey** | 0.9 ± 0.2 | 0.8 ± 0.4 | 1.5 ± 0.5 | 1.2 ± 0.3 | 1.1 ± 0.5 | 0.54 |
| **Herbaceous** | 1.1 ± 0.4 | 2.1 ± 0.6 | 1.3 ± 0.6 | 1.4 ± 0.5 | 9 ± 0.7 | 0.83 |
| **Vinous** | 2.5 ± 0.5 | 2.3 ± 0.7 | 2.5 ± 0.5 | 2.9 ± 0.6 | 2.8 ± 0.6 | 0.87 |
| **Lactic** | 0.2 ± 0.3 | 0.5 ± 0.3 | 0.5 ± 0.3 | 0.5 ± 0.3 | 0.5 ± 0.4 | 0.99 |
| **Musty** | 0.6 ± 0.3 | 1.1 ± 0.4 | 1.1 ± 0.6 | 0.9 ± 0.4 | 1.5 ± 0.5 | 0.86 |
| **Chemical character** * | $0.8 \pm 0.3^{1}$ | $2.1 \pm 0.3^{2}$ | $0.6 \pm 0.3^{1}$ | $1.0 \pm 0.3^{1}$ | $0.8 \pm 0.2^{1}$ | 0.04 |
| **Olfactory quality** * | $4.4 \pm 0.4^{3}$ | $2.5 \pm 0.6^{1}$ | $3.1 \pm 0.5^{1,2}$ | $4.5 \pm 0.5^{3}$ | $3.8 \pm 0.5^{2,3}$ | 0.05 |

* Values are statistically significant at $p < 0.05$. Different superscript numbers indicate that tasters perceived a significant different intensity.

Consequently, the E4 test was selected as the preferred fermenting conditions for the production of sweet wines from grapes dried in a climate chamber, namely with *Saccharomyces bayanus* yeast, at a controlled temperature not greater than 10 °C, no nitrogen added and non-pellicular maceration in the presence of pectolytic enzymes. The above fermenting conditions were used for the rest of the research.

### 3.3. Selecting the Ageing Method for Sweet Wines made from Grapes Dried in a Climate Chamber

The Muscat wine from V01 fermented under the preferred conditions (E4) was subjected to ageing for 1 year using oak chips (Ch), oak barrels (B) and stainless steel containers (S). A LDA where the ageing method was taken as the cluster variable was applied to the sensory profiles of the sampled wines. This would allow the identification of the descriptors that best differentiate them. Since a high percentage of the wines (86.9%) were correctly classified, a PCA was performed on such most significant descriptors (i.e., those with $p < 0.05$): oak, olfactory quality, vinous, raisin, citric, musty, honey, fruity and ripe fruit. Figure 2 shows PC1 and PC2, which explain respectively 39.5% and 22.8% of the initial variance. Where PC1 is correlated with the descriptors raisin and oak, and also with the olfactory quality, while PC2 is inversely related to the intensity of the citric notes.

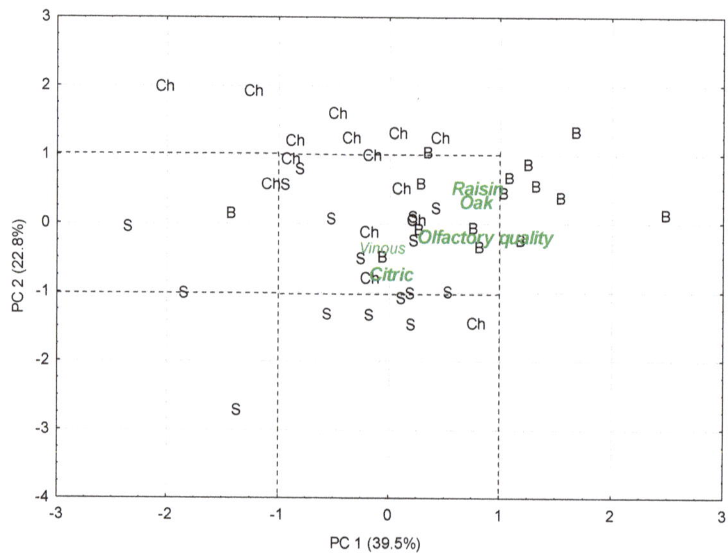


**Figure 2.** Principal components analysis of the sensory data of the same Muscat wine, vintage 01, aged under 3 different ageing methods (Ch: oak chips; B: oak barrels and S: stainless steel containers). Representation of both, the samples scores and the principal component loadings, onto the plane defined by the first two PCs.

When the wines are represented on this same plane, it can be seen that those wines, which were aged without any contact with oak (S) presented the lowest values for PC2, that is to say, they exhibited high intensity levels of citric notes, a typical character of Muscat wines when kept under these ageing conditions. Other authors [43] already confirmed that the degree to which these typical varietal attributes (or the compounds responsible for them) were maintained throughout the whole process from grape to the final wine depends on how the fermentation and ageing were conducted.

On the other hand, wines aged in oak barrels (B) were found on the right area of the graph (high values for PC1), which implies high intensity in oak notes and a very clearly perceived raisin note, both contributing to a highly valued olfactory quality.

With regard to wines aged in contact with oak chips (Ch), their position on the plane indicates low aromatic intensities of positive character notes and therefore it did not appear to be the best ageing system in principle.

This clear differentiation regarding ageing time corresponded to the discrimination observed for the same samples regarding the volatile composition [26]. Thus, the results obtained by means of sensory analysis seemed to confirm again the benefits of traditional ageing in oak casks as a method for ageing sweet wines made from raisins produced in a climate chamber.

### 3.4. Effect of the Length of The Ageing Period on the Aromatic Profile of Sweet Wines Made from Grapes Dried in a Climate Chamber

The previously selected vinification (E4) and ageing (B) methods were applied to the Muscat and Pedro Ximénez musts obtained from the V02 and V03 grapes dried in climate chambers, and the ageing process was monitored for 1 year. In the case of V01 grapes, only the wines from the Muscat variety were aged, since those from the Pedro Ximénez variety were accidentally contaminated by external causes at the beginning of the process and became unsuitable for this study.

In order to identify the variables that best discriminate wines according to their ageing time length, a LDA was performed on the initial (S0), intermediate (S6) and final (S12) samples, where the ageing time would be the cluster variable. The descriptors with the highest discrimination capacity were: aromatic intensity, fruity, citric, ripe fruit, floral, herbaceous, oak, vinous and together with olfactory quality (data not shown). The exclusive use of these sensory descriptors minimizes the amount of unnecessary data and is expected to improve the estimate and interpretation of the new components found by the PCA analysis, which resulted in three PCs (explaining 74% of the variance). Since standardized coefficients greater than 0.7 were considered as significant, it could be confirmed that the descriptors aromatic intensity, oak and olfactory quality correlated well with PC1 (which explained 28.7% of the variance), as well as herbaceous, although this latter correlated inversely, since it was a negative coefficient. With the same criterion, PC2 (explains 25.9% of the variance) correlated well with the citric and floral notes; while PC3 (19.4% of the variance explained) did so with the fruity and ripe fruit notes. Figure 3 represents the wines on the PC1–PC2 plane, although, for a clearer display, only the wines without any ageing (S0) and the samples from ageing wines that had been taken every other month (S01, S03, S05, S07, S09 and S11) are shown. As above mentioned, according to their loads, PC2 seems to be related to a greater aromatic complexity characterized by clear citric and floral notes, which would explain the higher values of this component in wines produced from Muscat grapes, with characteristic and intense primary aromas that are then enhanced during the fermentation phase (and which are located at the upper part of the graph), as opposed to the more neutral PX (at the lower area). It also seems logical that the scores for this PC decreased as the ageing time increases (from S0 to S11), given that the varietal typicity was partly eclipsed by the appearance of tertiary aromas.

It can be seen from the graph that for each vintage and variety the wines were sorted according to their ageing time along PC1 (from S0 to S11, from left to right). According to the loads of this component, the arrangement of the samples implies greater intensities of oak and aromatic intensity as the ageing process progresses, and what is more interesting, the perceived olfactory quality improved. The scores increased significantly over the first month, a behavior that has already been described by other authors [44] in relation to the volatile compounds found in sweet sherry wines during their oxidative ageing. It can also be seen that, based on the very different scores of the initial wines from the three vintages considered—which can be related to the different qualities of the musts as confirmed in previous sections—the wines that completed the ageing process presented similar PC1 values. This would attribute a certain buffering effect to the ageing process, contrary to what other authors concluded [45], who observed that different grape varieties have different wood extraction capacities, and that the ageing in wood barrels may enhance intrinsic varietal aroma differences.

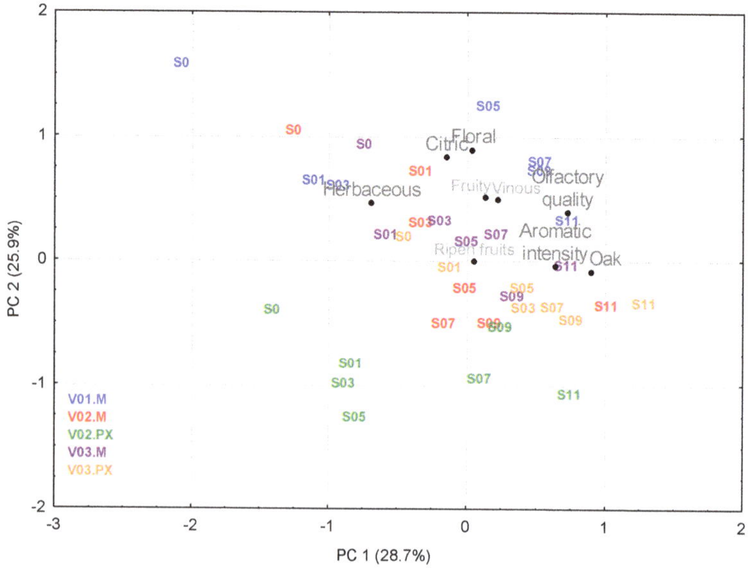


**Figure 3.** Principal components analysis of the sensory data of PX and Muscat wines from vintages V01, V02 and V03 fermented under the same conditions (E4) and aged for one year in oak barrels. Representation of both, the samples scores and the principal component loadings, onto the plane defined by the first two PCs.

It could be concluded that with regards to wood ageing and from the sensory point of view, ageing in oak barrels improved the olfactory perception of all the sweet wines made from grapes dried in a climate chamber, whether it be Muscat or Pedro Ximénez. Nevertheless, the effect became more noteworthy when the grapes were from less optimal vintages from the aromatic point of view.

## 4. Conclusions

In this study, sensory analysis was proven to be a reliable methodology to evaluate different process variables in the production of sweet wines, in which climate chambers to dehydrate the grapes were employed. The wine fermented at low temperature with *Saccharomyces bayanus* scored the highest in citric and floral notes, and this was preferred over all the other wines that were evaluated. Regarding the ageing stage, the wines aged in oak barrels were preferred to both, wines aged in the presence of oak chips as well as those aged without any wood contact. In addition, the use of climate chambers is a valuable alternative technique for the dehydration of grapes, compared to the traditional sun-drying method, because they allow a more precise control on the process and facilitate the solution to some production problems such as the musty off-odors associated to fungus contamination.

**Author Contributions:** Conceptualization, E.D.-G.; methodology, M.C.R.-D.; software, R.C.; validation, M.C.R.-D.; formal analysis, M.J.R.-B.; investigation, M.J.R.-B.; resources, C.G.B.; data curation, M.C.R.-D.; writing—original draft preparation, M.C.R.-D.; writing—review and editing, E.D.-G.; visualization, M.J.R.-B.; supervision, R.C.; project administration, R.C.; funding acquisition, C.G.B. All authors have read and agreed to the published version of the manuscript.

**Funding:** This research received no external funding.

**Conflicts of Interest:** The authors declare no conflict of interest. The funders had no role in the design of the study; in the collection, analyses or interpretation of data; in the writing of the manuscript, or in the decision to publish the results.

## References


1. Mencarelli, F.; Tonutti, P. (Eds.) *Sweet, Reinforced and Fortified Wines*; John Wiley & Sons, Ltd.: Oxford, UK, 2013; ISBN 9781118569184.
2. López, M.M. *Manual de Viticultura, Enología y Cata*; Almuzara: Córdoba, Spain, 2007.
3. Lea, A.G.H.; Piggott, J.R. *Fermented Beverage Production*; Blackie Academic & Professional: Glasgow, UK, 1997.
4. Franco, M.; Peinado, R.A.; Medina, M.; Moreno, J. Off-vine grape drying effect on volatile compounds and aromatic series in must from Pedro Ximénez grape variety. *J. Agric. Food Chem.* **2004**, *52*, 3905–3910. [CrossRef]
5. Valero, A.; Marín, S.; Ramos, A.J.; Sanchis, V. Survey: Ochratoxin A in European special wines. *Food Chem.* **2008**, *108*, 593–599. [CrossRef]
6. Ruíz Bejarano, M.J.; Rodríguez Dodero, M.C.; Barroso, C.G. Optimizing the process of making sweet wines to minimize the content of ochratoxin A. *J. Agric. Food Chem.* **2010**, *58*, 13006–13012. [CrossRef] [PubMed]
7. Serratosa, M.P.; Lopez-Toledano, A.; Millan, C.; Medina, M.; Merida, J. Changes of ochratoxin A in grapes inoculated with Aspergillus carbonarius and subjected to chamber-Drying under controlled conditions. *J. Agric. Food Chem.* **2010**, *58*, 11907–11912. [CrossRef]
8. Di Stefano, V.; Pitonzo, R.; Avellone, G.; Di Fiore, A.; Monte, L.; Ogorka, A.Z.T. Determination of Aflatoxins and Ochratoxins in Sicilian Sweet Wines by High-Performance Liquid Chromatography with Fluorometric Detection and Immunoaffinity Cleanup. *Food Anal. Methods* **2015**, *8*, 569–577. [CrossRef]
9. Bellincontro, A.; Matarese, F.; D'Onofrio, C.; Accordini, D.; Tosi, E.; Mencarelli, F. Management of postharvest grape withering to optimise the aroma of the final wine: A case study on Amarone. *Food Chem.* **2016**, *213*, 378–387. [CrossRef]
10. Corona, O.; Torchio, F.; Giacosa, S.; Río Segade, S.; Planeta, D.; Gerbi, V.; Squadrito, M.; Mencarelli, F.; Rolle, L. Assessment of Postharvest Dehydration Kinetics and Skin Mechanical Properties of "Muscat of Alexandria" Grapes by Response Surface Methodology. *Food Bioprocess Technol.* **2016**, *9*, 1060–1069. [CrossRef]
11. D'Onofrio, C.; Bellincontro, A.; Accordini, D.; Mencarelli, F. Malic acid as a potential marker for the aroma compounds of amarone winegrape varieties in withering. *Am. J. Enol. Vitic.* **2019**, *70*, 259–266. [CrossRef]
12. Bellincontro, A.; Prosperi, P.; De Santis, D.; Botondi, R.; Mencarelli, F. Control of environmental parameters in postharvest partial dehydration of wine grapes reduces water stress. *Postharvest Biol. Technol.* **2017**, *134*, 11–16. [CrossRef]
13. Modesti, M.; Petriccione, M.; Forniti, R.; Zampella, L.; Scortichini, M.; Mencarelli, F. Methyl jasmonate and ozone affect the antioxidant system and the quality of wine grape during postharvest partial dehydration. *Food Res. Int.* **2018**, *112*, 369–377. [CrossRef]
14. Petriccione, M.; Pagano, L.; Forniti, R.; Zampella, L.; Mastrobuoni, F.; Scortichini, M.; Mencarelli, F. Postharvest treatment with chitosan affects the antioxidant metabolism and quality of wine grape during partial dehydration. *Postharvest Biol. Technol.* **2018**, *137*, 38–45. [CrossRef]
15. Visconti, A.; Perrone, G.; Cozzi, G.; Solfrizzo, M. Managing ochratoxin a risk in the grape-wine food chain. *Food Addit. Contam. Part A Chem. Anal. Control. Expo. Risk Assess.* **2008**, *25*, 193–202. [CrossRef] [PubMed]
16. Serratosa, M.P.; Lopez-Toledano, A.; Medina, M.; Merida, J. Drying of Pedro Ximenez grapes in chamber at controlled temperature and with dipping pretreatments. Changes in the color fraction. *J. Agric. Food Chem.* **2008**, *56*, 10739–10746. [CrossRef]
17. Vega-Mercado, H.; Marcela Góngora-Nieto, M.; Barbosa-Cánovas, G.V. Advances in dehydration of foods. *J. Food Eng.* **2001**, *49*, 271–289. [CrossRef]
18. Serratosa, M.P.; Marquez, A.; Moyano, L.; Zea, L.; Merida, J. Chemical and morphological characterization of Chardonnay and Gewürztraminer grapes and changes during chamber-drying under controlled conditions. *Food Chem.* **2014**, *159*, 128–136. [CrossRef] [PubMed]
19. López de Lerma, N.; Moreno, J.; Peinado, R.A. Determination of the Optimum Sun-Drying Time for Vitis vinifera L. cv. Tempranillo Grapes by E-nose Analysis and Characterization of Their Volatile Composition. *Food Bioprocess Technol.* **2014**, *7*, 732–740. [CrossRef]
20. Martin-Gomez, J.; Varo, M.A.; Merida, J.; Serratosa, M.P. Bioactive Compounds of Chamber-Dried Blueberries at Controlled Temperature and Wines Obtained from Them. *J. Chem.* **2017**, *2017*. [CrossRef]
21. Márquez, R.; Castro, R.; Natera, R.; García-Barroso, C. Characterisation of the volatile fraction of Andalusian sweet wines. *Eur. Food Res. Technol.* **2008**, *226*, 1479–1484. [CrossRef]

22. Ruiz, M.J.; Zea, L.; Moyano, L.; Medina, M. Aroma active compounds during the drying of grapes cv. Pedro Ximenez destined to the production of sweet Sherry wine. *Eur. Food Res. Technol.* **2010**, *230*, 429–435. [CrossRef]
23. Ruiz, M.J.; Moyano, L.; Zea, L. Changes in aroma profile of musts from grapes cv. Pedro Ximenez chamber-dried at controlled conditions destined to the production of sweet Sherry wine. *LWT Food Sci. Technol.* **2014**, *59*, 560–565. [CrossRef]
24. López de Lerma, N.; García-Martínez, T.; Moreno, J.; Mauricio, J.C.; Peinado, R.A. Volatile composition of partially fermented wines elaborated from sun dried Pedro Ximénez grapes. *Food Chem.* **2012**, *135*, 2445–2452. [CrossRef] [PubMed]
25. Ruiz, M.J.; Moyano, L.; Zea, L. Sweet wines produced by an innovative winemaking procedure: Colour, active odorants and sensory profile. *South Afr. J. Enol. Vitic.* **2014**, *35*, 205–216. [CrossRef]
26. Ruiz-Bejarano, M.J.; Castro-Mejías, R.; del Carmen Rodríguez-Dodero, M.C.; García-Barroso, C. Volatile composition of Pedro Ximénez and Muscat sweet Sherry wines from sun and chamber dried grapes: A feasible alternative to the traditional sun-drying. *J. Food Sci. Technol.* **2016**, *53*, 2519. [CrossRef] [PubMed]
27. International Organization for Standardization (ISO). *Sensory Analysis—General Guidance for the Selection, Training and Monitoring of Assessors—Part 2: Expert Sensory Assessors*; ISO: Geneva, Switzerland, 2007.
28. International Organization for Standardization (ISO). *Sensory Analysis. General Guidance for the Design of Test Rooms*; ISO: Geneva, Switzerland, 2007.
29. International Organization for Standardization (ISO). *Sensory Analysis. Methodology. Flavour Profile Methods*; ISO: Geneva, Switzerland, 1985.
30. International Organization for Standardization (ISO). *Sensory Analysis. Methodology. Triangle Test*; ISO: Geneva, Switzerland, 2004.
31. International Organization for Standardization (ISO). *Sensory Analysis. Methodology. Ranking*; ISO: Geneva, Switzerland, 2006.
32. International Olive Council (IOC). *Standard Glass for Oil Tasting*; IOC: Madrid, Spain, 2004.
33. International Organization for Standardization (ISO). *Sensory Analysis. Guidelines for the Use of Quantitative Response Scales*; ISO: Geneva, Switzerland, 2003.
34. Serratosa, M.P.; Marquez, A.; Lopez-Toledano, A.; Merida, J. Sensory analysis of sweet musts in pedro ximenez cv. grapes dried using different methods. *South Afr. J. Enol. Vitic.* **2012**, *33*, 14–20. [CrossRef]
35. Ferreira, V.; Lopez, R. The actual and potential aroma of winemaking grapes. *Biomolecules* **2019**, *9*, 818. [CrossRef]
36. Schreier, P.; Jennings, W.G. Flavor composition of wines: A review. *Crit. Rev. Food Sci. Nutr.* **1979**, *12*, 59–111. [CrossRef]
37. Blank, M.; Hofmann, M.; Stoll, M. Seasonal differences in Vitis vinifera L. Cv. Pinot noir fruit and wine quality in relation to climate. *Oeno One* **2019**, *53*, 189–203. [CrossRef]
38. Niimi, J.; Tomic, O.; Næs, T.; Bastian, S.E.P.; Jeffery, D.W.; Nicholson, E.L.; Maffei, S.M.; Boss, P.K. Objective measures of grape quality: From Cabernet Sauvignon grape composition to wine sensory characteristics. *LWT* **2020**, *123*. [CrossRef]
39. Hernández-Orte, P.; Ibarz, M.J.; Cacho, J.; Ferreira, V. Effect of the addition of ammonium and amino acids to musts of Airen variety on aromatic composition and sensory properties of the obtained wine. *Food Chem.* **2005**, *89*, 163–174. [CrossRef]
40. Ugliano, M.; Bartowsky, E.J.; McCarthy, J.; Moio, L.; Henschke, P.A. Hydrolysis and Transformation of Grape Glycosidically Bound Volatile Compounds during Fermentation with Three Saccharomyces Yeast Strains. *J. Agric. Food Chem.* **2006**, *54*, 6322–6331. [CrossRef]
41. Aleixandre, J.L.; Padilla, A.I.; Navarro, L.L.; Suria, A.; García, M.J.; Álvarez, I. Optimization of Making Barrel-Fermented Dry Muscatel Wines. *J. Agric. Food Chem.* **2003**, *51*, 1889–1893. [CrossRef] [PubMed]
42. Selli, S.; Cabaroglu, T.; Canbas, A.; Erten, H.; Nurgel, C. Effect of skin contact on the aroma composition of the musts of Vitis vinifera L. cv. Muscat of Bornova and Narince grown in Turkey. *Food Chem.* **2003**, *81*, 341–347. [CrossRef]
43. Rapp, A. Volatile flavour of wine: Correlation between instrumental analysis and sensory perception. *Nahrung Food* **1998**, *42*, 351–363. [CrossRef]

44. Chaves, M.; Zea, L.; Moyano, L.; Medina, M. Changes in color and odorant compounds during oxidative aging of Pedro Ximenez sweet wines. *J. Agric. Food Chem.* **2007**, *55*, 3592–3598. [CrossRef] [PubMed]
45. Ortega-Heras, M.; González-Huerta, C.; Herrera, P.; González-Sanjosé, M.L. Changes in wine volatile compounds of varietal wines during ageing in wood barrels. *Anal. Chim. Acta* **2004**, *513*, 341–350. [CrossRef]

foods

MDPI

Article

# Chemical, Physical, and Sensory Effects of the Use of Bentonite at Different Stages of the Production of Traditional Sparkling Wines

Cristina Ubeda [1,2,*], María Ignacia Lambert-Royo [3], Mariona Gil i Cortiella [4], Rubén Del Barrio-Galán [3] and Álvaro Peña-Neira [3]

1 Departamento de Nutrición y Bromatología, Toxicología y Medicina Legal, Facultad de Farmacia, Universidad de Sevilla, C/Profesor García González 2, 41012 Sevilla, Spain
2 Instituto de Ciencias Biomédicas, Facultad de Ciencias, Universidad Autónoma de Chile, Santiago 8910060, Chile
3 Facultad de Ciencias Agronómicas, Universidad de Chile, Avenida Santa Rosa 11315, Santiago 8820808, Chile; milambertr@hotmail.com (M.I.L.-R.); rdelbarriogalan@gmail.com (R.D.B.-G.); apena@uchile.cl (Á.P.-N.)
4 Instituto de Ciencias Químicas Aplicadas, Facultad de Ingeniería, Universidad Autónoma de Chile, Santiago 8910060, Chile; mariona.gil@uautonoma.cl
* Correspondence: c_ubeda@us.es

**Abstract:** The addition of bentonite to wine to eliminate unstable haze-forming proteins and as a riddling adjuvant in the *remuage* is not selective, and other important molecules are lost in this process. The moment of the addition of bentonite is a key factor. Volatile profile (SPME-GC-MS), foam characteristics (Mosalux method), and sensory analyses were performed to study the effect of the distribution of the dosage of bentonite for stabilization of the wine among the addition on the base wine before the *tirage* (50%, 75%, and 100% bentonite dosage) and during the *tirage* (addition of the remaining dosage for each case). Results showed that the addition of 50% of the bentonite to the base wine (before the *tirage*) resulted in sparkling wines with the lowest quantity of volatile compounds, mainly esters and norisoprenoids. No significant differences were found among the sparkling wines after 9 months of aging in relation to foam properties measured by Mosalux, although higher foamability and crown's persistence were perceived in the sparkling wines with the addition of 75% and 100% of the bentonite dosage in sensory trials. The results of this study suggested that the amount of bentonite added as a fining agent in the *tirage* had greater effects than during the addition of this agent in the base wine.

**Keywords:** sparkling wines; bentonite; volatile compounds; foam properties; sensory analysis



**Citation:** Ubeda, C.; Lambert-Royo, M.I.; Gil i Cortiella, M.; Del Barrio-Galán, R.; Peña-Neira, Á. Chemical, Physical, and Sensory Effects of the Use of Bentonite at Different Stages of the Production of Traditional Sparkling Wines. *Foods* **2021**, *10*, 390. https://doi.org/10.3390/foods10020390

Academic Editor: Vito Michele Paradiso

Received: 31 December 2020
Accepted: 3 February 2021
Published: 10 February 2021



## 1. Introduction

The refusal of a wine by consumers can be driven by several reasons, some of which are subjective and others objective, such as defects in the product. Currently, haze formation in wines is an important concern for the industry because turbidity is one of the main causes of faulty perception by consumers and huge monetary losses due to the direct decrease of the quality of the wines and to the wine loss [1,2].

Different types of molecules such as polysaccharides, polyphenols, or proteins can contribute to the formation of haze owing to their instability [2]. Among them, proteins, and more specifically, proteins from grapes, seem to be the major contributors to the formation of haze [3]. The most employed strategy for the elimination of these unstable proteins is to treat wines with bentonite before bottling them.

Bentonite is a natural clay mineral with a high amount of montmorillonite, which is negatively charged at wine pH and interacts electrostatically with the positively charged wine proteins that are adsorbed on the bentonite surface, thus producing flocculation, and removing them from the wine [4].

During sparkling wine production, bentonite is generally used twice during the production process. First, it is added as a fining agent to achieve protein stabilization of the base wine, preventing the future formation of protein haze. Second, a small amount of bentonite is added as an adjuvant to the tirage solution to facilitate the flocculation of yeast strains during the process of riddling [5]. The lack of an available alternative for bentonite during sparkling wine production at the industrial scale, its high effectiveness, its simple application procedure, and its low cost explain its widespread use in wineries despite its negative effect on the foam quality and aromatic profile of wines [6–9]. Bentonite does not bind selectively to unstable proteins; thus, it also removes other positively charged species or aggregates. The loss of volatile compounds can occur directly, via the adsorption of these compounds onto bentonite [10], or indirectly, when the aromas are fixed by proteins or polysaccharides; moreover, some of the aromas are also discarded after the elimination of bentonite along with these macromolecules [6–8,11,12].

Furthermore, protein removal itself has its drawbacks, because the proteins and polysaccharides removed by bentonite affect the foamability of a sparkling wine [13,14]. Specifically, the addition of bentonite as a fining agent decreases foamability [15]; moreover, when added to facilitate the riddling process, bentonite significantly affects the foam quality, decreasing the maximum height and persistence of the wine foam [9,16].

Among the main types of proteins found in sparkling wine, i.e., those from grapes (chitinases and thaumatin-like proteins) and yeast (mainly mannoproteins), the ones from the fruit seem to play a major role in protein haze formation [17]. To overcome the disadvantages of the employment of bentonite, new additives are being sought to compensate for foam depreciation [18,19].

Despite all these negative effects that it has on the final product, a specific amount of bentonite is required to achieve protein stabilization of base wines, and it is still the most widely employed and effective agent in wine protein stabilization [20]. Hence, defining the appropriate application dosage of bentonite is extremely important for using enough to prevent haze. However, applying an excess of bentonite is not recommended, firstly to ensure the nitrogen quantity needed for fermentation (if it is added to the must or for the second fermentation) and secondly, as mentioned above, to prevent negative sensorial effects. It has been described that "matrix factors" modulating the removal of wine odor-active compounds during bentonite fining are the chemical nature of the clay, the hydrophobicity, the initial concentration of wine odor-active compounds, and the abundance and nature of wine proteins [7,10,12]. In this sense, ethyl esters seem to be the most affected volatiles, significantly decreasing their presence after bentonite treatment [7,10,12].

In addition to the quantity added, the moment of its addition is key to preserving the sensory characteristics of the wine. Some researchers have investigated the implications of bentonite treatments at different time points of the production process, especially before, during, and after fermentation [4,8,17,21]. Thus, Lambri et al. [8] concluded that a smaller dose of fining agent is needed when bentonite is added only to the must. However, other authors observed that the addition of bentonite during fermentation minimized both the dose amount required to allow wine protein stabilization and the negative sensory implications [4,22]. These results were in agreement with the ones of Lira et al. [17], who established that the best moment of addition of bentonite in Albariño wines was during alcoholic fermentation, particularly at the middle and at the end, giving rise to wines with higher aromatic intensity, being also preferred by the consumers in their sensory trials. Moreover, the application of bentonite at the middle and end of fermentation seems to provide better foaming properties to the wine [21].

However, despite the results found in these studies, the effects of the distribution of the needed dosage for stabilization during different phases of sparkling wine production have not been studied.

The aim of this study was to determine such effects on two essential quality parameters of sparkling wines: volatile composition and foam properties. A Chardonnay sparkling wine was tested, and the protein stabilization dosage of bentonite was distributed among the stages of fining of the base wine and before the second fermentation of the tirage liquor (as a riddling adjuvant) in different proportions. In addition, sensory analysis was conducted to corroborate and establish the effects/implications of a higher or lower bentonite dosage added at each stage.

## 2. Materials and Methods

### 2.1. Winemaking and Experimental Design

The study was carried out using a Chardonnay base wine variety of the 2017 vintage made in the San Pedro de Tarapacá winery, which is located in the Casablanca Valley region of Chile. This base wine had a 10.4% vol. and a pH of 3.42.

The stabilization dosage of bentonite for this base wine, which was determined using a fast heat test, was 17 g/hL [23]. For this study, activated sodium bentonite was used (SIHA® G, Eaton Industries, Dublin, Ireland). Figure 1 shows a detailed scheme of the experiment and production process of the sparkling wines of this study. Ninety liters of the base wine were distributed into three stainless steel tanks (STAGE 0) (Figure 1). Each tank was treated with 50%, 75%, and 100% of the stabilization dosage, i.e., 8.5, 12.75, and 17 g/hL, respectively. Bentonite was added as a 5% bentonite solution in water, and it was hydrated with cold water 24 h before the application. Bentonite acted for 72 h, after which the wine of every tank was racked off and transferred to clean stainless-steel tanks (B50, B75, and B100) (STAGE 1) (Figure 1). Following the traditional method (champenoise), the tirage was carried out. A preadapted yeast culture of Lalvin EC1118® *Saccharomyces cerevisiae bayanus* purchased from Lallemand (Chile) was used for a second fermentation in the bottle (750 mL green bottle Maipo type, Cristalchile, Chile). For the preadaptation of the yeast, 40 g of yeast were dissolved in 400 mL of water at 35 °C. After 30 min, this mixture was added to 1 L of water containing 200 g of sugar perfectly dissolved, and following this, 4.5 L of wine were incorporated slowly. This mixture was incubated overnight at 25–30 °C. Next day, a viable yeast cell counting, and density measuring were done. The addition of sugar, water, and base wine was repeated but gradually increasing the base wine volume until the *tirage* to force the yeast to adapt to the rough conditions of this matrix. At this point, a second addition of bentonite was carried out, adding to every bottle of base wine 24 g/L of sucrose, preadapted yeast, and the bentonite necessary for each treatment. Hence, the bottles of the base wines for the treatments, 50% and 75% (S50 and S75) were spiked with the dosage of hydrated 5% (m/v) bentonite needed to complete their stabilization; this was 8.5 and 4.25 g/hL, respectively. Moreover, 3 g/hL bentonite was added to all the bottles to facilitate riddling, avoid differences due to technological reasons, and to be able to assign the results and effects to the bentonite used as a clarifying agent. Fifteen days later, the second fermentation was complete, and samples were taken (S50, S75, and S100) (STAGE 2) (Figure 1). After 9 months of aging on lees at 16 °C, the remuage was carried out in one cycle with a Gyropalette® (Oenoconcept®, Epernay, Champagne, France), and sparkling wines were finished (A50, A75, and A100) (STAGE 3) (Figure 1). The resulting sparkling wines presented an alcoholic degree of 11.9 ± 0.1 and a pH of 3.31 ± 0.01. Stage 0 and Stage 1 wines were analyzed in triplicates (analytical replicates), and three bottles each of the wines from Stage 2 and Stage 3 were analyzed at each condition (biological replicates).

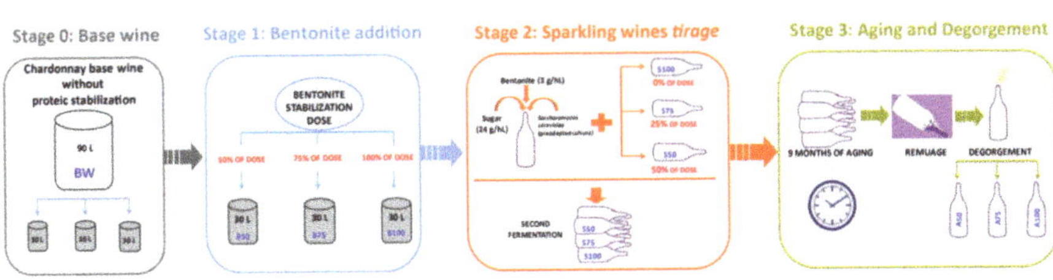


**Figure 1.** Schedule of the production process and the samples analyzed.

### 2.2. Reagents and Standards

The standard compounds employed in this study for the identification and quantification, i.e., ethyl butyrate, isoamyl acetate, ethyl hexanoate, hexyl acetate, ethyl lactate, ethyl octanoate, isoamyl hexanoate, ethyl nonanoate, methyl decanoate, ethyl decanoate, isoamyl octanoate, diethyl succinate, β-phenethyl acetate, isoamyl decanoate, isobutanol, isoamyl alcohol, hexanol, E-3-hexenol, phenylethyl alcohol, linalool, α-terpineol, citronellol, E-nerolidol, hexanoic acid, octanoic acid, and decanoic acid, were supplied by Sigma-Aldrich (Germany). Sodium chloride and 4-methyl-2-pentanol (internal standard) were purchased from Merck (Darmstadt, Germany).

### 2.3. Volatile Compound Analysis

Volatile compounds were extracted using headspace solid phase microextraction (HS-SPME), as described by Ubeda et al. [24]. For the extraction, a 2 cm 50/30 μm fiber made of carboxen/divinylbenzene/polydimethylsiloxane (Supelco, Bellefonte, PA, USA) was employed. For the identification of compounds, authentic reference standards were used, and matching was done with the 2.0 version of the standard NIST library and the linear retention indices (LRIs) from the literature (Pherobase; www.pherobase.com) and the NIST Mass Spectrometry Data Center; https://webbook.nist.gov/ (accessed on 20 November 2020)). LRIs were calculated using the retention times of *n*-alkanes (C6–C30) under identical conditions for each analysis temperature program. All data were expressed as concentrations (μg/L) obtained from calibration curves using the reference standards (relative area vs. concentration), except in the case of C13 norisoprenoids, for which the data were expressed as relative areas. The relative area was calculated by dividing the peak area of the major ion of each compound by the peak area of the major ion of the internal standard.

### 2.4. Determination of Foaming Properties

Foam properties were measured using the Mosalux procedure [25,26]. To carry out the measurement, the wines were degassed. Thereafter, a test tube with a porous piece of glass at the bottom and a $CO_2$ entry was filled with 100 mL of the sample, and a constant flow of $CO_2$ (10 L/h) was passed through the sample at a constant temperature of 16 °C. The parameters measured were HM, which is the maximum height reached by the foam and represents the foamability, and HS, which is the stable height of the foam that represents the ability of the wine to produce stable foam/persistence of the foam collar [25]. These analyses were performed in triplicates. The parameters, HM and HS, were expressed in millimeters.

### 2.5. Sensorial Analyses

Samples of sparkling wines after 9 months of aging were employed for sensory analysis: A50, A75, and A100 (addition of 50%, 75%, and 100% of the required dose of bentonite, respectively, to the base wines). They were evaluated by an expert panel of 17 tasters who are professional oenologists from the sparkling wine industry in Chile (six females and eleven males). The attributes selected were aromatic intensity, foamability, foam stability, and $CO_2$ integration. The last attribute provides information about foam aggressiveness in the mouth. For each evaluation, 50 mL of sparkling wine at 8 °C was served in each glass (Riedel®, Riedel Crystal America Inc. Kufstein, Austria). The selected attributes were indicated on a tasting card, and panelists were asked to rank each descriptor on a 15 cm unstructured scale (from unnoticeable to very strong).

### 2.6. Statistics

The InfoStat 2017p software (Free software. FCA-Universidad Nacional de Córdoba, Argentina. www.infostat.com.ar) was used for data analysis. The means were compared using ANOVA and a post hoc (Tukey test) ($\alpha$ = 0.05). Principal component analysis (PCA) was performed using IBM SPSS Statistics 26 software (IBM, Barcelona, Spain).

Sensory analysis data were processed using PanelCheck V1.4.2 (Free software, Norway. www.panelcheck.com).

## 3. Results and Discussion

### 3.1. *Effects on Volatile Compound Profile*

It has been demonstrated that the addition of bentonite to wine for fining purposes provokes an indirect removal of most of the fermentative aromatic compounds linked to the proteins removed, and only a few odor-active molecules are directly removed through adsorption [6,7,27]. Every chemical group studied (esters, alcohols, acids, terpenes, and norisoprenoids) among the 35 volatile compounds determined showed different tendencies; however, the most affected group by the bentonite treatment applied was the esters. The trend observed after the first addition of 75% (B75) and 100% (B100) of bentonite dosage seems to have caused the highest decrease in ester contents with respect to the base wines; however, the trend of these compounds after the addition of 50% of bentonite dosage (B50) was similar with respect to the base wine (Figure 2). The main compounds responsible for the strong decrease in ester contents after the addition of bentonite were ethyl butyrate, isoamyl acetate, and ethyl hexanoate (Table 1). These molecules are hydrophobic and easily adsorb on the clay of the fining agent [10]. This result agrees with that of Lambri et al. [7], who reported that the most affected esters after the application of bentonite to a white wine were ethyl butyrate, hexanoate, octanoate, isoamyl acetate, and phenylethyl acetate. In addition, hexyl acetatecontents decreased dramatically (between 73% and 82%) after the addition of bentonite in the three treatments. In contrast with these results, it was observed that stage 1 base wines presented an increase of ethyl octanoate and decanoate with respect to the stage 0 base wine (Table 1). This agrees with the results of Pozo-Bayón et al. [28], who also observed that the main changes affected ethyl octanoate and ethyl decanoate while testing the addition of bentonite in the tirage solution vs. non-addition. In contrast to the results found by Vincenzi et al. [12], we did not observe a correlation between the length of the hydrocarbon chain and the decrease in volatile compound contents.

The second fermentation and second addition of bentonite gave rise to sparkling wines that were not equally affected (stage 2). In the case of the addition of 50% of the dosage of bentonite in the tirage (S50), the resulting sparkling wines presented a significantly lower ester amount than B50 wines; however, S75 and S100 revealed an increase in the total amount of these compounds, which was probably due to the esters formed during the second fermentation. The increase in the ethyl butyrate and isoamyl acetate contents of S75 and S100 was statistically significant (Table 1). The loss of these two compounds (ethyl butyrate and isoamyl acetate) after the first addition and the increase after the second addition of bentonite could be explained by observing the macromolecular colloids present in the wine. In the first addition, i.e., a base wine without stabilization, proteins from grapes were present, and in the second addition, proteins from the yeast material involved in the second fermentation were present. Therefore, as suggested by Lambri et al. [7], these compounds may be easily attracted by proteins from the grape being mostly removed from the wine, and after the addition of bentonite in the tirage solution, they have more affinity for proteins from the yeast material, as they are not removed during the disgorgement.

Finally, as expected, the aging on lees produced a significant decrease in the total amount of esters (Stage 3), which was probably due to acid hydrolysis or even adsorption on the lees [29–31]. Although this was a massive loss of the ester content, the sparkling wines A75 and A100 preserved these volatile compounds more successfully than A50 (Figure 2). Isoamyl acetate, ethyl hexanoate, and ethyl octanoate were the compounds mainly responsible for this significant difference, because they were better preserved during the aging period in A75 and A100 (Table 1). This indicates that these compounds are easily bound to bentonite or the proteins that are removed with the fining agent, because their contents decreased dramatically after the first addition of bentonite and again when a high dosage of the fining agent was added to the tirage solution.

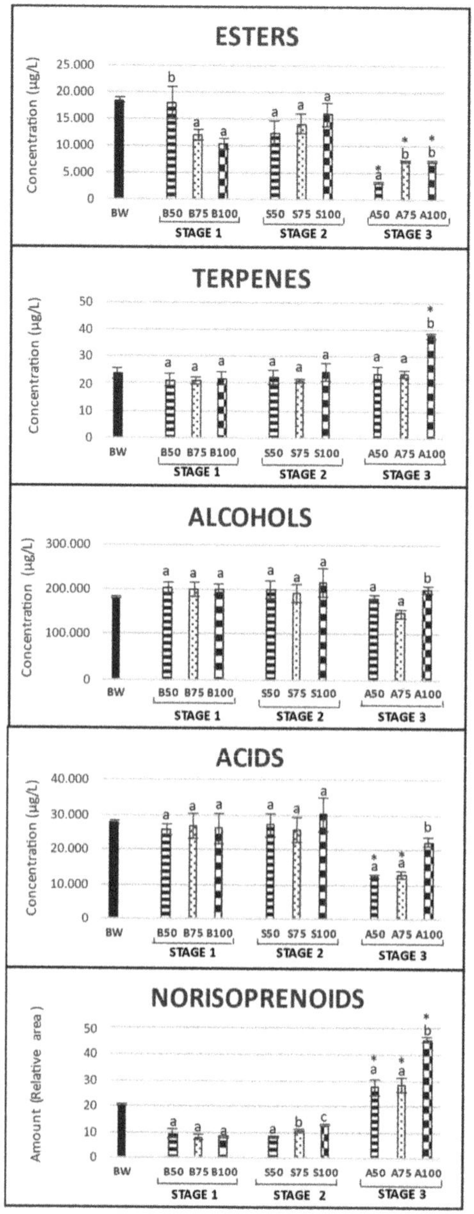


**Figure 2.** Total amounts of esters, alcohols, terpenes, acids, and norisoprenoids present in every Scheme 50. B75, B100 (wines before second fermentation with the first addition of bentonite); S50, S75, S100 (sparkling wines after second fermentation with the second addition of bentonite); A50, A75, A100 (sparkling wines after 9 months of aging on lees). Bars with different superscript letters indicate statistically significant differences ($p < 0.05$) by Tukey test among the samples belonging to the same stage *: statistically significant differences ($p < 0.05$) by Tukey test with the same treatment of the previous stage.

**Table 1.** Concentration of volatile compounds of Chardonnay base and sparkling wines along the production process.

| Volatile Compounds | LRI | ID | Stage 0 BW | B50 | Stage 1 B75 | B100 | S50 | Stage 2 S75 | S100 | A50 | Stage 3 A75 | A100 |
|---|---|---|---|---|---|---|---|---|---|---|---|---|
| ESTERS | | | | | | | | | | | | |
| Ethyl butyrate | 1055 | A | 798 ± 101 | 863 ± 245 $^{b}$ | 282 ± 56 $^{a}$ | 322 ± 88 $^{a}$ | 541 ± 139 $^{a}$ | 895 ± 221 $^{a,b*}$ | 1163 ± 80 $^{b,*}$ | 701 ± 59 $^{a}$ | 689 ± 62 $^{a}$ | 666 ± 88 $^{a,*}$ |
| Isoamyl acetate | 1122 | A | 10879 ± 19 | 8722 ± 1641 $^{b}$ | 4773 ± 1095 $^{a}$ | 3336 ± 829 $^{a}$ | 6019 ± 1166 $^{a}$ | 7311 ± 1187 $^{a,*}$ | 7873 ± 1093 $^{a,*}$ | 707 ± 12 $^{a,*}$ | 3677 ± 215 $^{b}$ | 3230 ± 144 $^{b}$ |
| Ethyl hexanoate | 1245 | A | 950 ± 70 | 988 ± 123 $^{b}$ | 597 ± 122 $^{a}$ | 653 ± 143 $^{a}$ | 836 ± 182 $^{a}$ | 804 ± 62 $^{a}$ | 963 ± 196 $^{a}$ | 6.47 ± 0.46 $^{a,*}$ | 758 ± 57 $^{b}$ | 844 ± 25 $^{b}$ |
| Hexyl acetate | 1285 | A | 608 ± 50 | 123 ± 21 $^{a,b}$ | 110 ± 19 $^{a}$ | 161 ± 35 $^{b}$ | 27.7 ± 3.8 $^{a,*}$ | 58.7 ± 5.5 $^{b,*}$ | 58.3 ± 14.2 $^{b,*}$ | 0.740 ± 0.062 $^{a}$ | 13.6 ± 3.0 $^{b,*}$ | 14.2 ± 1.0 $^{b,*}$ |
| Ethyl lactate | 1379 | A | 4.81 ± 1.40 | 4.66 ± 0.52 $^{a}$ | 4.91 ± 1.20 $^{a}$ | 5.13 ± 0.25 $^{a}$ | 5.11 ± 0.54 $^{a}$ | 5.93 ± 0.12 $^{a}$ | 5.41 ± 0.82 $^{a}$ | 103 ± 1 $^{b,*}$ | 6.11 ± 1.05 $^{a}$ | 10.1 ± 2.2 $^{a}$ |
| Ethyl octanoate | 1437 | A | 2440 ± 103 | 3166 ± 417 $^{a}$ | 2773 ± 146 $^{a}$ | 2596 ± 81 $^{a}$ | 2426 ± 419 $^{a}$ | 2449 ± 166 $^{a}$ | 2716 ± 397 $^{a}$ | 62 ± 23 $^{a,*}$ | 1516 ± 80 $^{b,*}$ | 1593 ± 37 $^{b,*}$ |
| Isoamyl hexanoate | 1468 | A | 3.73 ± 0.05 | 7.71 ± 0.58 $^{b}$ | 6.06 ± 1.11 $^{a,b}$ | 5.65 ± 0.92 $^{a}$ | 4.12 ± 0.56 $^{a,*}$ | 5.22 ± 0.89 $^{a}$ | 5.72 ± 1.73 $^{a}$ | 5.03 ± 1.81 $^{b}$ | 2.01 ± 0.09 $^{a,*}$ | 2.30 ± 0.10 $^{a,*}$ |
| Ethyl nonanoate | 1558 | A | 25.6 ± 0.7 | 43.9 ± 1.6 $^{b}$ | 35.4 ± 3.7 $^{a}$ | 32.1 ± 0.7 $^{a}$ | 35.3 ± 9.9 $^{a}$ | 36.7 ± 4.5 $^{a}$ | 38.7 ± 8.0 $^{a}$ | 49.9 ± 12.8 $^{b}$ | 4.51 ± 1.07 $^{a,*}$ | 6.54 ± 0.40 $^{a,*}$ |
| Methyl decanoate | 1600 | A | 13.8 ± 2.9 | 12.1 ± 1.5 $^{a}$ | 12.8 ± 0.6 $^{a}$ | 12.2 ± 2.4 $^{a}$ | 11.3 ± 2.2 $^{a}$ | 12.0 ± 0.3 $^{a}$ | 14.4 ± 3.8 $^{a}$ | 2.04 ± 0.07 $^{b,*}$ | 1.29 ± 0.29 $^{a,*}$ | 1.58 ± 0.21 $^{a,b*}$ |
| Ethyl decanoate | 1647 | A | 1548 ± 92 | 2773 ± 701 $^{a}$ | 2200 ± 252 $^{a}$ | 2141 ± 264 $^{a}$ | 1558 ± 393 $^{a}$ | 1608 ± 249 $^{a,*}$ | 1866 ± 485 $^{a}$ | 862 ± 74 $^{b}$ | 322 ± 54 $^{a,*}$ | 418 ± 18 $^{a,*}$ |
| Isoamy loctanoate | 1680 | A | 114 ± 31 | 205 ± 53 $^{b}$ | 146 ± 9 $^{a,b}$ | 117 ± 20 $^{a}$ | 123 ± 19 $^{a}$ | 93.3 ± 9.7 $^{a,*}$ | 137 ± 42 $^{a}$ | 6.15 ± 0.57 $^{a,*}$ | 13.5 ± 2.8 $^{b,*}$ | 16.2 ± 2.7 $^{b,*}$ |
| Diethy lsuccinate | 1675 | A | 11.5 ± 0.5 | 14.8 ± 2.6 $^{a}$ | 15.4 ± 3.2 $^{a}$ | 14.5 ± 3.7 $^{a}$ | 15.1 ± 1.9 $^{a}$ | 14.1 ± 1.9 $^{a}$ | 15.7 ± 2.7 $^{a}$ | nd $^{a,*}$ | 36.6 ± 3.2 $^{b,*}$ | 43.5 ± 2.0 $^{c*}$ |
| Ethyl 9-decenoate | 1698 | B | 11.4 ± 0.2 | 16.9 ± 3.5 $^{a}$ | 15.0 ± 1.2 $^{a}$ | 14.2 ± 0.6 $^{a}$ | 12.6 ± 1.8 $^{a}$ | 12.9 ± 0.8 $^{a}$ | 14.5 ± 1.6 $^{a}$ | 2.83 ± 0.37 $^{a,*}$ | 12.1 ± 0.9 $^{b}$ | 19.7 ± 1.4 $^{c*}$ |
| Propyl decanoate | 1725 | B | 4.88 ± 0.09 | 7.11 ± 1.22 $^{a}$ | 5.98 ± 1.03 $^{a}$ | 6.10 ± 0.67 $^{a}$ | 4.52 ± 0.77 $^{a}$ | 4.77 ± 0.28 $^{a}$ | 5.53 ± 0.96 $^{a}$ | 5.85 ± 0.73 $^{b}$ | 0.231 ± 0.031 $^{a,*}$ | 0.502 ± 0.031 $^{a,*}$ |
| Methyl dodecanoate | 1823 | B | 10.7 ± 2.4 | 4.55 ± 1.03 $^{a}$ | 6.57 ± 1.05 $^{a,b}$ | 8.51 ± 1.82 $^{b}$ | 7.39 ± 1.79 $^{a}$ | 7.63 ± 1.34 $^{a}$ | 8.90 ± 1.24 $^{a}$ | 2.06 ± 0.09 $^{c*}$ | 0.673 ± 0.091 $^{a,*}$ | 0.903 ± 0.032 $^{b,*}$ |
| β-phenethyl acetate | 1851 | A | 146 ± 7 | 108 ± 10 $^{a}$ | 120 ± 7 $^{a}$ | 120 ± 12 $^{a}$ | 78.4 ± 7.0 $^{a,*}$ | 89.3 ± 9.6 $^{a,*}$ | 95.2 ± 10.5 $^{a,*}$ | 1.91 ± 0.11 $^{a,*}$ | 21.7 ± 1.1 $^{b,*}$ | 38.9 ± 1.5 $^{c*}$ |
| Ethyl dodecanoate | 1864 | B | 769 ± 30 | 773 ± 72 $^{a}$ | 658 ± 100 $^{a}$ | 679 ± 65 $^{a}$ | 549 ± 108 $^{a}$ | 577 ± 86 $^{a}$ | 733 ± 111 $^{a}$ | 3.62 ± 1.20 $^{a,*}$ | 52.3 ± 6.8 $^{b,*}$ | 95.2 ± 6.4 $^{c*}$ |
| Isoamyl decanoate | 1888 | A | 10.2 ± 1.2 | 12.8 ± 2.4 $^{a}$ | 10.4 ± 2.1 $^{a}$ | 11.3 ± 1.6 $^{a}$ | 8.33 ± 2.71 $^{a}$ | 9.40 ± 1.33 $^{a}$ | 14.7 ± 2.7 $^{b}$ | 102 ± 10 $^{b,*}$ | 0.396 ± 0.015 $^{a,*}$ | 1.03 ± 0.18 $^{a,*}$ |
| Ethyl tetradecanoate | 2041 | B | 37.0 ± 8.7 | 37.6 ± 10.4 $^{a}$ | 34.2 ± 4.3 $^{a}$ | 25.1 ± 8.6 $^{a}$ | 29.7 ± 6.9 $^{a}$ | 31.6 ± 6.2 $^{a}$ | 32.8 ± 6.0 $^{a}$ | 18.3 ± 4.1 $^{b}$ | 7.26 ± 0.32 $^{a,*}$ | 19.8 ± 1.6 $^{b}$ |
| Ethyl hexadecanoate | 2235 | B | 29.6 ± 7.5 | 37.3 ± 10.8 $^{a}$ | 28.1 ± 7.4 $^{a}$ | 23.3 ± 4.6 $^{a}$ | 27.0 ± 7.5 $^{a}$ | 27.2 ± 7.3 $^{a}$ | 36.0 ± 2.0 $^{a,*}$ | 10.6 ± 3.6 $^{b}$ | 4.26 ± 0.60 $^{a,*}$ | 12.7 ± 0.9 $^{b,*}$ |
| ALCOHOLS | | | | | | | | | | | | |
| Isobutanol | 1074 | A | 23,739 ± 2359 | 34,191 ± 2111 $^{a}$ | 35,845 ± 14 $^{a}$ | 33,422 ± 5141 $^{a}$ | 35,867 ± 2617 $^{a}$ | 30,621 ± 5268 $^{a}$ | 36,070 ± 4798 $^{a}$ | 27,497 ± 639 $^{a,*}$ | 22,265 ± 3510 $^{a,*}$ | 27,906 ± 6129 $^{a}$ |
| Isoamyl alcohol | 1200 | A | 118,133 ± 865 | 125,616 ± 10,888 $^{a}$ | 121,876 ± 9627 $^{a}$ | 124,744 ± 4829 $^{a}$ | 120,297 ± 13,684 $^{a}$ | 119,670 ± 11,248 $^{a}$ | 131,238 ± 19,875 $^{a}$ | 122,046 ± 2687 $^{b}$ | 96,350 ± 5931 $^{a,*}$ | 127,663 ± 11,937 $^{b}$ |
| Hexanol | 1375 | A | 6115 ± 172 | 7088 ± 435 $^{a}$ | 7113 ± 1177 $^{a}$ | 7044 ± 1239 $^{a}$ | 7584 ± 1221 $^{a}$ | 7176 ± 952 $^{a}$ | 8037 ± 1607 $^{a}$ | 3612 ± 118 $^{a,*}$ | 4688 ± 1077 $^{a,b}$ | 6511 ± 232 $^{b}$ |
| E-3-Hexenol | 1366 | A | 9002 ± 41 | 9693 ± 150 $^{a}$ | 9558 ± 964 $^{a}$ | 9756 ± 1126 $^{a}$ | 10,312 ± 1212 $^{a}$ | 10,184 ± 485 $^{a}$ | 11,116 ± 1881 $^{a}$ | 7607 ± 2026 $^{a}$ | 7318 ± 1286 $^{a}$ | 9784 ± 522 $^{a}$ |
| Phenylethyl alcohol | 1940 | A | 24,449 ± 181 | 26,394 ± 612 $^{a}$ | 26,299 ± 3228 $^{a}$ | 25,880 ± 4477 $^{a}$ | 26,412 ± 2834 $^{a}$ | 24,905 ± 3354 $^{a}$ | 28,541 ± 4636 $^{a}$ | 19,282 ± 4706 $^{a,*}$ | 16,019 ± 996 $^{a,*}$ | 28,327 ± 2418 $^{b}$ |
| TERPENES | | | | | | | | | | | | |
| Linalool | 1555 | A | 5.86 ± 0.61 | 6.01 ± 0.58 $^{a}$ | 6.01 ± 0.79 $^{a}$ | 5.87 ± 1.32 $^{a}$ | 6.17 ± 0.51 $^{a}$ | 5.84 ± 0.57 $^{a}$ | 6.61 ± 0.94 $^{a}$ | 3.16 ± 0.08 $^{a,*}$ | 3.55 ± 0.30 $^{a,*}$ | 3.74 ± 0.17 $^{a,*}$ |
| α-terpineol | 1693 | A | 4.20 ± 0.20 | 3.11 ± 0.77 $^{a}$ | 3.17 ± 0.75 $^{a}$ | 3.49 ± 1.10 $^{a}$ | 3.80 ± 0.55 $^{a}$ | 3.74 ± 0.45 $^{a}$ | 4.19 ± 1.13 $^{a}$ | 3.80 ± 1.11 $^{a}$ | 4.25 ± 0.55 $^{a}$ | 6.49 ± 0.40 $^{b}$ |
| Citronellol | 1785 | A | 3.37 ± 0.48 | 3.19 ± 0.14 $^{a}$ | 3.23 ± 0.07 $^{a}$ | 3.41 ± 0.66 $^{a}$ | 3.01 ± 0.30 $^{a}$ | 2.91 ± 0.36 $^{a}$ | 3.42 ± 0.43 $^{a}$ | 8.82 ± 2.98 $^{a,*}$ | 7.47 ± 0.40 $^{a,*}$ | 13.0 ± 0.3 $^{b,*}$ |
| E-nerolidol | 2056 | A | 10.1 ± 0.6 | 8.70 ± 1.75 $^{a}$ | 8.54 ± 0.72 $^{a}$ | 9.13 ± 0.69 $^{a}$ | 9.12 ± 1.59 $^{a}$ | 8.44 ± 0.53 $^{a}$ | 9.82 ± 0.90 $^{a}$ | 7.90 ± 1.34 $^{a}$ | 8.53 ± 0.73 $^{a}$ | 14.4 ± 1.1 $^{b,*}$ |
| ACIDS | | | | | | | | | | | | |
| Hexanoic acid | 1880 | A | 10,775 ± 196 | 10,428 ± 337 $^{a}$ | 11,431 ± 1808 $^{a}$ | 11,225 ± 2013 $^{a}$ | 11,777 ± 1122 $^{a}$ | 11,158 ± 1593 $^{a}$ | 12,604 ± 1792 $^{a}$ | 6053 ± 350 $^{a,*}$ | 5132 ± 275 $^{a,*}$ | 10,959 ± 797 $^{b}$ |
| Octanoic acid | 2076 | A | 12,124 ± 220 | 11,470 ± 655 $^{a}$ | 11,689 ± 1355 $^{a}$ | 11,071 ± 1642 $^{a}$ | 11,839 ± 1702 $^{a}$ | 11,387 ± 1532 $^{a}$ | 13,122 ± 2352 $^{a}$ | 5481 ± 761 $^{a,*}$ | 6533 ± 996 $^{a,*}$ | 9236 ± 407 $^{b}$ |
| Decanoic acid | 2339 | A | 4889 ± 25 | 3620 ± 587 $^{a}$ | 3696 ± 376 $^{a}$ | 3708 ± 619 $^{a}$ | 3414 ± 369 $^{a}$ | 3375 ± 491 $^{a}$ | 4474 ± 741 $^{a}$ | 1040 ± 112 $^{a,*}$ | 1190 ± 41 $^{a,*}$ | 1972 ± 77 $^{b,*}$ |
| NORISOPRENOIDS | | | | | | | | | | | | |
| Vitispirane a | 1518 | B | 4.53 ± 0.40 | 2.13 ± 0.32 $^{a}$ | 1.86 ± 0.21 $^{a}$ | 1.81 ± 0.01 $^{a}$ | 2.21 ± 0.31 $^{a}$ | 2.59 ± 0.05 $^{b}$ | 3.14 ± 0.42 $^{b}$ | 9.93 ± 0.81 $^{a,*}$ | 8.73 ± 2.15 $^{a,*}$ | 8.56 ± 0.46 $^{a,*}$ |
| Vitispirane b | 1522 | B | 3.53 ± 0.82 | 1.50 ± 0.01 $^{a}$ | 1.65 ± 0.10 $^{a}$ | 1.44 ± 0.24 $^{a}$ | 1.50 ± 0.43 $^{a}$ | 1.93 ± 0.42 $^{a}$ | 2.27 ± 0.56 $^{a}$ | 6.62 ± 0.31 $^{a,*}$ | 6.88 ± 1.79 $^{a,*}$ | 5.77 ± 0.11 $^{a,*}$ |
| TDN | 1745 | B | 12.2 ± 0.1 | 5.69 ± 1.44 $^{a}$ | 4.53 ± 0.59 $^{a}$ | 4.82 ± 0.46 $^{a}$ | 4.69 ± 0.24 $^{a}$ | 6.04 ± 0.12 $^{a,b}$ | 7.43 ± 0.45 $^{b}$ | 10.9 ± 2.6 $^{a}$ | 12.4 ± 1.1 $^{a,*}$ | 16.2 ± 1.44 $^{b,*}$ |

BW (base wine); B50, B75, B100 (wines before second fermentation with the first addition of bentonite); S50, S75, S100 (sparkling wines after second fermentation with the second addition of bentonite); A50, A75, A100 (sparkling wines after 9 months of aging on lees). Values are expressed in μg/L except the norisoprenoids group which is expressed in relative area × 100. Values with different superscript letters indicate statistically significant differences among the samples of the same stage ($p < 0.05$) by Tukey test *: statistically significant difference with the same sample of the previous stage ($p < 0.05$) by Tukey test. nd: not detected; LRI: linear retention index; ID: reliability of identification: A: mass spectrum and LRI agreed with standards; B, mass spectrum agreed with mass spectral database and LRI agreed with the literature data (Pherobase: www.pherobase.com; NIST Mass Spectrometry Data Center: https://webbook.nist.gov/ (accessed on 20 November 2020)).

Unexpectedly, the amounts of ethyl lactate and isoamyl decanoate increased in wine A50 from the end of the second fermentation until the end of 9 months of bottle aging with lees.

Neither the first nor the second addition of bentonite produced significant differences in the total amount of alcohols in base wines at stage 1 (B50, B75, and B100) or sparkling wines at stage 2 (S50, S75, and S100) (Figure 2). Likewise, the second fermentation process did not give rise to significant differences among the total alcohols present in sparkling wines with respect to the base wines from which they were prepared. However, after 9 months of aging on lees (stage 3), the total amount of alcohols decreased significantly from stage 2 in A100 sparkling wines. The C6 alcohols, hexanol and E-3-hexenol, did not experience significant changes due to the first addition of bentonite, contrary to the results of Lambri et al. [7] but in accordance with those of Horvat et al. [22]. Nevertheless, the second addition plus the aging time decreased the contents of alcohols; however, this was not significant in almost all cases.

The terpenes group presented the same behavior as alcohols, without significant changes between the different treatments after the first and second addition of bentonite and between stages 1 and 2. The significant changes occurred after the aging period; the linalool content decreased and the citronellol content increased in all the sparkling wines studied, whereas the nerolidol content increased in A100. It is expected that during aging on lees, the effect of β-glucosidase enzyme activity releases the aglycone (odoriferous molecule) from the sugar in the volatile compounds present in their glycosidic form in the wine, thereby increasing their presence in the matrix during aging [32]. However, enzymes present in the wine matrix may disappear due to the addition of bentonite [21,33]. Hence, the treatments in which a higher dosage of bentonite was added to the tirage solution probably had less β-glucosidase enzyme available to act in the matrix because of its affinity to bentonite. Therefore, only A100, which had 3 g/hL of bentonite added as a riddling adjuvant, presented an increase in the contents of these compounds.

Similar to alcohols and terpenes, acids showed no significant differences between stages 1 and 2 and among the different treatments. Again, a decrease in the contents of all the acids determined in A50 and A75 was observed only after the aging period. In the case of A100, only the decanoic acid content significantly diminished (Table 1).

It is well known that some norisoprenoids such as 1,1,6-Trimethyl-1,2-dihydronaphthalene (TDN)and vitispiranes are aging markers [31,34]. These compounds tend to increase in content with aging time. Therefore, no significant differences were observed between stages 1 and 2; however, between stages 2 and 3 (end of second fermentation and after 9 months on lees), there was a significant increase in their amounts (Figure 2). The first addition of bentonite did not make any difference in the three treatments, but the addition in the tirage solution produced remarkable dissimilarities. The wines with the lower dosage added in this step, S75 and S100, presented slightly higher amounts of vitispirane A and TDN than S50. Although, after 9 months on lees, that difference reduced even more, and only the TDN content in A100 sparkling wines was significantly higher than that in A50 and A75.

The diversity of the effects experienced by the different volatile compounds after bentonite treatment may be explained by the fact that only a few odor-active compounds are directly adsorbed by bentonite, most of which are removed as an indirect effect of deproteinization [7]. Depending on the hydrophilic or hydrophobic characteristics of the volatile compounds, they are linked to the surface of the proteins through weak hydrogen bonds or to interior protein sites, respectively [7,12].

In general, the most effective treatment was the application of a 100% dosage of bentonite to the base wine before the second fermentation. These results agree with those obtained by several authors, indicating that bentonite fining could have a lower impact on the aroma quality when used before fermentation, i.e., when the fermentative aroma is yet to be produced [12].

### 3.2. Impact on Foaming Properties

Foaming properties of sparkling wines after 9 months on lees (A50, A75, and A100) were measured using the Mosalux methodology. Measurements taken using this methodology are quite heterogeneous; in this study, one measurement was taken for each bottle of the triplicates (biological replicates). Therefore, heterogenicity did not allow us to obtain significant differences among the different treatments (Figure 3). Instead, the results reflected a slightly non-significant higher maximum foam height in A100 than in A50 and A75 (Figure 3). All the sparkling wines analyzed received the same dosage of bentonite during the process but in two different stages of production. The addition of bentonite supposes the loss, among other molecules, of proteins from grapes and mannoproteins from yeast, which, as mentioned previously, are greatly responsible for sparkling wine foaming. Our results suggest that the molecules removed before the tirage are less responsible for the HM of the foam than the compounds released into the wine during aging. Previous studies have reported that glycoproteins, especially yeast mannoproteins, rather than grape proteins, more significantly affect the foaming properties of sparkling wines [35]. Reconstitution experiments performed by adding different molecular fractions isolated from wine to a model solution have pointed out the key role of mannoproteins in determining the capacity and stability of foam [14,36]. It has been reported that the glycosylated protein removal rate with sodium bentonite is low, as observed by Jaeckels et al. [37]. However, despite this, our results showed that the massive removal of thaumatin-like proteins (which play a major role in haze formation and the turbidity potential) from grapes in stage 1 after the addition of bentonite in the base wine seemed to affect the foam maximum height to a lesser extent than yeast proteins removed due to the addition of bentonite in stage 2. However, despite the absence of statistical significance, the HS results reflected that A50 showed slightly non-significant higher foam stability than A75 and A100 (Figure 3). This was not expected, since Kupfer et al. [38] described the key role of the yeast protein PAU5 in foam stability, showing that most of its removal occurred when bentonite fining of the wine was conducted before bottling. It might be that some compounds with foam stability properties from grapes are being removed; however, much research needs to be done to determine the effects of the stabilization of proteins from grapes.

### 3.3. Sensory Effects

A simple descriptive sensory analysis was performed with sparkling wines to assess whether their chemical and physical properties were perceived. Visual parameters were strongly influenced by the distribution of the bentonite dosages during the production process. Hence, A75 and A100 showed significantly higher foamability and persistence of the crown than A50 (Figure 4). The perceived foamability agreed with the non-significant Mosalux results, whereas the persistence did not. Sensory analysis reflected the expected results, which was probably due to the lack of significance of the Mosalux results, owing to the heterogeneity of the measurements. Aromatic intensity did not show significant differences among treatments. However, A75 was perceived to be more intense, followed by A100 and A50, which was probably because of the presence of a significantly lower quantity of isobutanol and isoamyl alcohol (Table 1, Figure 4). Higher quantities of these alcohols in red wine have been previously reported as blockers of the perception of fruity attributes [39,40]. It is possible that the lower concentrations of these alcohols in A75 allowed the perception of other nuances in the wine as more intense. Martínez-Rodríguez and Polo [41] observed that the addition of 3 g/100 L of sodium bentonite to the tirage solution increased the aroma intensity and quality of sparkling wines, unlike not adding bentonite to the tirage solution at all. Perhaps the addition of 25% of bentonite dosage to the base wine in the tirage step enhanced the intensity. However, as the dosage increased, less intensity was perceived for the convergence of the higher alcohol prevalence and the yeast protein-trapping effect.

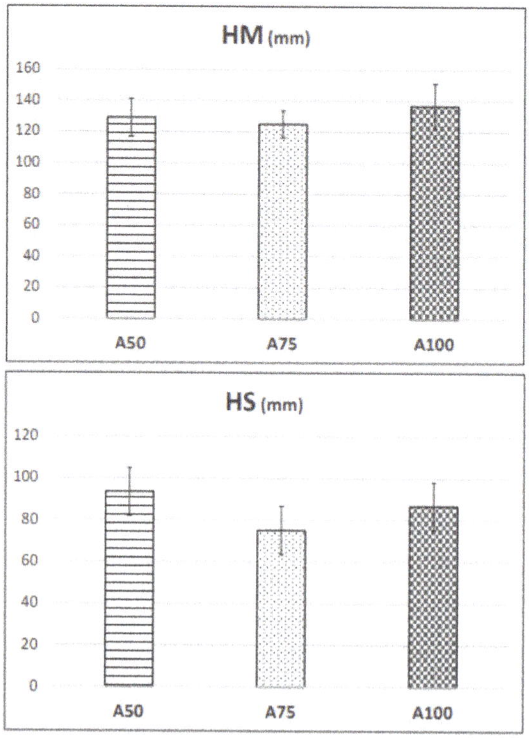


**Figure 3.** Foam properties of sparkling wines after 9 months of aging on lees from the STAGE 3 (A50, A75, A100) measured by the Mosalux method. HM: Foam maximum height; HS: Foam stability height.

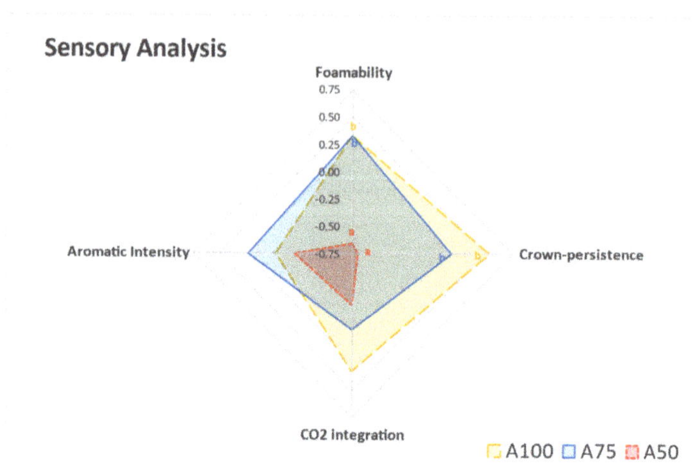


**Figure 4.** Sensory analysis of the sparkling wines after 9 months of aging on lees from STAGE 3 (A50, A75, A100). Different superscript letters in a sensorial attribute indicate statistically significant differences ($p < 0.05$) among the samples.

Finally, the samples did not present significant differences in in-mouth $CO_2$ integration; however, it seemed that as the quantity of bentonite added prior to aging time increased, the integration of $CO_2$ perceived by the panelists decreased. This parameter affects the dissolved carbon dioxide in the wine and directly influences the frequency of bubble formation in the glass, the growth rate of rising bubbles, the mouthfeel, and the aromatic perception [42].

### 3.4. Multivariate Analysis

Two different PCAs including all the volatile compounds, and the total sum of every group were performed as shown in Figure 5 (40 variables). One PCA (Figure 5a) comprised all the sparkling wine samples from stage 1 (B50, B75, and B100), stage 2 (S50, S75, and S100), and stage 3 (A50, A75, and A100). The analysis determined five principal components (PCs) which explained 90.9% of the total variance, with PC1 (Component 1) and PC2 (Component 2) accounting for 72.4% of the cumulative variance and permitting a significant separation of the samples. Thus, PC1 seemed to explain the effect of the 9 months of aging, discriminating among samples of stages 1 and 2 and those of stage 3. This indicated that the addition of bentonite to base wines did not allow differentiation among treatments and that the different dosages added did not cause major changes, even after the second fermentation. The samples of base wines from stage 1 and sparkling wines from stage 2 were mixed in the right side of the plane over the PC1 axis and the sparkling wines aged 9 months were located on the left side of PC1. Hence, typical aging markers such as diethyl succinate and ethyl lactate were placed on the left side joined to terpenes, which typically increase in concentration during aging due to acid hydrolysis of the glycosidic aroma precursors (Figure 5a). PC2 allowed the separation of the samples after 9 months of aging, depending on the bentonite treatment applied, showing that the fining agent added in the tirage caused greater effects among treatments than the addition of bentonite in the base wine. Figure 5b presents the PCA with only sparkling wines at stage 3, indicating that the first two PCs explained 87.6% of the variance. In this case, the sparkling wines A50 and A75 were grouped on the left side of the plane, and the sparkling wines A100 were grouped on the right side. The corresponding loading distribution clearly reflects the higher enrichment of volatile compounds of the sparkling wines with less quantity of bentonite added during the tirage (Figure 5b), which is in agreement with the results of the chemical and sensorial analyses (Table 1, Figure 2).

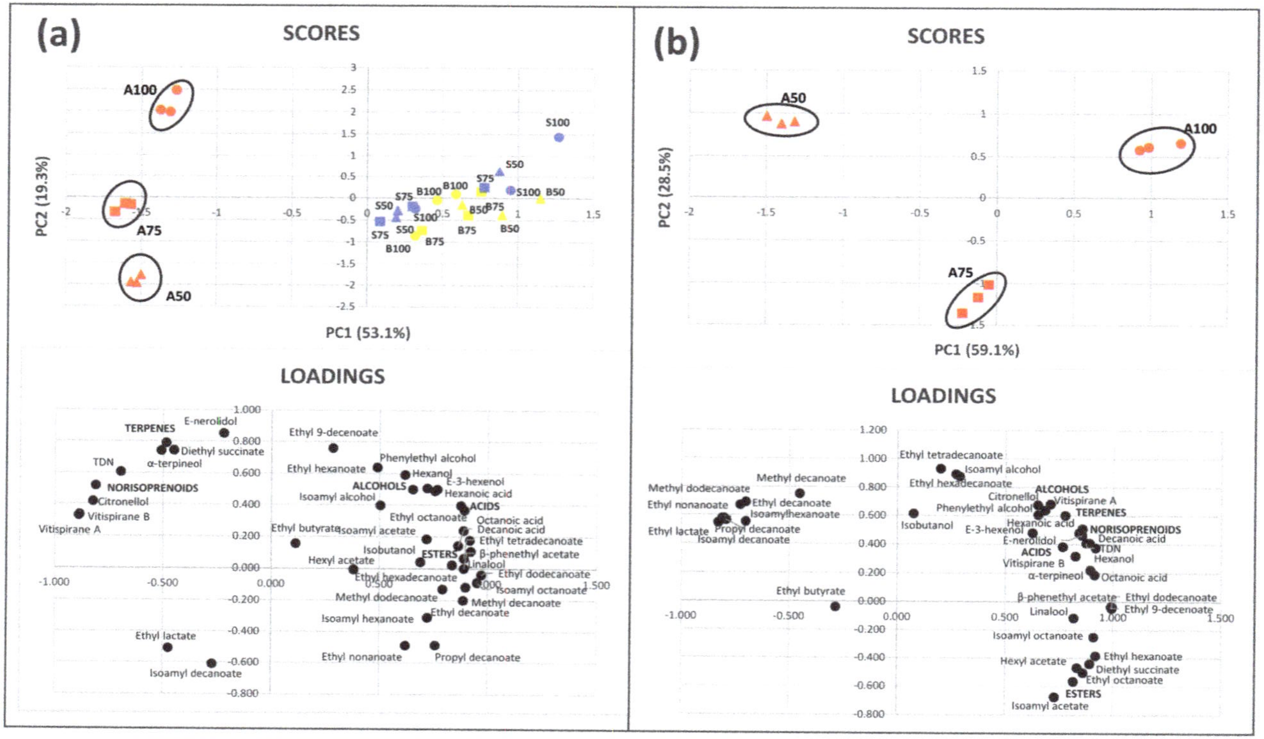


**Figure 5.** Data scores and loading biplot on the plane of the first two principal components (PC1 against PC2) of (**a**) Sparkling wines STAGE 1 (B50, B75, B100), STAGE 2 (S50, S75, S100), STAGE 3 (A50, A75, A100) (**b**) Sparkling wines from STAGE 3 (A50, A75, A100). B50, B75, and B100 (wines before second fermentation with the first addition of bentonite); S50, S75, and S100 (sparkling wines after second fermentation with the second addition of bentonite); A50, A75, and A100 (sparkling wines after 9 months of aging on lees).

## 4. Conclusions

The results of this study showed that the distribution of the dosage of bentonite needed for stabilization of the base wine before and the tirage significantly influences the volatile compounds profile and sensory perception of the sparkling wines. Our results suggest that the amount of bentonite added as a fining agent in the tirage causes greater effects during treatments than the addition of this agent in the base wine. The addition of 100% of the bentonite dosage to the base wine gives rise to wines with higher amounts of volatile compounds; however, the distribution of 75% of the bentonite before the tirage and 25% during it results in a diminution of higher alcohols contents, enhancing the perceived aromatic intensity. From a sensorial point of view, the addition of 50% of the bentonite dosage during the tirage has a negative effect on the foam and aromatic properties. These results reflect the state of the current procedures applied in most wineries; however, knowledge of the effects of the distribution of the dosage could help winemakers with highly unstable wines ensure protein stabilization (because the volatile profile seems to be mostly unaffected) or even enhance the aromatic intensity and complexity of sparkling wines.

**Author Contributions:** Conceptualization, C.U. and M.G.i.C.; methodology, C.U., M.G.i.C., R.D.B.-G. and M.I.L.-R.; data curation, C.U.; writing—original draft preparation, C.U.; writing—review and editing, C.U., M.G.i.C., M.I.L.-R., R.D.B.-G.; and Á.P.-N.; funding acquisition, C.U.; and Á.P.-N. All authors have read and agreed to the published version of the manuscript.

**Funding:** This research was funded by CONICYT-Chile [FONDECYT 1171227, FONDEQUIP EQM130129].

**Institutional Review Board Statement:** Not applicable.

**Informed Consent Statement:** Not applicable.

**Data Availability Statement:** Not applicable.

**Acknowledgments:** The authors are also grateful to San Pedro de Tarapacá winery for the base wine supply as well as to Santa Carolina winery for allowing us to use their facilities to develop the beginning of the study. Also, authors are grateful to Lallemand Chile for the essential help to organize the sensory trials in their facilities. In addition, Cristina Ubeda is thankful to the VI Plan Propio de Investigación y Transferencia of the University of Seville for the current contract (Contract number USE-18644-Z).

**Conflicts of Interest:** The authors declare no conflict of interest.

## References

1. Majewski, P.; Barbalet, A.; Waters, E.J. $1 billion hidden cost of bentonite fining. *Aust. N. Z. Grapegrow. Winemak.* **2011**, *569*, 58–62.
2. Van Sluyter, S.C.; McRae, J.M.; Falconer, R.J.; Smith, P.A.; Bacic, A.; Waters, E.J.; Marangon, M. Wine protein haze: Mechanisms of formation and advances in prevention. *J. Agric. Food Chem.* **2015**, *63*, 4020–4030. [CrossRef] [PubMed]
3. Waters, E.J.; Alexander, G.; Muhlack, R.; Pocock, K.F.; Colby, C.; O'neill, B.K.; Høj, P.B.; Jones, P. Preventing protein haze in bottled white wine. *Aust. J. Grape Wine Res.* **2005**, *11*, 215–225. [CrossRef]
4. Pocock, K.F.; Salazar, F.N.; Waters, E.J. The effect of bentonite fining at different stages of white winemaking on protein stability. *Aust. J. Grape Wine Res.* **2011**, *17*, 280–284. [CrossRef]
5. Poinsaut, P.; Hardy, G. Bentonites. III. Utillisation des bentonites en oenologie. *Revue des Oenologues* **1995**, *77*, 29–34.
6. Armada, L.; Falqué, E. Repercussion of the clarification treatment agents before the alcoholic fermentation on volatile composition of white wines. *Eur. Food Res. Technol.* **2007**, *225*, 553–558. [CrossRef]
7. Lambri, M.; Dordoni, R.; Silva, A.; De Faveri, D.M. Effect of bentonite fining on odor-active compounds in two different white wine styles. *Am. J. Enol. Vitic.* **2010**, *61*, 225–233.
8. Lambri, M.; Dordoni, R.; Giribaldi, M.; Violetta, M.R.; Giuffrida, M.G. Heat-unstable protein removal by different bentonite labels in white wines. *LWT Food Sci. Technol.* **2012**, *46*, 460–467. [CrossRef]
9. Kemp, B.; Condé, B.; Jégou, S.; Howell, K.; Vasserot, Y.; Marchal, R. Chemical compounds and mechanisms involved in the formation and stabilization of foam in sparkling wines. *Crit. Rev. Food Sci. Nutr.* **2019**, *59*, 2072–2094. [CrossRef]
10. Lambri, M.; Dordoni, R.; Silva, A.; De Faveri, D.M. Odor-active compound adsorption onto bentonite in a model white wine solution. *Chem. Eng. Trans.* **2013**, *32*, 1741–1746.
11. Lubbers, S.; Charpentier, C.; Feuillat, M.; Voilley, A. Influence of yeast walls on the behavior of aroma compounds in a model wine. *Am. J. Enol. Vitic.* **1994**, *45*, 29–33.

12. Vincenzi, S.; Panighel, A.; Gazzola, D.; Flamini, R.; Curioni, A. Study of combined effect of proteins and bentonite fining on the wine aroma loss. *J. Agric. Food Chem.* **2015**, *63*, 2314–2320. [CrossRef]
13. López-Barajas, M.; López-Tamames, E.; Buxaderas, S.; Suberbiola, G.; De la Torre-Boronat, M.C. Influence of wine polysaccharides of different molecular mass on wine foaming. *Am. J. Enol. Vitic.* **2001**, *52*, 146–150.
14. Vincenzi, S.; Crapisi, A.; Curioni, A. Foamability of Prosecco wine: Cooperative effects of high molecular weight glycocompounds and wine PR-proteins. *Food Hydrocoll.* **2014**, *34*, 202–207. [CrossRef]
15. Kemp, B.; Alexandre, H.; Robillard, B.; Marchal, R. Effect of production phase on bottle-fermented sparkling wine quality. *J. Agric. Food Chem.* **2015**, *63*, 19–38. [CrossRef]
16. Vanrell, G.; Canals, R.; Esteruelas, M.; Fort, F.; Canals, J.M.; Zamora, F. Influence of the use of bentonite as a riddling agent on foam quality and protein fraction of sparkling wines (Cava). *Food Chem.* **2007**, *104*, 148–155. [CrossRef]
17. Lira, E.; Rodríguez-Bencomo, J.J.; Salazar, F.N.; Orriols, I.; Fornos, D.; López, F. Impact of bentonite additions during vinification on protein stability and volatile compounds of Albariño wines. *J. Agric. Food Chem.* **2015**, *63*, 3004–3011. [CrossRef] [PubMed]
18. Colangelo, D.; Torchio, F.; De Faveri, D.M.; Lambri, M. The use of chitosan as alternative to bentonite for wine fining: Effects on heat-stability, proteins, organic acids, colour, and volatile compounds in an aromatic white wine. *Food Chem.* **2018**, *264*, 301–309. [CrossRef] [PubMed]
19. Apolinar-Valiente, R.; Salmon, T.; Williams, P.; Nigen, M.; Sanchez, C.; Marchal, R.; Doco, T. Improvement of the foamability of sparkling base wines by the addition of Acacia gums. *Food Chem.* **2020**, *313*, 126062. [CrossRef]
20. Chagas, R.; Monteiro, S.; Ferreira, R.B. Assessment of potential effects of common fining agents used for white wine protein stabilization. *Am. J. Enol. Vitic.* **2012**, *63*, 574–578. [CrossRef]
21. Lira, E.; Salazar, F.N.; Rodríguez-Bencomo, J.J.; Vincenzi, S.; Curioni, A.; López, F. Effect of using bentonite during fermentation on protein stabilisation and sensory properties of white wine. *Int. J. Food Sci. Technol.* **2014**, *49*, 1070–1078. [CrossRef]
22. Horvat, I.; Radeka, S.; Plavša, T.; Lukić, I. Bentonite fining during fermentation reduces the dosage required and exhibits significant side-effects on phenols, free and bound aromas, and sensory quality of white wine. *Food Chem.* **2019**, *285*, 305–315. [CrossRef]
23. Esteruelas, M.; Poinsaut, P.; Sieczkowski, N.; Manteau, S.; Fort, M.F.; Canals, J.M.; Zamora, F. Characterization of natural haze protein in sauvignon white wine. *Food Chem.* **2009**, *113*, 28–35. [CrossRef]
24. Ubeda, C.; Kania-Zelada, I.; del Barrio-Galán, R.; Medel-Marabolí, M.; Gil, M.; Peña-Neira, A. Study of the changes in volatile compounds, aroma and sensory attributes during the production process of sparkling wine by traditional method. *Int. Food Res. J.* **2019**, *119*, 554–563. [CrossRef]
25. Maujean, A.; Poinsaut, P.; Dantan, H.; Brissonnet, F.; Cossiez, E. Study of the performance and quality of the foam in sparkling wines. 2. Perfecting of a measuring technique for foaming ability, performance and stability of the foam in sparkling wines [measurement of foam properties, bentonite, vegetable black]. *Bull. de l'OIV* **1991**, *63*, 405–427.
26. Poinsaut, P. Le Mosalux, appareil de mesure du pouvoir moussant d'un vin. *Revue des Oenologues* **1991**, *59*, 35–43.
27. Di Gaspero, M.; Ruzza, P.; Hussain, R.; Vincenzi, S.; Biondi, B.; Gazzola, D.; Siligardi, G.; Curioni, A. Spectroscopy reveals that ethyl esters interact with proteins in wine. *Food Chem.* **2017**, *217*, 373–378. [CrossRef] [PubMed]
28. Pozo-Bayón, M.Á.; Hernández, M.T.; Martín-Álvarez, P.J.; Polo, M.C. Study of low molecular weight phenolic compounds during the aging of sparkling wines manufactured with red and white grape varieties. *J. Agric. Food Chem.* **2003**, *51*, 2089–2095. [CrossRef]
29. Makhotkina, O.; Kilmartin, P.A. Hydrolysis and formation of volatile esters in New Zealand Sauvignon blanc wine. *Food Chem.* **2012**, *135*, 486–493. [CrossRef] [PubMed]
30. Gallardo-Chacon, J.J.; Vichi, S.; Lopez-Tamames, E.; Buxaderas, S. Changes in the sorption of diverse volatiles by Saccharomyces cerevisiae lees during sparkling wine aging. *J. Agric. Food Chem.* **2010**, *58*, 12426–12430. [CrossRef] [PubMed]
31. Riu-Aumatell, M.; Bosch-Fusté, J.; López-Tamames, E.; Buxaderas, S. Development of volatile compounds of cava (Spanish sparkling wine) during long ageing time in contact with lees. *Food Chem.* **2006**, *95*, 237–242. [CrossRef]
32. Ganss, S.; Kirsch, F.; Winterhalter, P.; Fischer, U.; Schmarr, H.G. Aroma changes due to second fermentation and glycosylated precursors in Chardonnay and Riesling sparkling wines. *J. Agric. Food Chem.* **2011**, *59*, 2524–2533. [CrossRef]
33. Moio, L.; Ugliano, M.; Gambuti, A.; Genovese, A.; Piombino, P. Influence of clarification treatment on concentrations of selected free varietal aroma compounds and glycoconjugates in Falanghina (Vitis vinifera L.) must and wine. *Am. J. Enol. Vitic.* **2004**, *55*, 7–12.
34. Torrens, J.; Riu-Aumatell, M.; Vichi, S.; Lopez-Tamames, E.; Buxaderas, S. Assessment of volatile and sensory profiles between base and sparkling wines. *J. Agric. Food Chem.* **2010**, *58*, 2455–2461. [CrossRef]
35. Senée, J.; Robillard, B.; Vignes-Adler, M. Films and foams of Champagne wines. *Food Hydrocoll.* **1999**, *13*, 15–26. [CrossRef]
36. Coelho, E.; Rocha, S.M.; Coimbra, M.A. Foamability and foam stability of molecular reconstituted model sparkling wines. *J. Agric. Food Chem.* **2011**, *59*, 8770–8778. [CrossRef] [PubMed]
37. Jaeckels, N.; Tenzer, S.; Meier, M.; Will, F.; Dietrich, H.; Decker, H.; Fronk, P. Influence of bentonite fining on protein composition in wine. *LWT Food Sci. Technol.* **2017**, *75*, 335–343. [CrossRef]
38. Kupfer, V.M.; Vogt, E.I.; Siebert, A.K.; Meyer, M.L.; Vogel, R.F.; Niessen, L. Foam-stabilizing properties of the yeast protein PAU5 and evaluation of factors that can influence its concentration in must and wine. *Food Res. Int.* **2017**, *102*, 111–118. [CrossRef] [PubMed]

39. Cameleyre, M.; Lytra, G.; Tempere, S.; Barbe, J.C. Olfactory impact of higher alcohols on red wine fruity ester aroma expression in model solution. *J. Agric. Food Chem.* **2015**, *63*, 9777–9788. [CrossRef] [PubMed]
40. De-La-Fuente-Blanco, A.; Sáenz-Navajas, M.P.; Ferreira, V. On the effects of higher alcohols on red wine aroma. *Food Chem.* **2016**, *210*, 107–114. [CrossRef]
41. Martínez-Rodríguez, A.J.; Polo, M.C. Effect of the addition of bentonite to the tirage solution on the nitrogen composition and sensory quality of sparkling wines. *Food Chem.* **2003**, *81*, 383–388. [CrossRef]
42. Liger-Belair, G.; Polidori, G.; Zéninari, V. Unraveling the evolving nature of gaseous and dissolved carbon dioxide in Champagne wines: A state-of-the-art review, from the bottle to the tasting glass. *Anal. Chim. Acta* **2012**, *732*, 1–15. [CrossRef] [PubMed]

*foods*

MDPI

*Article*

# Chitosan in Sparkling Wines Produced by the Traditional Method: Influence of Its Presence during the Secondary Fermentation

**Antonio Castro Marín, Claudio Riponi and Fabio Chinnici ***

Department of Agricultural and Food Sciences, University of Bologna, Viale Fanin, 40, 40127 Bologna (BO), Italy; antonio.castromarin2@unibo.it (A.C.M.); claudio.riponi@unibo.it (C.R.)

* Correspondence: Fabio.chinnici@unibo.it; Tel.: +39-0512096012

Received: 31 July 2020; Accepted: 21 August 2020; Published: 25 August 2020

**Abstract:** Chitosan is a polysaccharide admitted in winemaking as clarifying, antimicrobial and chelating agent. In addition, evidence about its antioxidant and radical scavenging activities have been recently reported in wine conditions. As an insoluble adjuvant, chitosan efficacy also depends on the duration of its contact with the matrix. In the case of sparkling wines obtained following the traditional method, for instance, the addition of chitosan before the secondary fermentation would permit a prolonged contact of the polymer with wine and yeast lees. However, information on the effects of this practice on final products is totally unknown. In this work, the addition of chitosan during the secondary fermentation of a traditional sparkling wine production method has been investigated for its effects on both the physicochemical and sensory characteristics of the resulting wine. After 12 months of "sur lie" maturation, chitosan was found to increase the protein and amino acid content of wines up to about 50% and 9%, respectively, with limited change of phenolics and organic acids. Volatile compounds, particularly esters, were increased as well, which was reflected by higher values for fruity character and aroma intensity after sensory tests. Foaming features, evaluated by sensory and physical measurements, were also positively affected.

**Keywords:** chitosan; sparkling wine; foamability; sensory

## 1. Introduction

The traditional method for sparkling wines production is based on two consecutive alcoholic fermentations (AF). In a first step, the base wine is obtained by conventional white winemaking procedures. Next, selected yeasts and sugars (liqueur de tirage) are added to promote a second AF, carried out in sealed bottles, that results in the further formation of ethanol, dissolved $CO_2$ and volatile compounds, as main products [1,2]. Once secondary fermentation ends, bottled wines are subjected to a prolonged period in contact with dead yeast lees [3] during which autolysis of yeasts cells occurs. This leads to the release of various intracellular components such as nitrogen compounds, polysaccharides and some volatiles like terpenic and higher alcohols that impact the organoleptic properties of sparkling wines [2,4]. Secondary fermentation also affects the foaming properties of final product since peptides, amino acids and polysaccharides released during autolysis may have a positive effect on height and persistence of the foam itself, further contributing to the overall perceived wine quality [5].

Chitosan is the deacetylated derivative of chitin, the second most abundant biopolymer in the earth, extracted from shellfish, insects and fungus. Its structure, mainly constituted of glucosamine and *N*-acetylglucosamine units, confers it a great versatility with respect to several applications in food industries and interesting features including metal chelation, film forming properties or antimicrobial capacity [6–11]. Since its authorization in winemaking for metal and protein removal (maximum dose level 1 g/L) and microbial stabilization (maximum dose level 0.1 g/L) [12], the use of chitosan has aroused great interest in oenology. Colangelo and coworkers [13], for instance, reported a significant improvement of wines stability to heat test performed after fining treatments with chitosan. Other researches demonstrated that the presence of chitosan during fermentation can enhance the production of some volatile esters such as isoamyl acetate and phenethyl acetate together with medium-chain fatty acids and respective ethyl esters [14].

Chitosan can also act as an antioxidant in wine by means of various mechanisms, such as direct hydroxyl radical scavenging, prevention of the formation of 1-hydroxyethyl radical and metal chelation [15,16].

In principle, the traditional method for sparkling wine production could favour the action of chitosan as it permits both the presence of the polysaccharide during alcoholic fermentation and a prolonged permanence in the medium followed by a complete removal, as insoluble matter, during the degorgement step. However, information on the effects of the addition of chitosan during the prise de mousse are totally lacking up to now.

Therefore, the present study aimed to evaluate the effect of chitosan during the second fermentation and riddling stage of sparkling wines and to study the influence on sensory, foam and quality parameters of the finished product.

## 2. Materials and Methods

### 2.1. Chemicals

HPLC-grace acetonitrile, acetic acid, and methanol were obtained from Merck (Darmstadt, Germany). Milli-Q quality water was used for all HPLC experiments. Pure standards of volatile compounds, internal standard (2-octanol) and potassium metabisulphite were purchased from Sigma-Aldrich (Milano, Italy). Dichloromethane and methanol (SupraSolv) were supplied by Merck (Darmstadt, Germany), absolute ethanol (ACS grade) was obtained from Scharlau Chemie (Sentmenat, Spain) and pure water was obtained from a Milli-Q purification system (Millipore, Billerica, MA, USA). LiChrolut EN resin for solid-phase extraction (SPE) prepacked in 200 mg cartridges (3 mL total volume) were purchased from Merck (Darmstadt, Germany). Chitosan from *Aspergillus niger* (80–90% deacetylated, with average molecular weight of 10–30 kDa) was obtained from KitoZyme (Herstal, Belgium).

### 2.2. Samples Preparation

Base wines (75 mg/L total $SO_2$), obtained from cv Pinot gris and Pignoletto grapes, were filtered under nitrogen and 25 g/L of beet sucrose was added and arranged in two distinct trials, consisting of 50 bottles each, the first with no further additions (CTRL) and a second with addition of fungoid chitosan (250 mg/L) (KT). Before closing with bidules and crown caps, samples were inoculated with rehydrated active dried yeasts ($3 \times 10^6$ cells of *Saccharomyces cerevisiae* strain 1042 from University of Bologna—ESAVE collection) and ammonium phosphate (200 mg/L) was added. Six bottles (three each trial) were provided with manometer to monitor the internal pressure development. All the bottles were left at controlled temperature (18 °C) during the prise de mousse that lasted about 1 month during which the pressure increase was annotated daily, and the bottles were agitated to facilitate the chitosan resuspension. Samples were analysed as base wines, at the end of secondary fermentation (1 month) and after 12 months of sur lie maturation (degorgement).

### 2.3. Oenological Parameters

All the analyses were carried out according to OIV methods [17]. The pH was determined by using a pH meter (Mettler Toledo, Columbus, OH, USA). The alcoholic strength of wines was determined with an oenochemical distilling unit (Gibertini, Italy). Total phenolics (TPI) were spectrophotometrically determined (after wine filtration at 0.45 μm with cellulose filters) at 280 nm using an Uvidec 610 spectrophotometer (Jasco, Tokyo, Japan), and results were expressed as mg/L of gallic acid (GAE). All the analyses were carried out in triplicate. The browning development of the wines was followed measuring the absorbance at 420 nm (1 cm optical path) after filtration (0.45 μm, cellulose filters) at each sampling time.

### 2.4. Organic Acids

Quantification of organic acids, sugars and glycerol was conducted following the procedure described by Chinnici et al. [18]. The HPLC used was a Jasco apparatus (Tokyo, Japan) equipped with a binary pump (PU 1580), a 20 μL loop, a Rheodyne valve (Cotati, CA, USA), a photodiode detector (PU MD 910; Tokyo, Japan) and a column oven (Hengoed, Mid Glamorgan, UK). The column was a Bio-Rad Aminex HPX 87H (300 mm × 7.8 mm), thermostated at 35 °C. Isocratic elution was carried out with 0.005 N phosphoric acid at flow 0.4 mL/min. All the analyses were carried out in triplicate. Organic acids were quantified using external calibration curves obtained with standard compounds at known concentrations.

### 2.5. Phenolic Acids

Phenolic acid analysis was performed following a previous method after minor modifications [19]. A Jasco HPLC instrument (Tokyo, Japan), equipped with a quaternary gradient pump Jasco PU-2089, an autosampler Jasco AS-2057 Plus Intelligent Sampler and two detectors, a Jasco UV/Vis MD-910 PDA detector and a Jasco FP-2020 Plus Fluorescence detector, was used. The column was a C18 Poroshell 120 (Agilent technologies, Santa Clara, CA, USA), 2.7 μm, (4.6 mm × 150 mm), operating at 35 °C with a flow of 0.8 mL/min. Elution solvents were 2% acetic acid in HPLC grade water (Eluent A) and 2% acetic acid in HPLC grade acetonitrile (Eluent B). Gradient elution was as follow: from 98% to 95% A in 10 min, 95% to 90% A in 7 min, 90 to 82% A in 6 min, 82% to 80% A in 3 min, 80% to 70% A in 3 min, 70% to 50% A in 3 min, 50% to 0% A in 4 min and 98% A in 1 min. Quantification of phenolic compounds was carried out using an external calibration curve obtained by injecting solutions of standard compounds at known concentrations and plotting peak areas vs. concentrations. The amount of tartrate esters of caffeic, coumaric and ferulic acids and Grape reaction Product GRP were expressed as the respective hydroxycinnamic acid.

### 2.6. Total Protein Content

A protein assay kit TP0300 from Merck (Darmstadt, Germany) was used to quantify soluble proteins of sparkling wines. The procedure described is based on Peterson's modification of micro-Lowry method where known interferents (amino acid, peptide buffers and sucrose) were eliminated after protein precipitation with deoxycholate. Prior to analysis, wine samples were properly degasified and diluted 10 times with distilled water. Total protein concentrations are expressed in mg/L of BSA (bovine serum albumin).

### 2.7. Amino Acids and Amines

#### 2.7.1. Derivatization

A methodology proposed by Cejudo-Bastante et al. [20] was used. Briefly, 1.75 mL of borate buffer 1 M, 0.75 mL of methanol, 1 mL of sample and 20 μL of diethyl ethoxy methyl malonate (DEEMM) were left to react in a 10 mL screw-cap tube for 30 min in an ultrasound bath. Afterward, solution was warmed at 70 °C for 20 min in order to eliminate the excess of DEEMM. Once cooled, the samples were filtered with a 0.45 μm cellulose filter.

#### 2.7.2. HPLC Analysis

HPLC separation was performed on the instrument already cited in Section 2.5. A Waters (Milford, MA, USA) reversed-phase column Nova-Pak® C18 (3.9 mm × 300 mm; 4 μm), thermostated at 40 °C, was used. Mobile phases were A (25 mM acetate buffer pH = 5.65) and B (80:20 mixture of acetonitrile and methanol). Flow rate: 1.1 mL/min. HPLC gradient, for solvent A was: 0 min, 100%; 7 min, 96%; 18 min, 94%; 23 min, 92%; 25 min, 92%; 28 min, 85%; 50 min, 77%; 60 min, 55%; 65 min, 40%; 67 min, 20% and 70 min, 100%.

Detection was performed at 280 nm while quantification was based upon calibration curves obtained by plotting peak areas vs. concentration of solutions of standard amino acids and amines at known concentration.

### 2.8. Determination of Mannose

The content of mannoproteins (expressed as mg/L of mannose) was determined in wines after 12 months of ageing on yeast lees. A 10 mL of wine was first concentrated up to 2 mL under vacuum and then precipitated using cooled ethanol and HCl following the method of Segarra et al. [21]. After acid hydrolysis [22], samples were analysed with the HPLC apparatus cited in Section 2.4 equipped with a refraction index detector (Jasco 830-RI; Tokyo, Japan). The column was a Transgenomic CarboSep CHO-682 (300 mm × 7.8 mm) set at 80 °C. Elution was carried out using deionized water with a flow rate of 0.4 mL/min. Quantification of mannose was performed by means of a calibration curve of standard solutions of known concentrations.

### 2.9. Foamability

Analysis of foam quality was carried out by following a modified Mosalux method [23]. The instrument consisted of a glass column (400 mm × 24 mm), containing 50 mL of degasified wine to examine, with a porous septum (101–106 µm) at the base, which keeps the carbon dioxide separate from the wine, and a tap, necessary to block the flow of gas. A carbon dioxide cylinder was connected to the column, regulated at 1 bar and at a flow rate of 110 mL/min. Once the gas was opened, the evolution of the foam was recorded for 15 min. During this period, the height of the foam has been measured every 15 s. After 15 min, the cylinder and the column tap were closed, and the time required for the foam to disappear was measured. Three different parameters were measured: (i) HM, the maximum height reached by the foam after $CO_2$ injection, expressed in mm, (ii) HS, the foam height stability during 15 min of $CO_2$ injection, expressed in mm and (iii) TS, which is the foam stability time, expressed in seconds, once flow of $CO_2$ is interrupted.

### 2.10. Sensory Analysis

Sensory analysis was performed by 14 (8 men and 6 women aged from 27 to 64) well-experienced panelists recruited from the staff of the Department of Agricultural and Food Sciences, trained according to ISO 8586:2012. Wines sensory attributes were set based on testing cards already established by our research group for sparkling wines and further developed by asking the panellists to assess samples for appearance (foam in particular), aroma, flavour, mouthfeel and aftertaste. A total of 10 attributes were selected by consensus including 3 for the appearance, 3 for the aroma and 4 for the mouthfeel/tactile. A Quantitative Descriptive Analysis (QDA) test was performed on a continuous unstructured scale left anchored from absent to maximum. All sessions were performed in normalized room according to ISO 8589:2007. Wine samples were first individually served in the presence of each panellist to evaluate the foaming characteristics. In a second session, each of the wines were poured immediately before being served to perform the aroma and mouthfeel assessment. Coded and capped wines glasses and white trays were used (ISO 3591:1977). Data were elaborated by means of analysis of variance (ANOVA) and Friedman test to evaluate sample, panelists and replication variability of data.

### 2.11. Wine Volatile Compounds

A method already described and validated by Lopez et al. [24] was used for volatile extraction. A hundred microliter of a 2-octanol solution at 500 mg/L was added to 20 mL of degassed wine as internal standard and deposed on a previously activated LiChrolut EN cartridge. Analytes were eluted with 5 mL of dichloromethane and concentrated to 200 μL under a stream of nitrogen prior to GC-MS analysis. The Trace GC ultra-apparatus coupled with a Trace DSQ mass selective detector (Thermo Fisher Scientific, Milan, Italy) was equipped with a fused silica capillary column Stabilwax-DA (Restek, Bellefonte, PA, USA; 30 m, 0.25 mm i.d. and 0.25 μm film thickness). The carrier gas was He at a constant flow of 1.0 mL/min. The GC programmed temperature was 45 °C (held for 3 min) to 100 °C (held for 1 min) at 3 °C/min and then to 240 °C (held for 10 min) at 5 °C/min. Splitless mode injection (1 μL) was performed at 250 °C. Detection was carried out by electron ionization (EI) mass spectrometry in full scan mode, using ionization energy of 70 eV. Transfer line interface was set at 220 °C and ion source at 260 °C. Mass acquisition range was m/z 30–400. Compounds were identified by a triple criterion: (i) by comparing their mass spectra and retention time with those of authentic standards, (ii) compounds lacking of standards were identified after matching their respective mass spectra with those present in the commercial libraries NIST 08 and Wiley 7 and (iii) matching the linear retention index (LRI) obtained under our conditions, with already published LRI on comparable polar columns. Quantification of compounds was carried out via the respective total ion current peak areas after normalization with the area of the internal standard. Calibration curves were obtained by injections of standard solutions, subjected to the already cited extraction procedure, containing a mixture of commercial standard compounds at concentrations between 0.01 and 200 mg/L, and internal standard at the same concentration as in the samples. The calibration equations for each compound were obtained by plotting the peak area response ratio (target compound/internal standard) versus the corresponding concentration. For compounds lacking reference standards, the calibration curves of standards with similar chemical structure were used. Analyses were done in triplicate.

### 2.12. Statistical Analysis

Physicochemical data were given as mean ± SD. Evaluation of statistical significance was conducted by one-way analysis of variance (ANOVA) followed by a post hoc comparison Tukey test. Differences between groups were considered significant when $p < 0.05$. The univariate analysis (ANOVA) was performed using XLSTAT version 2016.02 (Addinsoft, Paris, France).

## 3. Results

### 3.1. Oenological Parameters

No significant differences between the treatments were recorded for the main oenological parameters (Table 1). Volatile acidity, pH and alcohol strength were adequate for this type of product. Yellow colour was subjected to little variations during the fining period, regardless the treatment adopted. After the end of the secondary fermentation, both samples showed a tendency to marginally increase the titratable acidity. This was followed by a subsequent reduction during the 12 months of ageing in the presence of yeast lees. This last evidence will be further discussed in the following section.

### 3.2. Organic Acids and Glycerol

HPLC quantification of organic acids after secondary fermentation showed similar values for both CTRL and KT samples (Table 2). As already reported by Pozo-Bayon et al. [25], glycerol content tended to slightly augment (by 0.2 g/L in our samples) after second fermentation because of yeast production, remaining unchanged for the following storage period. After 12 months of on lees ageing, concentration of tartaric acid was significantly decreased in both samples, which contributed to the reduction of titratable acidity reported in the previous section. The concentration of pyruvic acid, a secondary metabolite of alcoholic fermentation, increased in both samples after 12 months of ageing "sur lie" indicating its release from yeast cells autolysis.

**Table 1.** Oenological parameters and total protein concentration of the samples after secondary alcoholic fermentation (2nd AF) and after 12 months of ageing "sur lie." In the same row, different letters indicate significant differences according to Tukey's test ($p < 0.05$). $n = 3$. TPI = total phenolics; GAE = gallic acid equivalent.

| | Base Wine | 2nd AF | | 12 Months "Sur lie" | |
|---|---|---|---|---|---|
| | | CTRL | KT | CTRL | KT |
| Titratable acidity (g/L) | 5.75 ± 0.07 ab | 5.90 ± 0.14 a | 5.85 ± 0.07 a | 5.59 ± 0.06 ab | 5.51 ± 0.01 b |
| pH | 3.10 ± 0.01 a | 3.10 ± 0.01 a | 3.10 ± 0.02 a | 3.11 ± 0.01 a | 3.12 ± 0.01 a |
| Volatile acidity (g/L) | 0.29 ± 0.02 a | 0.28 ± 0.01 a | 0.30 ± 0.01 a | 0.32 ± 0.01 a | 0.28 ± 0.01 a |
| Alcohol (%v/v) | 10.32 ± 0.71 b | 11.42 ± 0.09 a | 11.36 ± 0.19 a | 11.40 ± 0.06 a | 11.30 ± 0.03 a |
| Optical Density 420 nm | 0.092 ± 0.001 a | 0.093 ± 0.01 a | 0.089 ± 0.012 a | 0.101 ± 0.001 a | 0.104 ± 0.001 a |
| TPI (GAE) | 172.3 ± 0.02 ab | 162.8 ± 0.07 b | 180.4 ± 0.28 a | 177.1 ± 0.03 ab | 176.9 ± 0.03 ab |
| Total proteins (mg/L) | 22.43 ± 0.45 d | 30.53 ± 1.28 b | 32.99 ± 1.37 b | 25.55 ± 0.85 c | 38.25 ± 1.01 a |

**Table 2.** Organic acids and glycerol amounts (g/L) after secondary alcoholic fermentation (2nd AF) and after 12 months of storage "sur lie" (shikimic and pyruvic acids as mg/L). In the same row, different letters indicate significant differences according to Tukey's test ($p < 0.05$). $n = 3$.

| | Base Wine | 2nd AF | | 12 Months "Sur lie" | |
|---|---|---|---|---|---|
| | | CTRL | KT | CTRL | KT |
| Tartaric acid | 3.57 ± 0.01 a | 3.58 ± 0.14 a | 3.66 ± 0.04 a | 2.53 ± 0.02 b | 2.61 ± 0.14 b |
| Pyruvic acid | 26.1 ± 0.23 b | 24.3 ± 2.19 b | 22.4 ± 1.46 b | 36.8 ± 0.42 a | 40.6 ± 1.54 a |
| Malic acid | 0.13 ± 0.01 a | 0.16 ± 0.04 a | 0.14 ± 0.01 a | 0.19 ± 0.01 a | 0.19 ± 0.01 a |
| Shikimic acid | 60.7 ± 0.35 a | 54.4 ± 2.15 a | 56.1 ± 0.75 a | 56.3 ± 1.22 a | 55.9 ± 1.22 a |
| Lactic acid | 2.37 ± 0.03 a | 2.30 ± 0.08 a | 2.35 ± 0.07 a | 2.35 ± 0.08 a | 2.36 ± 0.08 a |
| Acetic acid | 0.18 ± 0.01 ab | 0.16 ± 0.01 bc | 0.19 ± 0.02 a | 0.13 ± 0.01 d | 0.14 ± 0.01 cd |
| Succinic acid | 0.55 ± 0.01 a | 0.52 ± 0.04 a | 0.60 ± 0.03 a | 0.69 ± 0.06 a | 0.69 ± 0.06 a |
| Glycerol | 3.21 ± 0.05 a | 3.39 ± 0.12 a | 3.40 ± 0.11 a | 3.40 ± 0.12 a | 3.35 ± 0.10 a |
| Sum | 6.80 ± 0.03 a | 6.70 ± 0.22 a | 6.93 ± 0.08 a | 5.88 ± 0.16 b | 5.98 ± 0.16 b |

## 3.3. Protein Content

If compared with base wine, after secondary fermentation total proteins increased in both CTRL and KT samples to the same extent (Table 1). This was somehow expected since, as already reported, yeast metabolism and initial autolysis favour the release of proteins and peptides from cell cytoplasm to the wine since the very beginning of the ageing [5].

However, at 12 months, protein content further increased in KT sparkling wines while in CTRL samples, a decrease was observed. Untreated samples followed the common pattern already observed by Nunez and coworkers [5] where late reduction of protein content during "sur lies" ageing could be attributed to both the residual cells protease activity and the presence of alcohol [26,27].

Nevertheless, in KT samples, interactions between positively charged amine groups of the polymer and negatively charged components of cell wall may occur [14,28], which promotes an increased cell permeability, further speeding up the process of yeast autolysis and the release of proteins.

## 3.4. Phenolic Acids

Evolution of phenolic compounds after secondary alcoholic fermentation and 12 months of ageing on lees is presented in Table 3. A total of 18 compounds were identified in both the sparkling wines. Generally, treatments with chitosan did not affect the polyphenolic profile of wines compared to control samples, with the exception of (+)-catechin, which was present in significantly lower amounts ($p < 0.05$) after 12 months of storage in the presence of the biopolymer. This is due to the affinity of chitosan for flavanols present in wines, leading to its absorptive removal [16,29]. Overall, after 12 months of permanence on yeast lees, phenols slightly diminished or, in some cases, remained unchanged with respect to the base wine. As already evidenced elsewhere [25,30], at reducing conditions like those of sparkling wines, phenolic acids concentration tends not to be considerably modified, because of the scarcity of dissolved oxygen and the protective role of $CO_2$ against phenolic oxidation. Table 3 also evidences a temporary diminution of almost all the phenolic compounds just after the secondary fermentation. This has been often observed, due to absorption of phenolics onto yeast cells [25,31]. During the subsequent period of lees ageing, two concurrent phenomena are then expected to be occurred: (i) the partial release of those phenols into the wine, following the cell disorganization and (ii) the hydrolysis of hydroxycinnamates esters that promotes the increase of the corresponding phenolic acids [32].

**Table 3.** Concentrations (mg/L) of phenolic acids after secondary alcoholic fermentation (2nd AF) and after 12 months of storage "sur lie." In the same row, different letters indicate significant differences according to Tukey's test ($p < 0.05$). $n = 3$. GRP = Grape reaction Product.

| | Base Wine | 2nd AF | | 12 Months "Sur lie" | |
|---|---|---|---|---|---|
| | | CTRL | KT | CTRL | KT |
| *Hydroxybenzoic acids and flavanols* | | | | | |
| Gallic | 21.79 ± 0.26 ᵃ | 21.21 ± 0.50 ᵃ | 21.05 ± 1.04 ᵃ | 23.17 ± 0.16 ᵃ | 22.69 ± 0.07 ᵃ |
| Syringic | 0.74 ± 0.05 ᵃ | 0.85 ± 0.07 ᵃ | 1.04 ± 0.03 ᵃ | 1.18 ± 0.03 ᵃ | 0.94 ± 0.49 ᵃ |
| *p*-Hydroxybenzoic | 1.15 ± 0.01 ᵃ | 0.15 ± 0.02 ᶜ | 0.09 ± 0.09 ᶜ | 0.77 ± 0.22 ᵇ | 0.62 ± 0.04 ᵇ |
| (+)-Catechin | 3.58 ± 0.07 ᵃ | 3.53 ± 0.12 ᵃ | 3.54 ± 0.21 ᵃ | 3.16 ± 0.03 ᵃ | 2.60 ± 0.04 ᵇ |
| *Hydroxycinnamic acids* | | | | | |
| *t*-Caftaric acid | 5.39 ± 0.08 ᵃ | 4.14 ± 0.05 ᶜ | 4.12 ± 0.08 ᶜ | 4.64 ± 0.01 ᵇ | 4.49 ± 0.03 ᵇᶜ |
| GRP | 5.81 ± 0.09 ᵃ | 3.47 ± 0.10 ᵇ | 3.17 ± 0.05 ᶜ | 5.87 ± 0.03 ᵃ | 5.75 ± 0.06 ᵃ |
| *t*-Coutaric acid | 1.92 ± 0.06 ᵃ | 1.87 ± 0.02 ᵃ | 1.87 ± 0.01 ᵃ | 1.89 ± 0.08 ᵃ | 1.83 ± 0.01 ᵃ |
| *c*-Coutaric acid | 2.46 ± 0.01 ᵃ | 1.37 ± 0.01 ᵇᶜ | 1.36 ± 0.01 ᶜ | 1.69 ± 0.05 ᵇ | 1.63 ± 0.02 ᵇ |
| Fertaric acid | 4.13 ± 0.07 ᵃ | 3.11 ± 0.02 ᶜ | 3.07 ± 0.06 ᶜ | 3.60 ± 0.01 ᵇ | 3.65 ± 0.07 ᵇ |
| Caffeic acid | 1.72 ± 0.01 ᵃ | 0.81 ± 0.11 ᵇᶜ | 0.78 ± 0.16 ᶜ | 1.16 ± 0.07 ᵇ | 1.11 ± 0.03 ᵇ |
| *p*-Coumaric acid | 1.57 ± 0.05 ᵃ | 0.56 ± 0.10 ᶜ | 0.51 ± 0.04 ᶜ | 0.89 ± 0.01 ᵇ | 0.92 ± 0.04 ᵇ |
| Ferulic acid | 1.74 ± 0.02 ᵃ | 0.77 ± 0.08 ᶜ | 0.76 ± 0.05 ᶜ | 0.98 ± 0.05 ᵇ | 0.97 ± 0.04 ᵇ |
| *Flavonols* | | | | | |
| Quercetin | 0.11 ± 0.01 ᵃ | 0.12 ± 0.01 ᵃ | 0.09 ± 0.01 ᵃ | 0.11 ± 0.01 ᵃ | 0.10 ± 0.01 ᵃ |
| *Other* | | | | | |
| Tyrosol | 3.20 ± 0.2 ᵇ | 3.81 ± 0.04 ᵃ | 3.72 ± 0.13 ᵃ | 3.94 ± 0.05 ᵃ | 3.81 ± 0.05 ᵃ |

### 3.5. *Amino Acids and Amines*

The data relative to amino acids (Table 4) illustrate the typical decrease in their total amount following the second fermentation because of the assimilation by yeasts [26,33]. By comparing the concentrations in base and refermented wines it appears, in fact, that apart from asparagine and glutamine, all the amino acids where metabolized by yeasts to various extent. It should, however, be considered that at the end of fermentation, residual nitrogen composition of wines depends on a balance between initial depletion by yeasts and successive excretion or passive exsorption, these occurring latter during the last phases of fermentation [23,29]. In addition, it is worth noting that when compared to untreated wines (CTRL), KT seemed to elicit a generalized lower consumption (or higher excretion) of amino acids, particularly with respect to glycine, arginine and lysine, that drove to significantly higher final amounts of amino acids for chitosan treated wines, at the end of secondary fermentation. After ageing on lees, amino acids significantly increased (Table 4). This evidence is in accordance with that obtained in previous works [33,34] where the cellular pool of amino acids has been claimed to be released to the medium by exsorption after yeast cell degradation. During the permanence on lees, both the treated and untreated samples evolved in a very similar way, maintaining the differences already recorded after the second fermentation, being the KT samples richer in these compounds with respect to CTRL. For what concern amines, their total amount did not change noticeably during the distinct production phases (Table 4).

**Table 4.** Concentrations (mg/L ± STD) of amino acids, ammonium ion and amines after secondary alcoholic fermentation (2nd AF) and after 12 months of storage "sur lie." In the same row, different letters indicate significant differences according to Tukey's test ($p < 0.05$). $n = 3$.

| | Base Wine | 2nd AF | | 12 Months Storage | |
|---|---|---|---|---|---|
| | | CTRL | KT | CTRL | KT |
| Aspartic acid | 7.45 ± 0.02 [a] | 1.34 ± 0.03 [e] | 1.67 ± 0.04 [d] | 3.29 ± 0.05 [c] | 3.65 ± 0.03 [b] |
| Glutamic acid | 10.52 ± 0.88 [a] | 4.24 ± 0.20 [d] | 5.82 ± 0.00 [cd] | 6.40 ± 0.05 [bc] | 8.01 ± 0.09 [b] |
| Serine | 6.41 ± 0.56 [a] | 0.97 ± 0.03 [b] | 1.02 ± 0.03 [b] | 1.95 ± 0.01 [b] | 1.94 ± 0.04 [b] |
| Asparagine | 4.70 ± 0.23 [a] | 5.27 ± 0.00 [a] | 5.11 ± 0.17 [a] | 5.56 ± 0.52 [a] | 5.71 ± 0.14 [a] |
| Glutamine | 25.65 ± 1.72 [b] | 34.38 ± 0.44 [a] | 34.27 ± 0.75 [a] | 36.34 ± 0.79 [a] | 34.31 ± 1.76 [a] |
| Glycine | 55.04 ± 2.17 [a] | 43.37 ± 0.05 [c] | 49.08 ± 1.04 [b] | 51.89 ± 0.25 [ab] | 56.97 ± 1.77 [a] |
| Histidine | 17.57 ± 0.19 [a] | 7.55 ± 0.10 [c] | 7.95 ± 0.19 [c] | 10.10 ± 0.12 [b] | 10.58 ± 0.23 [b] |
| Threonine | 2.20 ± 0.76 [a] | 1.00 ± 0.06 [a] | 0.91 ± 0.05 [a] | 1.47 ± 0.06 [a] | 1.55 ± 0.08 [a] |
| Arginine | 17.94 ± 0.04 [b] | 14.43 ± 0.09 [d] | 16.47 ± 0.22 [c] | 16.45 ± 0.08 [c] | 19.23 ± 0.28 [a] |
| Alanine | 6.52 ± 0.09 [a] | 2.23 ± 0.07 [c] | 3.03 ± 0.11 [bc] | 3.23 ± 0.30 [b] | 3.42 ± 0.30 [b] |
| Tyrosine | 4.84 ± 0.03 [b] | 3.71 ± 0.01 [d] | 4.18 ± 0.07 [c] | 4.80 ± 0.00 [b] | 5.21 ± 0.01 [a] |
| Ammonium | 28.43 ± 0.43 [c] | 45.98 ± 0.51 [b] | 45.61 ± 0.67 [b] | 47.34 ± 0.59 [ab] | 48.56 ± 0.20 [a] |
| Ethanolamine | 15.36 ± 0.12 [c] | 15.43 ± 0.20 [c] | 15.80 ± 0.07 [bc] | 16.4 ± 0.23 [ab] | 16.45 ± 0.08 [a] |
| Valine | 10.43 ± 0.89 [a] | 3.32 ± 0.04 [c] | 4.62 ± 0.06 [bc] | 4.79 ± 0.03 [bc] | 6.01 ± 0.06 [b] |
| Methionine | 5.59 ± 0.07 [a] | 1.85 ± 0.15 [b] | 1.93 ± 0.20 [b] | 1.90 ± 0.18 [b] | 1.98 ± 0.26 [b] |
| Isoleucine | 9.61 ± 0.69 [a] | 0.94 ± 0.00 [c] | 1.30 ± 0.16 [c] | 2.14 ± 0.17 [bc] | 2.65 ± 0.02 [b] |
| Leucine | 18.64 ± 0.32 [a] | 3.94 ± 0.03 [e] | 4.92 ± 0.12 [d] | 5.87 ± 0.02 [c] | 6.83 ± 0.05 [b] |
| Phenylalanine | 7.10 ± 0.08 [a] | 2.57 ± 0.07 [d] | 3.45 ± 0.07 [c] | 3.34 ± 0.02 [c] | 4.38 ± 0.15 [b] |
| Ornithine | 2.46 ± 0.14 [d] | 4.05 ± 0.06 [bc] | 3.90 ± 0.06 [c] | 4.52 ± 0.04 [a] | 4.36 ± 0.13 [ab] |
| Lysine | 35.61 ± 0.27 [a] | 11.38 ± 0.60 [d] | 14.29 ± 1.14 [c] | 15.63 ± 0.05 [c] | 18.6 ± 0.48 [b] |
| Putrescine | 18.93 ± 0.42 [a] | 14.48 ± 0.84 [b] | 15.07 ± 0.44 [b] | 15.38 ± 0.44 [b] | 16.38 ± 0.58 [b] |
| SUM amino acids | 248.3 ± 3.46 [a] | 146.5 ± 0.32 [e] | 163.9 ± 3.04 [d] | 179.7 ± 1.94 [c] | 195.4 ± 4.31 [b] |
| SUM amines | 36.75 ± 0.15 [a] | 34.96 ± 1.43 [a] | 35.77 ± 1.58 [a] | 36.3 ± 0.39 [a] | 37.19 ± 0.10 [a] |

Individual changes were found for putrescine which diminished after secondary fermentation in all the samples, partially counterbalanced by little and progressive increase in ornithine amounts independently of the treatments.

### 3.6. Foamability Parameters

A notable portion of the perceived quality of sparkling wines is linked to foam features. For this reason, foamability was analysed on samples after 12 months of ageing. Foam profile and related parameters are reported in Figure 1. Results show higher values for foam height (HM) and stability time (TS) in KT samples when compared to CTRL. This could be correlated to the higher content of proteins in wines aged in the presence of the polysaccharide (Table 1) as already commented above. The pivotal role of proteins on foam quality has been studied by several researchers [5,35,36].

Those authors demonstrated that released proteins from yeast cell autolysis would improve foam development and stability in wines by reducing surface tension and increasing viscosity. Furthermore, in addition to proteins, amino acids have also been considered as foaming agents [37]. Their action is associated with the positive charge that these molecules carry in acidic wine conditions, resulting in the presence of both hydrophilic and hydrophobic groups. As with proteins, this favours the retention of amino acids in the air–liquid interphase, improving wine foamability [38]. Amines have been found to behave in a similar way [37].

Mannoproteins are another wine component consistently reported to positively affect foam height and stability [37]. In our samples, however, after 12 months of ageing on lees, we did not find significant differences in mannose content of wines (114 and 124 mg/L for KT and CTRL, respectively), suggesting that such polysaccharides could not be the reason for the better foam quality in chitosan-treated sparkling samples.

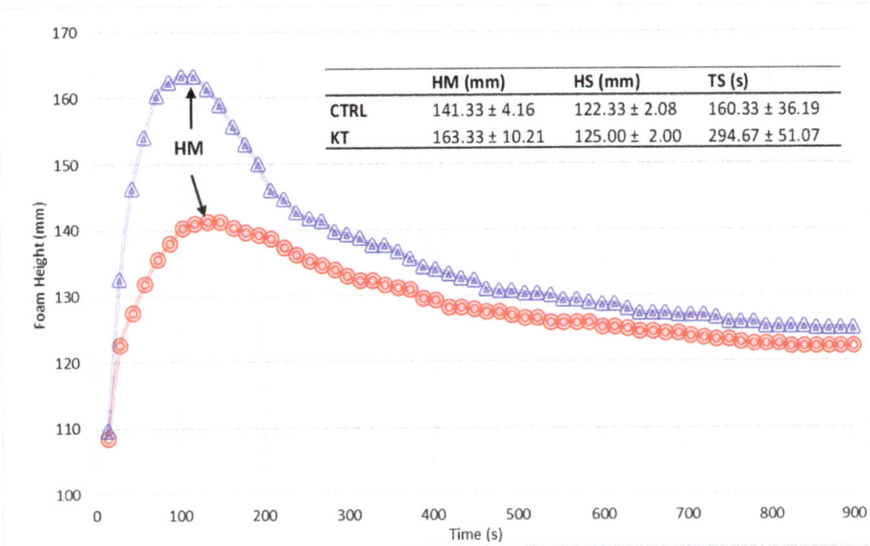


**Figure 1.** Evolution of the foam height during 15 min of measurement of wine sampled after 12 months of "sur lie" ageing. In the inset are outlined the recorded foam parameters HM = maximum foam height; HS = stability height; TS = stability time. Control (CTRL) (-O-) Chitosan (KT) (-Δ-).

### *3.7. Evolution of Volatile Compounds during Traditional Sparkling Winemaking Process*

The most significant volatile compounds identified in sparkling wines after the secondary fermentation and after 12 months of maturation on yeasts lees are reported in Table 5. Figure 2 also shows the sum of volatile compounds grouped by chemical families in order to be separately discussed.

**Table 5.** List of identified compounds, HMF = 5-hydroxymethylfurfural. [a] Identification assignment: Std = comparing mass spectra, linear retention index (LRI) and retention times with pure compounds, MS = by comparing mass spectra with NIST 08 and Wiley 7 spectral database, LRI = matching LRI on comparable polar columns (taken from the following publicly available databases: [39,40]).

| Compound | tR (min) | LRI | Identification [a] |
|---|---|---|---|
| Isobutyl alcohol | 5.70 | 1106 | Std, MS, LRI |
| Isoamyl acetate | 6.74 | 1133 | Std, MS, LRI |
| n-butanol | 7.19 | 1145 | Std, MS, LRI |
| 3-penten-2-ol | 7.80 | 1149 | Std, MS, LRI |
| 3-methyl-1-butanol | 8.92 | 1190 | Std, MS, LRI |
| Ethyl n-caproate | 9.86 | 1218 | Std, MS, LRI |
| Ethyl pyruvate | 11.33 | 1267 | Std, MS, LRI |
| 2-hexanol | 12.47 | 1304 | MS, LRI |
| 3-methyl-1-pentanol | 13.51 | 1331 | Std, MS, LRI |
| Ethyl lactate | 13.86 | 1340 | Std, MS, LRI |
| n-hexanol | 14.19 | 1349 | Std, MS, LRI |
| 2-hydroxy-3-pentanone | 14.63 | 1360 | Std, MS, LRI |
| 3-ethoxy-1-propanol | 15.10 | 1372 | Std, MS, LRI |
| 3-hexen-1-ol | 15.37 | 1379 | Std, MS, LRI |
| Ethyl octanoate | 17.40 | 1432 | Std, MS, LRI |
| Linalool oxide | 18.60 | 1463 | SMS, LRI |
| Furfural | 18.78 | 1467 | Std, MS, LRI |
| *c*-5-hydroxy-2-methyl-1,3-dioxane | 20.19 | 1503 | MS, LRI |
| Ethyl-3-hydroxybutyrate | 21.05 | 1524 | Std, MS, LRI |
| 2-methyl-3-thiolannone | 21.36 | 1531 | MS, LRI |

**Table 5.** *Cont.*

| Compound | tR (min) | LRI | Identification [a] |
|---|---|---|---|
| 2,3-butanediol | 23.08 | 1572 | Std, MS, LRI |
| Ethyl 3-hydroxypropionate | 23.93 | 1584 | MS, LRI |
| *t*-4-hydroxymethyl-2-methyl-1,3 dioxolane | 24.35 | 1606 | MS, LRI |
| 2-furancarboxylic acid, ethyl ester | 24.55 | 1616 | MS, LRI |
| n-butyric acid | 24.71 | 1624 | Std, MS, LRI |
| Decanoic acid, ethyl ester | 25.38 | 1659 | Std, MS, LRI |
| Pentanoic acid | 25.87 | 1689 | MS, LRI |
| Furfuryl alcohol | 26.02 | 1695 | Std, MS, LRI |
| Diethyl succinate | 26.44 | 1710 | Std, MS, LRI |
| 3-methylthio-1-propanol | 27.48 | 1746 | Std, MS, LRI |
| 1,3-propanediol diacetate | 28.03 | 1766 | MS, LRI |
| Ethyl 4-hydroxybutanoate | 29.79 | 1840 | Std, MS, LRI |
| 2-phenylethyl-acetate | 30.01 | 1851 | Std, MS, LRI |
| *t*-5-hydroxy-2-methyl-1,3-dioxane | 30.11 | 1856 | MS, LRI |
| Hexanoic acid | 30.39 | 1870 | Std, MS, LRI |
| Benzyl alcohol | 31.16 | 1905 | Std, MS, LRI |
| 2-phenylethanol | 31.92 | 1931 | Std, MS, LRI |
| Benzothiazole | 32.96 | 1966 | MS, LRI |
| 2,3-dihydroxypyrazine | 33.99 | 2001 | Std, MS, LRI |
| Diethyl Malate | 34.70 | 2038 | MS, LRI |
| Octanoic acid | 34.96 | 2052 | Std, MS, LRI |
| Diethyl-2-hydroxypentanedioate | 37.32 | 2197 | Std, MS, LRI |
| 4-vinyl-2-methoxyphenol | 37.78 | 2220 | Std, MS, LRI |
| Ethyl 5-oxotetrahydrofuran-2-furancarboxylate | 38.82 | 2270 | MS, LRI |
| decanoic acid | 39.31 | 2293 | Std, MS, LRI |
| Ethyl 2-hydroxy-3-phenylpropanoate | 39.39 | 2297 | MS, LRI |
| Glycerol | 40.20 | 2328 | Std, MS, LRI |
| Diethyl tartrate | 40.33 | 2182 | MS, LRI |
| Ethyl hydrogen succinate | 40.94 | 2356 | MS, LRI |
| 4-vinyl phenol | 41.26 | 2368 | Std, MS, LRI |
| Benzoic acid | 41.85 | 2390 | Std, MS, LRI |
| 3-furoic acid | 42.08 | 2399 | MS, LRI |
| Dodecanoic acid | 42.78 | 2444 | Std, MS, LRI |
| HMF | 43.12 | 2467 | Std, MS, LRI |
| Acetovanillone | 47.96 | 2662 | MS, LRI |
| n-hexadecanoic acid | 49.08 | 2803 | Std, MS, LRI |
| 4-hydroxy-benzenethanol | 51.20 | 2917 | Std, MS, LRI |
| Octadecanoic acid | 53.25 | 2998 | Std, MS, LRI |

### 3.7.1. Fatty Acids

Our results suggested that the presence of chitosan during the secondary fermentation generally enhanced the release of volatile fatty acids (Figure 2), likely impacting the aromatic profile of wines [41]. Similar results were reported in a previous work where chitosan was added in white musts during alcoholic fermentation [14]. Fatty acids are important constituents of cell membranes. Electrostatic interactions between chitosan amine groups and negatively charged cell surface components may induce an increase of permeability of yeast cell membranes, energetic unbalance and augmented excretion of fatty acids synthesised inside the cell [28,42,43]. Regarding the ageing period, a slight rising of some fatty acids was observed in both samples, with amounts of medium chain fatty acids such as 3-hydroxybutanoic, hexanoic, octanoic and decanoic increasing with time (Figure 2 and Table S1). This trend that could lead to an impact on the sensory attributes of final sparkling wines will be further discussed in a following section.

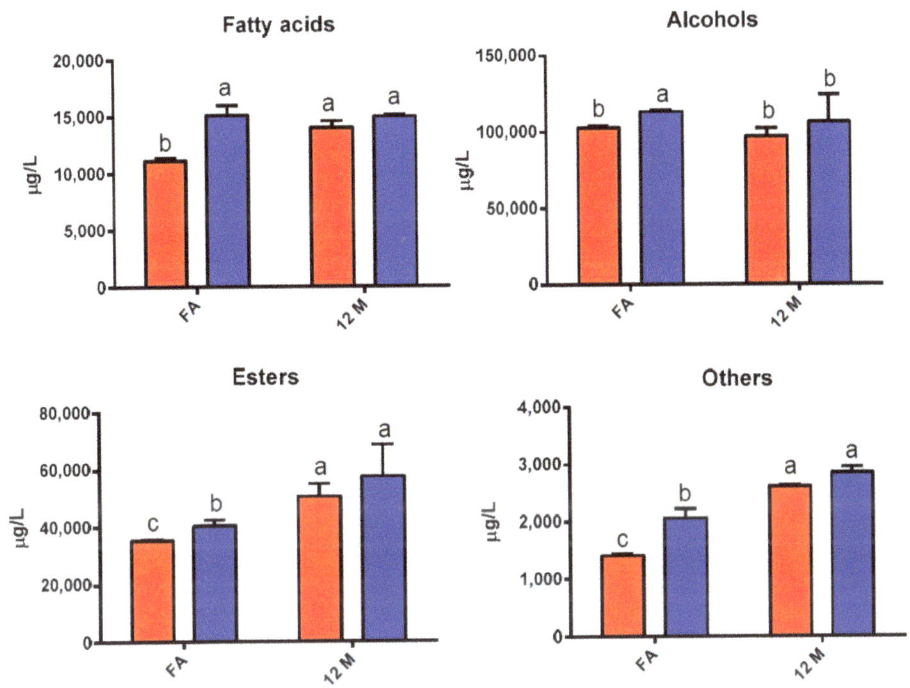


**Figure 2.** Concentrations (µg/L) of volatile compounds grouped by chemical family after secondary alcoholic fermentation (2nd AF) and after 12 months of storage "sur lie." Different letters indicate significant differences according to Tukey's test ($p < 0.05$). $n = 3$.

#### 3.7.2. Alcohols

Alcohols are related to an intense odour and play and important role in wine aroma. At concentrations lower than 300 mg/L, for instance, higher alcohols can impart wine complexity, but, at higher amounts, their intense odour could harm wine finesse [44]. None of our samples exceed the critical threshold (Figure 2), all reaching concentration levels around 100 mg/L as a sum.

Interestingly, after the second fermentation, the formation of volatile alcohols seemed to be slightly, though not significantly, higher in KT samples (10 mg/L higher as a sum). Some of these compounds are synthesised by yeast metabolism of sugars or amino acids by means of the Ehrlich pathway [45]. Isobutanol, in particular, was found at higher amounts after second fermentation in the presence of chitosan (Table S1). This alcohol comes from valine degradation by *Saccharomyces cerevisiae*, via the sequential formation of a α-ketoacid (ketoisovalerate), which is then reduced to isovaleraldehyde [46]. This latter can either be reduced to isobutyl alcohol or oxidized to isobutyric acid, which, also, was found at higher amounts in KT wines (Table S1). The reason of this metabolic expression in the presence of chitosan remains unclear.

Total amount of alcohols substantially did not change after 12 months of ageing on lees, but changes did occur for some compounds, independently from the sample considered.

A major variation in content was found for 2-phenylethanol, which at the end of ageing was reduced by about 10 mg/L with respect the initial amount (Table S1). This would impact the sensory features of the wines, considering the rose-reminiscent note of this alcohol.

#### 3.7.3. Esters

The presence and evolution of volatile esters in winemaking is of great interest since they play a fundamental role in the sensory properties of wines, imparting pleasant aromatic character such as candy, perfume-like and fruitness flavour [47]. Evolution of volatile esters in both samples during traditional sparkling winemaking is shown in Figure 2. Again, generation of these compounds was favoured by the presence of chitosan when compared to the control samples. However, this evidence was only significative at the end of secondary fermentation and seemed not to be further present after on-lees ageing.

Esters are generated from the reaction between alcohols and acids [45]. Therefore, an enhancement of the esterification reaction due to the greater availability of some volatile alcohols and fatty acids on KT wines after secondary fermentation (see Section 3.7.2) could be the origin of the increased content of esters in samples treated with chitosan. For example, isoamyl acetate, one of the most important acetate esters in wines, known for its distinctive banana aroma, was produced in higher concentrations in KT samples after secondary fermentation with *S. cerevisiae* (Table S1). As expected, some esters (acetates in particular) tended to decrease with time, with notable exceptions for the ethyl esters of some carboxylic acids (succinic, tartaric and lactic), which are usually regarded as markers of aged sparkling wines [4] and altogether contribute to the overall increase of this chemical class after 12 months of ageing (Figure 2).

#### 3.7.4. Other Compounds

The combined sum of some compounds, such as heterocyclic dioxane and dioxolane (generated from the acetylation between acetaldehyde and glycerol) or furans and pyrazines produced after the Maillard reaction between monosaccharides and amino acids, is also shown in Figure 2. This graph also comprises some carbonyl compounds (ketones and aldehydes) included in Table S1 under the common name of "others." As displayed in Figure 2, the presence of chitosan generally led to higher levels of these compounds, especially just after the secondary fermentation. Specifically, the major contributors to this higher level on KT samples after second fermentation are acetoin, 2-hydroxy-3-pentanone, ethyl-5-oxotetrahydro-2-furancarboxylate and 2,3-dihydroxypyrazine (Table S1). These compounds may contribute to pleasant, buttery and nutty nuances.

Further, after the ageing period, an overall increase of these compounds was observed, where samples treated with chitosan continue to show greater richness in these volatile compounds, mainly due to the presence of acetovanillone and 2,3-dihydroxypyrazine.

### *3.8. Sensory Profile of Sparkling Wines after Secondary Fermentation and after 12 Months of Ageing Sur Lie*

Sensory analysis was carried out after fermentation and after 12 months of ageing "sur lie" (Figure 3). As depicted on Figure 3A, no significant differences were appreciated at the end of secondary fermentation except for perlage persistence, which was higher in CTRL wines.

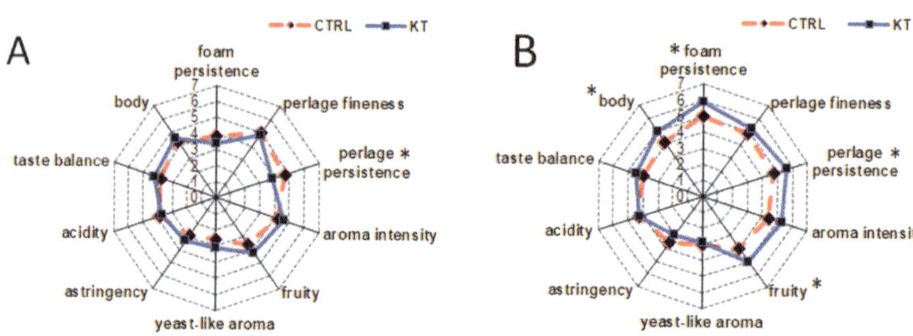


**Figure 3.** Sensory evaluation of sparkling wines after secondary fermentation (**A**) and 12 months of ageing "sur lie" (**B**). * Indicates significant differences according to Tukey's test ($p < 0.05$).

However, after 12 months of ageing in the presence of lees, the judges did find differences in the aromatic profile and foamability. Regarding the former, the richness in volatile compounds after ageing period (See Section 3.7) seemed to determine some impact to the wines, and KT samples were judged as the ones with higher aromatic intensity and richer fruity character. Despite the lack of significant differences between the distinct classes of volatiles of aged wines, in fact, the overall higher contents of aromatic compounds, especially some acids and esters (Table S1), has certainly contributed to this result. Sensory analysis also confirmed the data reported on Section 3.6 regarding foaming properties, as both perlage and foam persistence were significantly higher in wines added with chitosan because of the enhanced content of proteins and amino acids. Treated wines, in addition, were rated as more bodied and structured.

## 4. Conclusions

Based on our results, it was confirmed that the use of chitosan in traditional sparkling wines production may result in a higher content of fixed (mainly proteins and amino acids) and volatile compounds. This evidence could be associated to the ability of chitosan to interact with both the wall and the membrane of yeasts cells by electrostatic interactions at wine pH. This would eventually lead to the increase of permeability and the augmentation in the release of the cited compounds. Furthermore, this trend had an impact on the overall quality of wines, by increasing foamability and aromatic profile, making chitosan an interesting tool for the production of sparkling wines.

**Supplementary Materials:** The following are available online at http://www.mdpi.com/2304-8158/9/9/1174/s1, Table S1: Volatile compounds (mg/L) detected after second alcoholic fermentation (2nd AF) and after 12 months of storage "sur lie" in control (CTRL) and chitosan treated (KT) samples. In the same row, different letters indicate significant differences according to Tukey's test ($p < 0.05$). $n = 3$.

**Author Contributions:** Conceptualization, A.C.M. and F.C.; methodology, F.C.; formal analysis, A.C.M.; investigation, A.C.M.; resources, F.C.; data curation, F.C.; writing—original draft preparation, A.C.M. and F.C.; writing—review and editing, F.C.; validation, F.C.; visualization, F.C. and C.R.; supervision, F.C.; funding acquisition, C.R. All authors have read and agreed to the published version of the manuscript.

**Funding:** This research received no external funding.

**Acknowledgments:** The authors gratefully acknowledge Andrea Curioni and Simone Vincenzi (University of Padova) for their assistance in conducting the wine foamability measurements.

**Conflicts of Interest:** The authors declare no conflict of interest.

## References

1. Berbegal, C.; Polo, L.; García-esparza, M.J.; Lizama, V.; Ferrer, S.; Pardo, I. Immobilisation of yeasts on oak chips or cellulose powder for use in bottle-fermented sparkling wine. *J. Food Microbiol.* **2019**, *78*, 25–37. [CrossRef] [PubMed]
2. Martínez-García, R.; García-Martínez, T.; Puig-Pujol, A.; Mauricio, J.C.; Moreno, J. Changes in sparkling wine aroma during the second fermentation under CO2 pressure in sealed bottle. *Food Chem.* **2017**, *237*, 1030–1040. [CrossRef] [PubMed]
3. Kemp, B.; Alexandre, H.; Robillard, B.; Marchal, R. Effect of Production Phase on Bottle-Fermented Sparkling Wine Quality. *J. Agric. Food Chem.* **2015**, *63*, 19–38. [CrossRef] [PubMed]
4. Alexandre, H.; Guilloux-Benatier, M. Yeast autolysis in sparkling wine—A review. *Aust. J. Grape Wine Res.* **2006**, *12*, 119–127. [CrossRef]
5. Nunez, Y.P.; Carrascosa, A.V.; González, R.; Polo, M.C.; Martínez-Rodríguez, A.J. Effect of accelerated autolysis of yeast on the composition and foaming properties of sparkling wines elaborated by a champenoise method. *J. Agric. Food Chem.* **2005**, *53*, 7232–7237. [CrossRef] [PubMed]
6. Friedman, M.; Juneja, V.K. Review of antimicrobial and antioxidative activities of chitosans in food. *J. Food Prot.* **2010**, *73*, 1737–1761. [CrossRef]
7. Luo, Y.; Wang, Q. Recent Advances of Chitosan and Its Derivatives for Novel Applications in Food Science. *J. Food Process. Beverages* **2013**, *1*, 13.
8. No, H.K.; Meyers, S.P.; Prinyawiwatkul, W.; Xu, Z. Applications of chitosan for improvement of quality and shelf life of foods: A review. *J. Food Sci.* **2007**, *72*, R87–R100. [CrossRef]

9. Nunes, C.; Maricato, É.; Cunha, Â.; Rocha, M.A.M.; Santos, S.; Ferreira, P.; Silva, M.; Rodrigues, A.; Amado, O.; Coimbra, J.; et al. Chitosan–genipin film, a sustainable methodology for wine preservation. *Green Chem.* **2016**, *18*, 5331–5341. [CrossRef]
10. Rinaudo, M. Chitin and chitosan: Properties and applications. *Prog. Polym. Sci.* **2006**, *31*, 603–632. [CrossRef]
11. Rocha, M.A.M.; Coimbra, M.A.; Nunes, C. Applications of chitosan and their derivatives in beverages: A critical review. *Curr. Opin. Food Sci.* **2017**, *15*, 61–69. [CrossRef]
12. Commission Regulation (EU) 53/2011 of 21 January 2011. *Off. J. Eur. Union* **2011**, *L19*, 1–6.
13. Colangelo, D.; Torchio, F.; De Faveri, D.M.; Lambri, M. The use of chitosan as alternative to bentonite for wine fining: Effects on heat-stability, proteins, organic acids, colour, and volatile compounds in an aromatic white wine. *Food Chem.* **2018**, *264*, 301–309. [CrossRef] [PubMed]
14. Castro-Marin, A.; Gabriela Buglia, A.; Riponi, C.; Chinnici, F. Volatile and fixed composition of sulphite-free white wines obtained after fermentation in the presence of chitosan. *LWT—Food Sci. Technol.* **2018**, *93*, 174–180. [CrossRef]
15. Castro, A.; Culcasi, M.; Cassien, M.; Stocker, P.; Thétiot-laurent, S.; Robillard, B.; Chinnici, F.; Pietri, S. Chitosan as an antioxidant alternative to sulphites in oenology: EPR investigation of inhibitory mechanisms. *Food Chem.* **2019**, *285*, 67–76. [CrossRef] [PubMed]
16. Chinnici, F.; Natali, N.; Riponi, C. Efficacy of Chitosan in Inhibiting the Oxidation of (+)-Catechin in White Wine Model Solutions. *J. Agric. Food Chem.* **2014**, *62*, 9868–9875. [CrossRef] [PubMed]
17. International Organisation of Vine and Wine (OIV). *Compendium of International Methods of Wine and Must Analysis*; International Organisation of Vine and Wine (OIV): Paris, France, 2015.
18. Chinnici, F.; Spinabelli, U.; Riponi, C.; Amati, A. Optimization of the determination of organic acids and sugars in fruit juices by ion-exclusion liquid chromatography. *J. Food Compos. Anal.* **2005**, *18*, 121–130. [CrossRef]
19. Raposo, R.; Chinnici, F.; Ruiz-Moreno, M.J.; Puertas, B.; Cuevas, F.J.; Carbú, M.; Guerrero, R.F.; Ortíz-Somovilla, V.; Moreno-Rojas, J.M.; Cantos-Villar, E. Sulfur free red wines through the use of grapevine shoots: Impact on the wine quality. *Food Chem.* **2018**, *243*, 453–460. [CrossRef]
20. Cejudo-Bastante, M.J.; Sonni, F.; Chinnici, F.; Versari, A.; Perez-Coello, M.S.; Riponi, C. Fermentation of sulphite-free white musts with added lysozyme and oenological tannins: Nitrogen consumption and biogenic amines composition of final wines. *LWT—Food Sci. Technol.* **2010**, *43*, 1501–1507. [CrossRef]
21. Segarra, I.; Lao, C.; López-Tamames, E.; De La Torre-Boronat, M.C. Spectrophotometric Methods for the Analysis of Polysaccharide Levels in Winemaking Products. *Am. J. Enol. Vitic.* **1995**, *46*, 564–570.
22. Albersheim, P.; Nevins, D.J.; English, P.D.; Karr, A. A method for the analysis of sugars in plant cell-wall polysaccharides by gas-liquid chromatography. *Carbohydr. Res.* **1967**, *5*, 340–345. [CrossRef]
23. Maujean, A.; Poinsaut, P.; Dantan, H.; Brissonnet, F.; Cossiez, E. Étude de la tenue et de la qualité de mousse des vins effervescents II—Mise au point d'une technique de mesure de la moussabilité, de la tenue et de la stabilité de la mousse des vins effervescents. *Bull. OIV* **1990**, *63*, 405–427.
24. López, R.; Aznar, M.; Cacho, J.; Ferreira, V. Determination of minor and trace volatile compounds in wine by solid-phase extraction and gas chromatography with mass spectrometric detection. *J. Chromatogr. A* **2002**, *966*, 167–177. [CrossRef]
25. Pozo-Bayón, M.A.; Hernández, M.T.; Martín-Álvarez, P.J.; Polo, M.C. Study of Low Molecular Weight Phenolic Compounds during the Aging of Sparkling Wines Manufactured with Red and White Grape Varieties. *J. Agric. Food Chem.* **2003**, *51*, 2089–2095. [CrossRef]
26. Dizy, M.; Polo, M.C. Changes in concentration of nitrogenous compounds during fermentation of white grape musts at pilot plant scale/Cambios en la concentración de compuestos nitrogenados durante la fermentación de mosto a escala piloto. *Food Sci. Technol. Int.* **2007**, *2*, 87–93. [CrossRef]
27. Kelly-Treadwell, P.H. Protease activity in yeast: Its relationship to autolysis and Champagne character. *Aust. Grapegrow. Winemak.* **1988**, *292*, 58–66.
28. Lopez-Moya, F.; Lopez-Llorca, L. Omics for Investigating Chitosan as an Antifungal and Gene Modulator. *J. Fungi* **2016**, *2*, 11. [CrossRef]
29. Spagna, G.; Pifferi, P.G.; Rangoni, C.; Mattivi, F.; Nicolini, G.; Palmonari, R. The stabilization of white wines by adsorption of phenolic compounds on chitin and chitosan. *Food Res. Int.* **1996**, *29*, 241–248. [CrossRef]

30. Sartor, S.; Toaldo, I.M.; Panceri, C.P.; Caliari, V.; Luna, A.S.; de Gois, J.S.; Bordignon-Luiz, M.T. Changes in organic acids, polyphenolic and elemental composition of rosé sparkling wines treated with mannoproteins during over-lees aging. *Food Res. Int.* **2019**, *124*, 34–42. [CrossRef]
31. Del Barrio-Galán, R.; Úbeda, C.; Gil, M.; Sieczkowski, N.; Peña-Neira, Á. Different application dosages of a specific inactivated dry yeast (SIDY): Effect on the polysaccharides, phenolic and volatile contents and color of Sauvignon blanc wines. *OENO One* **2018**, *52*, 333–346.
32. Mazauric, J.P.; Salmon, J.M. Interactions between yeast lees and wine polyphenols during simulation of wine aging: II. Analysis of desorbed polyphenol compounds from yeast lees. *J. Agric. Food Chem.* **2006**, *54*, 3876–3881. [CrossRef] [PubMed]
33. Fornairon-Bonnefond, C.; Caramasa, C.; Moutounet, M.; Salmon, J.M. New trends on yeast autolysis and wine ageing on lees: A bibliographic review. *J. Int. Sci. Vigne Vin* **2002**, *36*, 49–69. [CrossRef]
34. Leroy, M.J.; Charpentier, M.; Duteurtre, B.; Feuillat, M.; Charpentier, C. Yeast Autolysis during Champagne Aging. *Am. J. Enol. Vitic.* **1990**, *41*, 21–28.
35. Martínez-Rodriguez, A.J.; Polo, M.C. Effect of the addition of bentonite to the tirage solution on the nitrogen composition and sensory quality of sparkling wines. *Food Chem.* **2003**, *81*, 383–388. [CrossRef]
36. Vanrell, G.; Canals, R.; Esteruelas, M.; Fort, F.; Canals, J.M.; Zamora, F. Influence of the use of bentonite as a riddling agent on foam quality and protein fraction of sparkling wines (Cava). *Food Chem.* **2007**, *104*, 148–155. [CrossRef]
37. Martínez-Lapuente, L.; Ayestarán, B.; Guadalupe, Z. Influence of Wine Chemical Compounds on the Foaming Properties of Sparkling Wines. In *Grapes and Wines*; Jordão, A.M., Cosme, F., Eds.; IntechOpen: Rijeka, Croatia, 2018.
38. Martínez-Lapuente, L.; Guadalupe, Z.; Ayestarán, B.; Pérez-Magariño, S. Role of major wine constituents in the foam properties of white and rosé sparkling wines. *Food Chem.* **2015**, *174*, 330–338. [CrossRef]
39. PubChem. Available online: https://pubchem.ncbi.nlm.nih.gov/srd (accessed on 22 January 2020).
40. Flavornet. Available online: http://www.flavornet.org/flavornet.html (accessed on 22 January 2020).
41. Ferreira, V.; Felipe, J.; Juan, F. Flavor of Wine. In *Food Flavors*; Henryk, J., Ed.; CRC Press: Boca Raton, FL, USA, 2011; pp. 269–300.
42. Hadwiger, L.A.; Kendra, D.F.; Fristensky, B.W.; Wagoner, W. Chitosan Both Activates Genes in Plants and Inhibits RNA Synthesis in Fungi. In *Chitin in Nature and Technology*; Springer: Boston, MA, USA, 1986; pp. 209–214.
43. Wakil, S.J.; Stoops, J.K.; Joshi, V.C. Fatty Acid Synthesis and its Regulation. *Ann. Rev. Biochem.* **1983**, *52*, 537–579. [CrossRef]
44. Rapp, A.; Versini, G. Influence of nitrogen compounds in grapes on aroma compounds of wines. In *Developments in Food Science*; Elsevier: Amsterdam, The Netherlands, 1995; Volume 37, pp. 1659–1694.
45. Ribereau-Gayon, P.; Glories, Y.; Maujean, A.; Dubordieu, D. *Handbook of Enology, Volume 2: The Chemistry of Wine Stabilization and Treatments*, 2nd ed.; John Wiley Sons, Ltd.: West Sussex, UK, 2001; Volume 2.
46. Hazelwood, L.A.; Daran, J.M.; Van Maris, A.J.A.; Pronk, J.T.; Dickinson, J.R. The Ehrlich pathway for fusel alcohol production: A century of research on Saccharomyces cerevisiae metabolism. *Appl. Environ. Microbiol.* **2008**, *74*, 2259–2266. [CrossRef]
47. Saerens, S.M.G.; Delvaux, F.R.; Verstrepen, K.J.; Thevelein, J.M. Production and biological function of volatile esters in Saccharomyces cerevisiae. *Microb. Biotechnol.* **2010**, *3*, 165–177. [CrossRef]

foods

MDPI

Article

# Bee Pollen Role in Red Winemaking: Volatile Compounds and Sensory Characteristics of Tintilla de Rota Warm Climate Red Wines

**Antonio Amores-Arrocha *, Pau Sancho-Galán, Ana Jiménez-Cantizano and Víctor Palacios**

Department of Chemical Engineering and Food Technology, Vegetal Production Area, Faculty of Sciences, University of Cadiz, Agrifood Campus of International Excellence (ceiA3), IVAGRO, P.O. Box 40, 11510 Puerto Real, Cadiz, Spain; pau.sancho@uca.es (P.S.-G.); ana.jimenezcantizano@uca.es (A.J.-C.); victor.palacios@uca.es (V.P.)

* Correspondence: antonio.amores@uca.es

Received: 3 July 2020; Accepted: 22 July 2020; Published: 23 July 2020

**Abstract:** One of the main aspects that define wine quality is its aromatic profile. Nutritional deficiencies in musts can lead to olfactory defects and a decline in quality. Commercial activators and nutrients are usually added to the must in these cases. The natural composition of bee pollen can provide all the necessary nutrients for yeasts. This investigation aims to analyze the impact of pollen addition on the profile of volatile and sensory compounds in Tintilla de Rota warm climate red wines. Volatile compounds were measured by Gas Chromatography–Mass Spectrometry, Odorant Activity Values analysis to find out each compound's fragrant participation, and sensorial analysis was conducted for a qualified panel of wine-tasters. As a result of the chromatographic analysis, 80 volatile compounds of different chemical families were identified and quantified. Bee pollen increased mainly isoamyl alcohol, esters, and terpenes compounds families in wines. Odorant Activity Values analysis showed an increase in fruity odorant series mainly, followed by floral, for all wines with pollen addition. The sensory analysis showed that low pollen doses (0.1 g/L and 0.25 g/L) increased tasting notes of fruit and floral attributes and fruit and floral odorant series as well, highlighting an increase in red and black fruit notes mainly. On the other hand, high doses deviated the sensory profile towards fleshy stone fruit, and raisin fruit, mostly. In addition, high bee pollen doses produce an increase in the odorant category responsible for the chemical, fatty, and grassy aromas mainly, and high and intermediate dose (1 g/L) an increase in the earthy notes in the aromas. Therefore, low bee pollen doses (0.1 and 0.25 g/L) can improve both the aromatic compound profile, as well as the Odorant Activity Values levels and the sensory profile in Tintilla de Rota red wines.

**Keywords:** bee pollen; Tintilla de Rota; alcoholic fermentation; warm climate; volatile compounds; sensory profile; fermentative activator; red winemaking; red wines

## 1. Introduction

The present tendency in wine consumption focuses on well-structured wines and full bodied in the mouth [1]. In addition, it is remarkable that the aroma of wine constitutes an important factor in consumer preference [2]. The compounds involved in the aroma can be derived from many sources: alcoholic fermentation, from biosynthesis, and from the conversion from neutral grape compounds to active components [3]. The majority of esters, as well as higher alcohols, volatile acids, and compounds within the thiol and terpene families, which are varietal compounds, are produced during alcoholic fermentation by yeast [4].

A complex and varied nutrient composition rich in amino acids, as well as fatty acids and vitamins in grape musts [5,6] ensure an adequate alcoholic fermentation development. In order to obtain them, weather conditions should be suitable during the grapes ripening stage [7]. Unfortunately, the current context of global warming is giving rise to problematic ripening processes directly responsible of changes in grape must composition [8,9], generating nutrient deficiencies for the yeasts. In this regard, potential difficulties may appear during alcoholic fermentation and, as a consequence, sensory profile defects in wines [10]. In warm climate areas, such as Southern Andalusia (Spain), wines may experience a loss in aromatic and sensory expression [11]. In order to confront these new environmental conditions in traditional winegrowing regions, several authors suggest cultivating autochthonous varieties better adapted to them. However, currently the main solution used by winemakers to resolve these problems is to employ commercial synthetic fermentative activators and nutrients [12,13].

Bee pollen is a natural product with a rich composition mainly composed by proteins, vitamins, minerals, and carbohydrates, but also of amino acids, fatty acids, sterols, phospholipids, carotenoids and polyphenols [14–19]. Amores-Arrocha et al. [20,21] stated bee pollen as a "Green nutrient activator", as they observed improvements in fermentation kinetics (increased fermentation rate, reduction of the yeast lag phase and increased cell multiplication), both in white and red winemaking processes, although its use is currently not legally authorized for industrial processing. In addition, at low doses, bee pollen has not affected the red wine's physicochemical composition or color parameters [20] Other published studies showed how bee pollen use improved both volatile and sensory compounds profiles of young white wines [22] as well as the aging kinetics and sensory profile of white wines undergoing biological aging [23]. Therefore, the aim of this research is to explore the effect of bee pollen use on the profile of both volatile compounds and sensory profile in red wines elaborated with an autochthonous grape variety: Tintilla de Rota.

## 2. Materials and Methods

### 2.1. Experimental Layout

Tintilla de Rota clusters of grapes were collected from vineyards of the privately-owned winery Luis Pérez, located in Jerez de la Frontera (southern Andalusia, Spain) (36°42′00.6′′ N, 6°11′34.0′′ W, 100 m above sea level). Vineyard soil was mainly limestone, known locally as "albariza". Grapes were destemmed and crushed. To avoid oxidations, the skins and grape musts mixture was sulphited with 25 mg/kg of $K_2O_5S_2$ (Sigma-Aldrich Chemical S.A., Madrid, Spain), tempered (20 °C) and no pH correction. A control (0 g/L) and six different doses (0.1, 0.25, 1, 5, 10 and 20 g/L) of commercial grounded bee pollen (Valencia, Spain) were studied. Bee pollen was added to each fermenter and homogenized with the paste, before adding the yeast inoculum. All vinifications assays were carried out in duplicate using temperature controlled glass fermenters (V = 5 L). Commercial yeast, *Saccharomyces cerevisiae* Lalvin 71B® strain (Lallemand, Barcelona, Spain), 10 g/hL inoculum was used to perform the alcoholic fermentation (AF). For malolactic fermentation (MLF), once AF was over, a Lactic Acid Bacteria (LAB) inoculum (1 g/hL) *Oenococcus oeni* S11B P2 Instant (Laffort, Bordeaux, France) was added.

### 2.2. Analysis of Volatile Compounds and Odor Activity Values (OAV)

Volatile compounds and their corresponding OAVs were analyzed following the methodology indicated by several authors [22,24]. Odor series were assigned to each component based on the main odor descriptor according to Peinado et al. [25], to obtain quantitative information from chemical analysis based on target criteria.

### 2.3. Sensory Wines Evaluation

A 10-member panel of trained and instructed experts experienced in wine tasting performed the evaluations. The sensory analysis was carried out in individual cabins equipped with a lighting control system. 50 mL of wine were provided to each taster in a regular wine-tasting glasses (ISO 3591, 1997) [26], topped with glass to avoid volatile compounds evaporation. Each sample was encoded using a random three-digit code to be tasted according to the order indicated. The wines were then served at a temperature of 20 ± 2 °C. In order not to overload judges, each of them performed the same tasting on a two different days in order to carry out the sensory analysis more than once presenting wine glasses randomly each time. Specific tasting notes were given to each taster and each attribute was scored according to an increase in intensity, based on a 10-point rating scale. All the sensorial characteristics employed in the tasting sheets have been selected taking into account the commercial wine style sensorial profile and following Jackson [27] tasting descriptors for red wines. Fruity, floral, and spicy aromas, as well as acidity, astringency, bitterness, sweetness, and milky notes were evaluated as generic attributes. In addition, the global judgment was added, as an attribute which encompasses the balance of all the attributes in conjunction with each other. Fruits groups (red, black, white, tropical, citrus, fruits with bone, raisin, and nuts), flowers groups (white, red, blue), vegetable, spices, woody, toast, balsamic, minerals, animals, and chemical, were the specific attributes evaluated by the judges.

### 2.4. Data Treatment

Means and standard deviations with significant differences were determined by bidirectional ANOVA and Bonferroni's multiple range (BSD) test; $p < 0.05$ was considered significant (GraphPad Prism version 6.01 for Windows, GraphPad Software, San Diego, CA, USA). For statistical significance, all tests were conducted in triplicate ($n = 3$). Principal Components Analysis (PCA) was performed using the SPSS 24.0 statistical computer package (SPSS Inc., Chicago, IL, USA).

## 3. Results and Discussion

### 3.1. Evaluation of the Effects of the Addition of Pollen on Wine Volatile Compounds and their Corresponding Odorant Activity Values (OAV)

#### 3.1.1. Higher Alcohols and Methanol

Higher alcohols total concentrations were found in order of 200 mg (Table 1). In general, higher alcohols are not affected by the use of pollen, except for the 0.25 g/L dose, where isoamyl alcohol is slightly higher, without exceeding 400 mg/L. Taking into account the red wines, Yeast Assimilable Nitrogen (YAN) levels [20] were higher than those of white wines [21], it could be expected higher alcohol levels, however, this does not occur. These results imply that there is no relationship between YAN and increased alcohol production, and skin presence is buffering the effect of pollen.

#### 3.1.2. Aldehydes

In general, the aldehyde content of red wines (3066–5699 μg/L) was lower than white wines [22]. No correlation was observed between aldehyde formation and pollen dose, and their contents fluctuate between the different doses. As might be expected, acetaldehyde was the major compound in this family, followed by benzeneacetaldehyde. Both compounds contributed to the wines sensory profile with nutty and floral notes. Additionally, nonanal and 3-methyl-butanal were identified, which, due to low perception thresholds, contribute to the wines' aromatic profile.

#### 3.1.3. Alcohols

Alcohol content was representing between 4–5.5% of the total volatile compounds (Table 1). Its values showed fluctuations in all samples, without following any correlation. This behavior is mainly marked by phenylethyl alcohol, which is the main alcohol, alongside with 1-pentanol and 1H-Indole-3-ethanol. Both phenylethyl alcohol and 1H-Indole-3-ethanol showed fluctuations while 1-pentanol together with most alcohols tended to increase significantly with respect to the control.

**Table 1.** Comparative effect of the addition of bee pollen (0.1, 0.25, 1, 5, 10 and 20 g/L) on concentrations of volatile compounds (μg/L) and control elaboration of Tintilla de Rota red wines.

| Compounds | Tintilla de Rota Red Wines Pollen Doses | | | | | | |
|---|---|---|---|---|---|---|---|
| | Control | 0.1 g/L | 0.25 g/L | 1 g/L | 5 g/L | 10 g/L | 20 g/L |
| *Higher alcohols* | | | | | | | |
| 2-Propanol | 34,953.9 ± 1276.1 $^{a}$ | 36,052.5 ± 2210.1 $^{a}$ | 38,842.7 ± 4907.9 $^{a}$ | 34,985.5 ± 1788.9 $^{a}$ | 36,902.1 ± 799.7 $^{a}$ | 34,768.3 ± 1495 $^{a}$ | 33,171.9 ± 815.5 $^{a}$ |
| *n*-Propyl alcohol | 27,691.4 ± 2165.4 $^{a}$ | 28,078.3 ± 3292.1 $^{a,b}$ | 27,692.6 ± 1445 $^{a}$ | 33,332.2 ± 3069.1 $^{b,c}$ | 29,588.4 ± 720.7 $^{a,c}$ | 30,273.1 ± 2148.2 $^{a,c}$ | 34,846.2 ± 2039.4 $^{c}$ |
| Isobutanol | 31,385.1 ± 1539.9 $^{a,b}$ | 31,248.3 ± 1983.1 $^{a}$ | 39,121 ± 683.8 $^{b}$ | 38,397.8 ± 1745.5 $^{a,b}$ | 34,307.7 ± 358.1 $^{a,b}$ | 36,460.3 ± 1599.1 $^{a,b}$ | 50,569 ± 1463.5 $^{c}$ |
| Isoamyl alcohols | 354,323.6 ± 16,785.8 $^{a,b}$ | 350,155.6 ± 21,853 $^{a,b}$ | 393,105.9 ± 5534 $^{a}$ | 354,546.7 ± 16,445 $^{a,b}$ | 316,247.7 ± 3734.7 $^{b}$ | 347,556.9 ± 15,377.3 $^{a,b}$ | 355,663.7 ± 10,652.6 $^{a,b}$ |
| Total | 448,353.9 ± 21,767.1 | 445,534.8 ± 29,338.3 | 498,762.1 ± 12,570.7 | 461,262.2 ± 23,048.6 | 417,045.8 ± 5613.1 | 449,058.5 ± 20,619.7 | 474,250.8 ± 14,971 |
| % higher alcohols | 77.53% | 76.12% | 77.82% | 74.73% | 70.49% | 71.89% | 67.65% |
| *Methanol* | 32,708.5 ± 3833 $^{a,b}$ | 34,834.6 ± 3461.2 $^{a,c,d}$ | 27,651.5 ± 141.1 $^{b}$ | 40,294 ± 5164 $^{c,d,e}$ | 40,450.6 ± 1741.7 $^{d,e}$ | 43,161.7 ± 5621 $^{e}$ | 33,131.9 ± 4624.9 $^{a,b}$ |
| Total | 32,708.5 ± 3833 | 34,834.6 ± 3461.2 | 27,651.5 ± 141.1 | 40,294 ± 5164 | 40,450.6 ± 1741.7 | 43,161.7 ± 5621 | 33,131.9 ± 4624.9 |
| % methanol | 5.66% | 5.95% | 4.31% | 6.53% | 6.84% | 6.91% | 4.73% |
| *Acids* | | | | | | | |
| Butanoic acid | 30 ± 0.2 $^{a,b}$ | 33 ± 0.7 $^{b}$ | 34.2 ± 0.8 $^{c,d}$ | 29.3 ± 0.8 $^{a}$ | 32 ± 2.5 $^{b,c,d}$ | 29.8 ± 0.4 $^{a,b}$ | 29.4 ± 0.7 $^{a,b}$ |
| 3-Methyl-butanoic acid | 189 ± 1.6 $^{a}$ | 190.4 ± 2.7 $^{a}$ | 182.2 ± 10.7 $^{a}$ | 188.8 ± 13.5 $^{a}$ | 193.4 ± 6.8 $^{a}$ | 189.9 ± 14.1 $^{a}$ | 184.2 ± 7.3 $^{a}$ |
| Hexanoic acid | 556.3 ± 42.4 $^{a}$ | 651.6 ± 5.8 $^{b}$ | 466.5 ± 26.1 $^{c,e,f}$ | 390.4 ± 4.5 $^{d}$ | 463.5 ± 29.4 $^{e,f}$ | 477.2 ± 3.3 $^{f}$ | 770.4 ± 23 $^{g}$ |
| Heptanoic acid | 20.7 ± 0.7 $^{a}$ | 24 ± 0.6 $^{a}$ | 24.7 ± 0.5 $^{a}$ | 25.6 ± 0.7 $^{a}$ | 27.9 ± 0.9 $^{a}$ | 74.8 ± 0.9 $^{b}$ | 144.4 ± 10.7 $^{c}$ |
| 2-Hexenoic acid | 34 ± 0.6 $^{a}$ | 39.4 ± 0.6 $^{b,c}$ | 39.9 ± 0.3 $^{c}$ | 44.6 ± 1.6 $^{d,e}$ | 45.7 ± 1.7 $^{e}$ | 53.7 ± 4.5 $^{f}$ | 60 ± 0.2 $^{g}$ |
| Octanoic acid | 624.4 ± 19.8 $^{a}$ | 366.6 ± 23.9 $^{b,c}$ | 443.7 ± 30.9 $^{c}$ | 648.1 ± 37.2 $^{a}$ | 1424.5 ± 14 $^{d,e}$ | 1443.3 ± 30.6 $^{e}$ | 1253.8 ± 39.3 $^{f}$ |
| Nonanoic acid | 87.7 ± 2.1 $^{a}$ | 76.4 ± 3.4 $^{a}$ | 90.5 ± 3.7 $^{a}$ | 116.2 ± 1.4 $^{a}$ | 556.2 ± 9.7 $^{b}$ | 182.6 ± 0.4 $^{c}$ | 184.4 ± 11.1 $^{c}$ |
| *n*-Decanoic acid | 0.19 ± 0.01 $^{a}$ | 398.5 ± 26.5 $^{b,c,f}$ | 412.4 ± 1.3 $^{c,f}$ | 311.9 ± 6.9 $^{d}$ | 866.4 ± 46.2 $^{e}$ | 440.9 ± 3.0 $^{f}$ | 450.3 ± 39.6 $^{f}$ |
| 9-Decenoic acid | 21.4 ± 1.5 $^{a}$ | 154.3 ± 12.4 $^{b}$ | 33.8 ± 1.9 $^{c,d}$ | 39.1 ± 0.8 $^{d}$ | 80.1 ± 1.3 $^{e,g}$ | 52.7 ± 2.7 $^{f}$ | 87.9 ± 0.6 $^{g}$ |
| Benzoic acid | 84.2 ± 5.9 $^{a}$ | 200.8 ± 0.1 $^{b}$ | 108.7 ± 1.4 $^{a,c}$ | 125 ± 5.8 $^{c}$ | 328.8 ± 29.8 $^{d}$ | 274 ± 3.5 $^{e}$ | 371.3 ± 12.2 $^{f}$ |
| Benzeneacetic acid | 40.8 ± 0.4 $^{a}$ | 42.8 ± 0.8 $^{a,b,c}$ | 41.9 ± 1.1 $^{a}$ | 43.9 ± 0.5 $^{a,b,c}$ | 46 ± 0.7 $^{b,c}$ | 46.4 ± 0.7 $^{c}$ | 49.5 ± 2.2 $^{c}$ |
| Total | 1688.6 ± 75.2 | 2177.7 ± 77.5 | 1878.5 ± 78.7 | 1962.8 ± 73.8 | 4064.5 ± 143.1 | 3265.3 ± 64.1 | 3585.5 ± 147 |
| % acids | 0.29% | 0.37% | 0.29% | 0.32% | 0.69% | 0.52% | 0.51% |
| *C-6 alcohols* | | | | | | | |
| 1-Hexanol | 493.9 ± 9.5 $^{a}$ | 431.1 ± 1.7 $^{b,c,d}$ | 436.6 ± 19.2 $^{c,d}$ | 433 ± 10.2 $^{d}$ | 434.2 ± 20.2 $^{e}$ | 520.3 ± 1.4 $^{a}$ | 539.5 ± 53.1 $^{a}$ |
| (E)-3-Hexen-1-ol | 16.7 ± 0.2 $^{a}$ | 29.1 ± 0.5 $^{b,c}$ | 30.9 ± 0.9 $^{c,d}$ | 33.1 ± 0.8 $^{d}$ | 35.7 ± 1.7 $^{e}$ | 36.1 ± 0.6 $^{e}$ | 37.7 ± 1.4 $^{e}$ |
| (Z)-3-Hexen-1-ol | 25.3 ± 2.4 $^{a}$ | 30 ± 1.4 $^{a,d}$ | 31.8 ± 1.0 $^{a}$ | 91.2 ± 0.4 $^{b}$ | 101.1 ± 6.8 $^{c}$ | 6.2 ± 0.3 $^{d}$ | 9.5 ± 0.1 $^{a,d}$ |
| Total | 535.9 ± 12.1 | 490.3 ± 3.6 | 499.2 ± 21 | 557.2 ± 11.4 | 571 ± 28.7 | 562.6 ± 2.3 | 586.7 ± 54.6 |
| % C-6 alcohols | 0.09% | 0.08% | 0.08% | 0.09% | 0.10% | 0.09% | 0.08% |
| *Alcoholes* | | | | | | | |
| 3-Penten-2-ol | 103.6 ± 4.9 $^{a}$ | 45.1 ± 2.4 $^{b,c}$ | 50.6 ± 0.1 $^{c}$ | 63.6 ± 4.7 $^{d,e,f}$ | 67.2 ± 3.9 $^{e,f}$ | 70.9 ± 0.1 $^{f}$ | 129.5 ± 2.0 $^{g}$ |
| 1-Pentanol | 1358.2 ± 53.1 $^{a}$ | 1812.7 ± 73.5 $^{b,c,d,e}$ | 1331.8 ± 123.8 $^{a}$ | 1746.8 ± 14.0 $^{c,d,e}$ | 1847.5 ± 17.9 $^{d,e}$ | 1992.8 ± 0.5 $^{e}$ | 2925.1 ± 187.4 $^{f}$ |
| 3-Ethyl-2-pentanol | 10.1 ± 0.3 $^{a}$ | 12.7 ± 0.7 $^{b,c}$ | 13.9 ± 1.0 $^{c}$ | 16.6 ± 0.1 $^{d,e}$ | 17.8 ± 0.1 $^{e}$ | 20 ± 0.2 $^{f}$ | 21.3 ± 1.9 $^{f}$ |
| 4-Methyl-1-pentanol | 13 ± 0.7 $^{a}$ | 16 ± 0.8 $^{b,c,d}$ | 17.3 ± 0.6 $^{c,d}$ | 18.3 ± 0.7 $^{d,e}$ | 20.5 ± 0.4 $^{e,f}$ | 20.7 ± 1.1 $^{f}$ | 25.8 ± 1.7 $^{g}$ |
| 3-Methyl-1-pentanol | 290.4 ± 8.0 $^{a}$ | 278.7 ± 2.9 $^{a}$ | 184.1 ± 7.2 $^{b}$ | 320.1 ± 20.5 $^{c}$ | 288.6 ± 15.5 $^{a}$ | 273.6 ± 4.3 $^{a}$ | 110.2 ± 8.7 $^{d}$ |
| 3-Ethoxy-1-Propanol | 99.3 ± 0.7 $^{a}$ | 100.7 ± 0.7 $^{a}$ | 112.4 ± 2.8 $^{b,c,e}$ | 118.9 ± 1.1 $^{c,d,e,f}$ | 125 ± 0.8 $^{d,e,f}$ | 120 ± 1.6 $^{e,f}$ | 125.9 ± 0.6 $^{f}$ |
| 1-Octanol | 20.8 ± 1.0 $^{a}$ | 23 ± 0.7 $^{a,b}$ | 24.7 ± 0.4 $^{b,c,d,e}$ | 25.5 ± 0.1 $^{c,d,e,f}$ | 25.9 ± 0.1 $^{d,e,f}$ | 25.8 ± 1.6 $^{e,f}$ | 27.6 ± 0.7 $^{f}$ |
| 1-Nonanol | 3.8 ± 0.3 $^{a}$ | 2.4 ± 0.2 $^{b,c,d,e}$ | 2.7 ± 0.02 $^{c,d,f}$ | 2.4 ± 0.2 $^{d,e}$ | 2.2 ± 0.2 $^{e}$ | 2.8 ± 0.2 $^{f,g}$ | 3.1 ± 0.02 $^{g}$ |
| Benzyl alcohol | 84.5 ± 7.1 $^{a}$ | 93.7 ± 0.6 $^{a}$ | 133 ± 10.6 $^{a,b}$ | 179.7 ± 3.8 $^{b}$ | 603.9 ± 58.2 $^{c}$ | 248.8 ± 2.3 $^{d}$ | 247.7 ± 18.8 $^{d}$ |
| Phenylethyl Alcohol | 23,942.5 ± 1414.2 $^{a}$ | 23,512 ± 235.9 $^{a}$ | 28,318.7 ± 302.5 $^{b,c}$ | 24,110.5 ± 1481.8 $^{a}$ | 27,830.4 ± 2039.4 $^{c}$ | 22,209.8 ± 331.3 $^{a}$ | 23,346.3 ± 2021.2 $^{a}$ |
| 1 H-Indol-3-ethanol | 862.1 ± 25 $^{a}$ | 1313.5 ± 10.6 $^{b,c}$ | 1452 ± 118.6 $^{c}$ | 1823.4 ± 2.6 $^{d,e}$ | 1866.8 ± 64.1 $^{e}$ | 452 ± 28.7 $^{f}$ | 551.7 ± 6.5 $^{f}$ |
| 1-Butanol | 31.2 ± 1.6 $^{a}$ | 22.7 ± 0.9 $^{a,b}$ | 15.3 ± 0.9 $^{b,e}$ | 23.3 ± 1.7 $^{a,e}$ | 114.3 ± 5.8 $^{c}$ | 44.3 ± 1.0 $^{d}$ | 29.7 ± 0.8 $^{a}$ |
| 3-Metil-2-buten-1-ol | 41.5 ± 1.7 $^{a}$ | 25.8 ± 1.2 $^{b}$ | 46.9 ± 1.4 $^{a,c}$ | 49.3 ± 4.3 $^{c}$ | 57.1 ± 2.8 $^{d}$ | 65.5 ± 4.1 $^{e}$ | 77.9 ± 1.1 $^{f}$ |
| Total | 26,861 ± 1518.6 | 27,259 ± 331.2 | 31,703.5 ± 570 | 28,498.3 ± 1535.6 | 32,867.2 ± 2209 | 25,547 ± 377 | 27,622 ± 2251.4 |
| % Alcoholes | 4.64% | 4.66% | 4.95% | 4.62% | 5.56% | 4.09% | 3.94% |

**Table 1.** *Cont.*

| Compounds | Tintilla de Rota Red Wines Pollen Doses | | | | | | |
|---|---|---|---|---|---|---|---|
| | Control | 0.1 g/L | 0.25 g/L | 1 g/L | 5 g/L | 10 g/L | 20 g/L |
| *Phenols* | | | | | | | |
| 2,6-di-terc-butil-4-ethylfenol | 23.1 ± 0.8 [a] | 29.1 ± 0.5 [b,c] | 29.8 ± 0.4 [c] | 20.1 ± 0.1 [d,e,f] | 18.6 ± 1.4 [e,f] | 20.8 ± 0.9 [a,f] | 23.3 ± 0.1 [a] |
| 4-Etilfenol | 6.7 ± 0.2 [a] | 8.2 ± 0.4 [b,e] | 9.4 ± 0.7 [c,d,e] | 10.1 ± 0.1 [d] | 8.6 ± 0.2 [e] | 4.9 ± 0.5 [f] | 7.2 ± 0.1 [a] |
| 4-Vinilguaiacol | 33.4 ± 1.1 [a] | 55.7 ± 5.3 [b] | 76.6 ± 3.7 [c,d] | 74.9 ± 6.3 [d] | 23.1 ± 0.8 [e,f] | 23 ± 1.4 [f] | 41.5 ± 0.3 [g] |
| Acetovainillin | 19.7 ± 0.7 [a] | 32.7 ± 0.8 [b,d,f] | 46.2 ± 0.5 [c] | 32.6 ± 0.7 [d,f] | 26.3 ± 0.7 [e] | 20.5 ± 1.4 [a] | 36.3 ± 0.7 [f] |
| Total | 82.9 ± 2.9 | 125.6 ± 7 | 162 ± 5.2 | 137.7 ± 7.2 | 76.6 ± 3.1 | 69.2 ± 4.1 | 108.3 ± 1.2 |
| % Phenols | 0.01% | 0.02% | 0.03% | 0.02% | 0.01% | 0.01% | 0.02% |
| *Terpenes* | | | | | | | |
| Linalool oxide | 50.5 ± 0.9 [a] | 74.2 ± 1.8 [b] | 86.3 ± 8.1 [c,f] | 59.1 ± 0.8 [d] | 50.1 ± 3.5 [a] | 38.2 ± 0.6 [e] | 88.3 ± 2.7 [f] |
| Linalool | 11.5 ± 0.5 [a] | 12.9 ± 0.8 [a,b] | 13.5 ± 0.1 [b] | 16.7 ± 0.3 [c,e] | 11.8 ± 0.5 [a] | 18.4 ± 0.8 [d] | 17 ± 0.5 [e,d] |
| α-Terpieol | 9.7 ± 0.2 [a] | 10.8 ± 0.1 [a] | 15.3 ± 0.1 [b] | 19.1 ± 0.9 [c,e] | 23.2 ± 1.4 [d] | 18.9 ± 1.4 [e] | 34 ± 2.4 [f] |
| (R)-(+)-β-Citronellol | 8.7 ± 0.2 [a] | 9.7 ± 0.2 [a,b] | 10.2 ± 0.1 [b] | 13.8 ± 0.9 [c,e] | 16.8 ± 0.9 [d,f] | 12.7 ± 0.8 [e] | 15.7 ± 1.4 [f] |
| 2,6-dimetil-3,7-Octadiene-2,6-diol, | 29 ± 0.1 [a] | 32.8 ± 0.4 [b,e,f,g] | 38.9 ± 0.2 [c,d,e] | 41.2 ± 1.5 [d] | 36.2 ± 0.7 [e,f] | 33.3 ± 0.6 [f,g] | 32.6 ± 0.7 [g] |
| 8-Hydroxylinalool | 49.4 ± 2.1 [a] | 100.4 ± 0.9 [b,d,g] | 202.4 ± 0.3 [c] | 91.1 ± 7.9 [d] | 387.1 ± 8.9 [e] | 142.2 ± 11.8 [f] | 130.3 ± 9.8 [f,g] |
| Total | 158.8 ± 4.1 | 240.8 ± 4.3 | 366.7 ± 8.9 | 240.9 ± 12.4 | 525.2 ± 15.8 | 263.7 ± 16 | 318 ± 17.4 |
| % Terpenes | 0.03% | 0.04% | 0.06% | 0.04% | 0.09% | 0.04% | 0.05% |
| *Esters* | | | | | | | |
| Ethyl acetate | 60,290.1 ± 5264.2 [a] | 65,998.8 ± 4222.7 [a,b] | 68,955 ± 3516.9 [a,b] | 75,224.6 ± 6073.5 [b,c] | 85,400.4 ± 1651.2 [c,d] | 94,930.8 ± 4988.6 [d] | 151,421.2 ± 10,126 [e] |
| Ethyl butyrate | 0.15 ± 0.01 [a] | 27.2 ± 0.7 [b,g] | 37.5 ± 2.8 [c] | 55.3 ± 1.8 [d] | 127.1 ± 2.4 [e] | 16.7 ± 0.1 [f] | 23 ± 1.9 [g] |
| Ethyl isovalerate | 5.6 ± 0.3 [a] | 14.5 ± 0.2 [b] | 17.7 ± 0.1 [c] | 20.9 ± 0.8 [d,g] | 25.1 ± 0.7 [e] | 27.6 ± 0.1 [f] | 19.9 ± 2.1 [g] |
| Isoamyl acetate | 29.4 ± 0.9 [a] | 45.5 ± 1.6 [b,c] | 50.5 ± 0.3 [c] | 75.5 ± 1.9 [d] | 180.1 ± 3.8 [e] | 107.6 ± 0.1 [f] | 126.9 ± 1.4 [g] |
| Ethyl hexanoate | 60 ± 0.1 [a] | 107.6 ± 2.1 [b,c,g] | 105.1 ± 0.7 [c,g] | 124.9 ± 1.8 [d,e,f] | 118.9 ± 0.3 [e,h] | 133.6 ± 1.4 [f] | 115 ± 4.3 [g,h] |
| Hexyl acetate | 26.3 ± 0.01 [a] | 29.4 ± 0.01 [a,b] | 30.3 ± 0.01 [b,c,d,e] | 32.8 ± 0.7 [c,g] | 28.3 ± 1.4 [a,e] | 33.4 ± 0.01 [d,g] | 42.6 ± 2.1 [f] |
| Ethyl 2-hydroxy-3-methyl butanoate | 13.2 ± 1.3 [a] | 15.8 ± 0.9 [a] | 17.7 ± 0.7 [a] | 29.6 ± 0.8 [b,e] | 63.3 ± 2.7 [c] | 37 ± 1.4 [d] | 33.4 ± 3.0 [e] |
| Ethyl octanoate | 204.5 ± 6.1 [a] | 349.9 ± 32.6 [b,c,g] | 372 ± 0.6 [c,g] | 244.8 ± 22.1 [d,e,f] | 261.2 ± 26.5 [e,f] | 245.4 ± 21.8 [f] | 346.7 ± 3.1 [g] |
| Ethyl nonanoate | 10.9 ± 0.9 [a] | 12.2 ± 1.0 [b,c,g] | 11 ± 0.6 [a,g] | 13 ± 0.8 [c] | 16.3 ± 0.6 [d,e,f] | 15.8 ± 1.4 [e] | 17.1 ± 0.5 [f] |
| Ethyl 2-hydroxy-4-methylpentanoate | 36.8 ± 1.6 [a] | 36.2 ± 3.1 [a] | 37.4 ± 2.7 [a] | 82.6 ± 0.7 [b] | 112.9 ± 7.6 [c] | 39.3 ± 1.9 [a] | 47.8 ± 2.0 [d] |
| Isoamyl lactate | 43.3 ± 1.8 [a] | 139 ± 0.6 [b,f] | 229.3 ± 3.1 [c] | 101.2 ± 1.0 [d] | 255.5 ± 7.1 [e] | 146.6 ± 2.1 [f] | 127.8 ± 0.4 [f] |
| Ethyl decanoate | 112.1 ± 7.1 [a] | 122.4 ± 0.3 [a] | 154.8 ± 3.6 [b] | 111.8 ± 5.3 [a] | 270.2 ± 7.5 [c,d] | 127.1 ± 1.6 [a] | 273.9 ± 5.8 [d] |
| Diethyl succinate | 509.3 ± 12.2 [a] | 681.5 ± 2.5 [b,d,e] | 950.9 ± 48 [c,f] | 636.5 ± 51.2 [d,e] | 633.3 ± 44.2 [e] | 907.9 ± 43.2 [f] | 1310.2 ± 125.7 [g] |
| Ethyl 9-decenoate | 59.7 ± 0.8 [a,c] | 63.9 ± 2.3 [a,b] | 66.1 ± 0.8 [b,d] | 64.3 ± 1.3 [a,b,d] | 59.4 ± 0.7 [a] | 56.5 ± 1.3 [c] | 62.8 ± 0.8 [a,b,d] |
| Ethyl phenylacetate | 1.27 ± 0.01 [a] | 2.23 ± 0.01 [b,c,d,f,g] | 2.27 ± 0.03 [c,d,f,g] | 2.39 ± 0.07 [d,e,f] | 2.55 ± 0.04 [e] | 2.27 ± 0.08 [f,g] | 2.2 ± 0.08 [g] |
| Phenethyl acetate | 33.8 ± 0.1 [a] | 97.4 ± 6.8 [b,e,d,f] | 104.5 ± 4.1 [c,d] | 94.9 ± 2.6 [d,f] | 279.1 ± 5.1 [e] | 80.9 ± 2.6 [f] | 251 ± 3.1 [g] |
| Diethyl malate | 24.4 ± 0.7 [a] | 32.9 ± 0.9 [b,c,d] | 34.9 ± 0.9 [c,d,e] | 36 ± 0.9 [d,e] | 37 ± 0.6 [e] | 40.6 ± 0.7 [f] | 45.1 ± 0.9 [g] |
| Methyl vanillate | 242.7 ± 0.5 [a] | 287.4 ± 5.2 [a] | 610.1 ± 8.2 [b,c] | 550.3 ± 45.8 [c,e,f] | 1541.4 ± 18.6 [d] | 454 ± 44.7 [e,f] | 470.4 ± 62.1 [f] |
| Ethyl lactate | 121.4 ± 1.5 [a] | 173.8 ± 5.3 [b,c] | 155.9 ± 4.3 [c] | 204.6 ± 21.0 [d,e] | 199 ± 3.2 [e] | 271.1 ± 25.7 [f] | 256 ± 4.0 [f] |
| Butanoic acid 3-hydroxy ethyl ester | 51.8 ± 2.5 [a] | 63 ± 1.3 [a,b] | 57 ± 5.2 [a] | 76 ± 3.5 [b,d] | 231.3 ± 3.8 [c] | 82.7 ± 1.3 [d] | 105.1 ± 8.0 [e] |
| Ethyl (Z)-4-decenoate | 60.9 ± 1.2 [a] | 165.9 ± 3.7 [b,d] | 203.7 ± 17.7 [c] | 191.3 ± 6.1 [d,c] | 391.5 ± 6.4 [e] | 31.9 ± 1.6 [a] | 57.4 ± 0.6 [a] |
| Ethyl dodecanoate | 68.2 ± 2.6 [a] | 35.6 ± 2.0 [b] | 143.5 ± 14.1 [c,f] | 172 ± 2.8 [d] | 303.9 ± 8.1 [e] | 121.1 ± 1.0 [f] | 91.8 ± 0.3 [g] |
| Methyl tetradecanoate | 34.8 ± 1.5 [a] | 66.4 ± 7.2 [b,c,d,f] | 88.5 ± 6.6 [c,d,f] | 72.8 ± 3.8 [d,f] | 297.9 ± 18.1 [e] | 80.5 ± 0.8 [f] | 87.5 ± 1.9 [f] |
| Succinoic acid 2-hydroxy-3-methyl diethyl ester | 93.3 ± 8.1 [a] | 110.3 ± 10.4 [a,b,e] | 126.9 ± 3.4 [b,c,e,f] | 144.2 ± 4.8 [c,e,f] | 450.8 ± 36.4 [d] | 131.8 ± 1.5 [e] | 147.2 ± 2.3 [f] |
| Methyl hexadecanoate | 81.4 ± 4.6 [a] | 100.9 ± 5.2 [b,e] | 147.6 ± 0.7 [c] | 219.3 ± 10.1 [d,g] | 106.3 ± 2.1 [e] | 186.4 ± 4.3 [f] | 227.6 ± 5.2 [g] |
| Hexadecanoic acid, ethyl ester | 310.8 ± 7.4 [a,f,g] | 420.9 ± 33.5 [b,d] | 732.6 ± 20.6 [c,e] | 435.7 ± 10.9 [d] | 719.8 ± 61.2 [e] | 300.3 ± 2.5 [f] | 356.1 ± 14.5 [g] |
| Propanoic acid 2-methyl-propyl ester | 84.7 ± 0.8 [a] | 116.1 ± 11.1 [b] | 160 ± 2.4 [c,d,f] | 154.5 ± 4.5 [d,f] | 181.7 ± 0.1 [e] | 147.7 ± 8.1 [f] | 155.2 ± 9.2 [f] |

**Table 1.** *Cont.*

| Compounds | Tintilla de Rota Red Wines Pollen Doses | | | | | | | | | | | | | | | | | | | | |
|---|---|---|---|---|---|---|---|---|---|---|---|---|---|---|---|---|---|---|---|---|---|
| | Control | | | 0.1 g/L | | | 0.25 g/L | | | 1 g/L | | | 5 g/L | | | 10 g/L | | | 20 g/L | | |
| Ethyl 8-nonenoato | 205.6 | ± | 3.5 $^{a}$ | 216.4 | ± | 3.1 $^{a,b}$ | 235.9 | ± | 0.8 $^{b,f}$ | 209.4 | ± | 2.2 $^{a,f}$ | 293.9 | ± | 6.5 $^{c,d}$ | 313 | ± | 17.6 $^{d}$ | 363.5 | ± | 20.5 $^{e}$ |
| Total | 62,816.5 | ± | 5332.3 | 69,533.2 | ± | 4366.3 | 73,838.8 | ± | 3669.9 | 79,381.1 | ± | 6282.9 | 92,587.9 | ± | 1927.1 | 99,069.5 | ± | 5177.4 | 156,584.3 | ± | 10,411.7 |
| % esters | 10.86% | | | 11.88% | | | 11.52% | | | 12.86% | | | 15.65% | | | 15.86% | | | 22.34% | | |
| *Aldehydes* | | | | | | | | | | | | | | | | | | | | | |
| Acetaldehyde | 4890.4 | ± | 279.7 $^{a,b}$ | 4810.7 | ± | 204.3 $^{a}$ | 5520.7 | ± | 263.1 $^{b}$ | 4482.1 | ± | 472.6 $^{a}$ | 2887.5 | ± | 294.1 $^{c,d}$ | 3292.9 | ± | 322.7 $^{d}$ | 4422.9 | ± | 417.9 $^{a}$ |
| Benzeneacetaldehyde | 66.6 | ± | 1.9 $^{a}$ | 74.3 | ± | 5.7 $^{a,e}$ | 131.9 | ± | 10.4 $^{b,c,d}$ | 145.3 | ± | 11.7 $^{c}$ | 119.8 | ± | 7.0 $^{d}$ | 80.5 | ± | 3.8 $^{a,e}$ | 85.1 | ± | 2.6 $^{e}$ |
| Nonanal | 10.5 | ± | 0.6 $^{a,e}$ | 13.4 | ± | 0.4 $^{b,d}$ | 16.2 | ± | 1.0 $^{c}$ | 14.2 | ± | 0.5 $^{d}$ | 10.7 | ± | 0.6 $^{a}$ | 9.9 | ± | 0.1 $^{a,e}$ | 8.6 | ± | 0.7 $^{e}$ |
| 3-methyl-butanal | 15.1 | ± | 0.5 $^{a}$ | 28.5 | ± | 1.5 $^{b,c}$ | 30.1 | ± | 0.5 $^{c,d}$ | 35.1 | ± | 2.6 $^{d}$ | 48.8 | ± | 2.6 $^{e}$ | 10 | ± | 0.3 $^{a}$ | 15.4 | ± | 0.4 $^{a}$ |
| Total | 4982.6 | ± | 282.7 | 4926.8 | ± | 212 | 5698.9 | ± | 275 | 4676.8 | ± | 487.4 | 3066.8 | ± | 304.4 | 3393.3 | ± | 326.9 | 4532.1 | ± | 421.6 |
| % Aldehydes | 0.86% | | | 0.84% | | | 0.89% | | | 0.76% | | | 0.52% | | | 0.54% | | | 0.65% | | |
| *Thiols* | | | | | | | | | | | | | | | | | | | | | |
| 3-(methylthio)-1-Propanol | 21.4 | ± | 1.8 $^{a}$ | 49.3 | ± | 3.8 $^{b}$ | 148.9 | ± | 7.5 $^{c}$ | 95.7 | ± | 3.7 $^{d,f}$ | 198.5 | ± | 3.2 $^{e}$ | 88.1 | ± | 8.6 $^{f}$ | 111.7 | ± | 2.4 $^{g}$ |
| Total | 21.4 | ± | 1.8 | 49.3 | ± | 3.8 | 148.9 | ± | 7.5 | 95.7 | ± | 3.7 | 198.5 | ± | 3.2 | 88.1 | ± | 8.6 | 111.7 | ± | 2.4 |
| % thiols | 0.004% | | | 0.01% | | | 0.02% | | | 0.02% | | | 0.03% | | | 0.01% | | | 0.02% | | |
| *Acetals* | | | | | | | | | | | | | | | | | | | | | |
| 1-(1-etoxietoxi)-pentano | 2.1 | ± | 0.1 $^{a}$ | 2 | ± | 0.1 $^{a}$ | 1.8 | ± | 0.1 $^{a}$ | 3.2 | ± | 0.1 $^{b}$ | 4.4 | ± | 0.2 $^{c}$ | 5.4 | ± | 0.1 $^{d}$ | 5.7 | ± | 0.2 $^{e}$ |
| Total | 2.1 | ± | 0 | 2 | ± | 0 | 1.8 | ± | 0.1 | 3.2 | ± | 0 | 4.4 | ± | 0.2 | 5.4 | ± | 0.1 | 5.7 | ± | 0.2 |
| % acetals | 0.0004% | | | 0.0003% | | | 0.0003% | | | 0.0010% | | | 0.0010% | | | 0.0010% | | | 0.0010% | | |
| *Norisoprenoids* | | | | | | | | | | | | | | | | | | | | | |
| 3-Oxo-α-ionol | 7.6 | ± | 0.1 $^{a}$ | 13.3 | ± | 0.4 $^{b,c,d,e}$ | 13.1 | ± | 0.1 $^{c,d,e}$ | 14.1 | ± | 0.2 $^{d,e}$ | 14.7 | ± | 0.1 $^{e}$ | 42.8 | ± | 0.6 $^{f}$ | 44.9 | ± | 1.0 $^{f}$ |
| Total | 7.6 | ± | 0.1 | 13.3 | ± | 0.4 | 13.1 | ± | 0.1 | 14.1 | ± | 0.2 | 14.7 | ± | 0.1 | 42.8 | ± | 0.6 | 44.9 | ± | 1 |
| % Norisoprenoids | 0.001% | | | 0.002% | | | 0.002% | | | 0.002% | | | 0.002% | | | 0.010% | | | 0.010% | | |
| *Lactones* | | | | | | | | | | | | | | | | | | | | | |
| Dihydro-5-pentyl-2-(3 H)-Furanona | 59.6 | ± | 0.7 $^{a}$ | 55.8 | ± | 4.8 $^{a}$ | 107.2 | ± | 2.8 $^{b,d}$ | 78.8 | ± | 0.5 $^{c}$ | 106.8 | ± | 0.3 $^{d}$ | 66.8 | ± | 0.2 $^{a}$ | 120.6 | ± | 8.4 $^{e}$ |
| 2,3-dihydro-benzofuran | 50.6 | ± | 1.1 $^{a}$ | 50 | ± | 4.0 $^{a}$ | 50.6 | ± | 1.1 $^{a}$ | 42.3 | ± | 0.8 $^{b}$ | 49.5 | ± | 0.8 $^{a}$ | 48.8 | ± | 1.5 $^{a}$ | 50.1 | ± | 1.6 $^{a}$ |
| Total | 110.2 | ± | 1.7 | 105.8 | ± | 8.8 | 157.8 | ± | 4 | 121.1 | ± | 1.4 | 156.2 | ± | 1.1 | 115.6 | ± | 1.7 | 170.7 | ± | 10 |
| % lactones | 0.02% | | | 0.02% | | | 0.02% | | | 0.02% | | | 0.03% | | | 0.02% | | | 0.02% | | |

Different superscript letters indicate that there are significant differences between the samples ($p < 0.05$, ANOVA, BSD test).

#### 3.1.4. Acids

Acid compounds content of red wines is much lower than that of white wines [22], which could be attributed to varietal character and also to skins presence during alcoholic fermentation. Some authors have found that grape skins provide fatty acids (oleic and linoleic), reducing the of volatile acids synthesis by yeasts [28,29]. However, analyzing the acid profile, two very different behaviors can be distinguished (Table 1). On the one hand, both control as well as low and intermediate doses (0.1–1 g/L) presented concentrations between 1688.54 and 2177.68 µg/L. On the other hand, the high doses (5–20 g/L) showed a greater range of concentration, with values between 3265.30 and 4064.52 µg/L. This could be due to autolysis phenomena produced between the end of AF and MLF, since volatile fatty acids from the cell membranes (hexanoic, octanoic, decanoic, and dodecanoic fatty acids) may be transferred into the medium via cell lysis [28,30].

#### 3.1.5. Esters

Esters family percentages of representation in red wines were observed between 10.86–22.34% (Table 1). Therefore, with the bee pollen addition, the formation of esters is favored in Tintilla de Rota red wines. A linear correlation is also observed between esters concentration and pollen doses ($R^2$ = 0.96). In most cases, esters increase with the dose of pollen: ethyl acetate ($R^2$ = 0.96), isoamyl acetate, ethyl octanoate, diethyl succinate ($R^2$ = 0.68), phenethyl acetate, ethyl lactate ($R^2$ = 0.64), methyl hexadecanoate, hexadecanoic acid, ethyl ester, ethyl 8-nonanoate ($R^2$ = 0.90), hexyl acetate ($R^2$ = 0.78), ethyl nonanoate ($R^2$ = 0.71) and diethyl malate ($R^2$ = 0.67). According to some authors, after MLF, an increase in esters concentration, ethyl acetate, ethyl hexanoate, ethyl lactate, and ethyl octanoate can be observed [31–36]. It should be noted that the main ester is ethyl acetate, whose descriptor is acrylic flavor, so it could be one of the compounds responsible for providing unpleasant aromas [37], especially for wines elaborated with the highest bee pollen doses (10 g/L and 20 g/L). This effect could be promoted because of YAN content and not because of pollen.

#### 3.1.6. C6-Alcohols

C6-alcohols were found in a range between 490 and 587 µg/L (Table 1). These compounds have fresh herbs and vegetables flavors and could be formed prior to AF via enzymatic action on their major precursors (linoleic acid and linolenic acid) [38]. During AF, these compounds are reduced to alcohols by yeast, mainly hexanol and hexenol, being the second most fragrant, but it was found in lower concentration in wines [38,39]. There are also these compounds present in grapes in their glycol form, but in lower concentration than in the pre-fermentation stages. Furthermore, the presence of antioxidant polyphenolic substances from the skins allows a low oxidation of final red wines.

#### 3.1.7. Terpenes

Terpenes represented very small percentages within the different volatile compound families of the obtained red wines (<0.1%) (Table 1). These compounds are typical of some grape varieties, and provide important floral notes in wine aroma [40]. The main terpenes able to contribute to wine aroma are naturally found in grape skins. In parallel to enzymatic actions, winemaking operations favors the extraction of these compounds from the grape must (macerations, pump-over, or head dipping) [39]. As can be seen in Table 1, most of the terpenes increased with pollen addition, with 8-hydroxylinalool standing out. However, there was no direct correlation between the applied dose and these family of compounds. It could be suggested that in addition to the grape skins, bee pollen was directly providing terpenes to wines.

### 3.1.8. Phenols

Most of the phenolic compounds are formed during AF, by decarboxylation of hydroxycinnamic acids carried out by *Saccharomyces cerevisiae*. Phenols were found in very low concentrations in red wines. 4-vinylguaiacol and acetovainillone were detected, providing spicy notes as typical varietal aromas of Tintilla de Rota wines. Phenol contents fluctuated with pollen doses unable to establish any pattern in this regard.

### 3.1.9. Thiols, Acetals, and Norisoprenoids

Thiols, acetals, and norisoprenoids represented the three minor compounds families of the volatile compound profile of red wines samples. Thiols were represented by 3-(methylthio)-1-propanol, whose concentration increased by bee pollen addition. This compounds formation has its origin in cysteine precursors present in grape must [39] which are degraded by yeasts to give rise to thiols. Thiols increase also could be justified by the natural richness of bee pollen in cysteine [41–43]. Acetals are compounds that can come from acetaldehyde and glycerol, however they are more commonly found at high levels in fortified wines [39,44]. Acetals family was represented by 1-(1-Ethoxyethoxy)-pentane, whose concentration showed a rising trend from the 5 g/L dose. Norisoprenoid family was represented by 3-oxo-α-ionol, whose concentration increased with pollen addition especially at high doses (10 and 20 g/L pollen). Some norisoprenoid, such as 3-oxo-α-ionol, have been found in several types of honey [45]. Considering bee pollen as the main raw material of honey, it is possible these compounds are being released by pollen during winemaking.

### 3.1.10. Lactones

Certain lactones have fermentation origins and are able to take part in wine aroma [46]. Both compounds identified in red wines studied were dihydro-5-pentylglycol-2(3H)-furanone and 2.3-dihydro-benzofuranone, representing under 0.1% of the total amount of volatile compounds. Dihydro-5-pentyl-2-(3H)-furanone showed a fluctuating behavior, while 2.3-dihydro-benzofuranone remained constant at all doses.

### 3.1.11. Principal Component Analysis (PCA) of Volatile Compounds

Table 2 shows the results of the loadings of the factors extracted in the analysis of the Principal Components Analysis (PCA). Concentration of total volatile compounds by aromatic families: higher alcohols, methanol, acids, C6-alcohols, alcohols, phenols, terpenoids, esters, aldehydes, acetals, norisoprenoids, and lactones, were variables included. PCA analysis extracted three factors that represent more than 91% of the total variance. Factor 1 (F1) correlated positively with acids, C6-alcohols, esters, acetals and norisoprenoids, the latter two represented by 1-(1-ethoxyethoxy)-pentane and norisoprenoids by 3-oxo-α-ionol. All families whose concentration can be influenced by the presence of pollen are represented in this factor, being the esters the ones with the highest load (0.958). Factor 2 (F2) represented families of compounds not related to the influence of pollen. The decrease of some of these families could be related to ester formation. These could be considered as intermediate compounds or precursors of other compounds and therefore their concentration may fluctuate with the pollen dose. F2 could explain all the fluctuations in all families resulting from consumption of some compounds in order to form others. In some cases, it is observed families where at certain pollen doses, their concentrations are lower than the control. This behavior is possibly a reflection of the consumption of these compounds to form others. Factor 3 (F3) belongs to pollen response on the aromatic profile of wines. This factor involves volatile compound families displaying higher concentration increases with pollen presence, not correlated with doses.

**Table 2.** Loadings of the main components of the volatile compounds in Tintilla de Rota dosed with bee pollen and control wines.

| | F1 | F2 | F3 |
|---|---|---|---|
| Higher alcohols | 0.179 | 0.936 | −0.059 |
| Methanol | 0.193 | −0.862 | −0.082 |
| Acids | 0.648 | −0.533 | 0.492 |
| C-6 alcohols | 0.744 | −0.449 | 0.151 |
| Alcohols | −0.282 | 0.102 | 0.948 |
| Phenols | −0.284 | 0.831 | 0.152 |
| Terpenes and derivatives | 0.129 | −0.196 | 0.961 |
| Esters | 0.958 | 0.056 | 0.164 |
| Aldehydes | −0.355 | 0.895 | −0.242 |
| Thiols | 0.190 | −0.080 | 0.955 |
| Acetals | 0.895 | −0.420 | 0.121 |
| Norisoprenoids | 0.945 | −0.079 | −0.140 |
| Lactones | 0.513 | 0.336 | 0.769 |
| Explained variance (%) | 32.75 | 30.37 | 28.80 |

Rotated component matrix loadings of Principal Components Analysis of volatile compounds in Tintilla de Rota red wines, using varimax with Kaiser normalization.

F1 increased with the pollen dose, reaching higher levels for high doses (10 and 20 g/L). This suggests that the effect of ester formation is the main one, taking into account that it is one of the compound families with high participation in the aromatic profile of wines (10.86–22.34%). F2 positioned control and dose of 0.1 g/L with similar behaviors, while those doses that generate an increase mainly of higher alcohols, aldehydes and phenols advanced towards positive values. However, this factor is offset by the effect of the methanol. As previously observed, this compound exhibited fluctuations in its concentration, which means that F2 had no correlation with the pollen dose. Factors F3, together with F1, showed an increasing trend of both pollen-formed and pollen-produced compounds.

Therefore, it could be pointed out that pollen use in Tintilla de Rota grapes variety vinification favors to produce fruity and floral aromas compounds in wines.

3.1.12. Odorant Activity Values (OAV) Analysis.

Odorant Activity Values (ΣOAV) of all volatile compounds involved in the aromatic profile of wines are shown in Table 3. Widely, there was an OAV increase between 30.7 and 63.6% in all pollen dosage wines without correlation. The most important odorant series in wines was fruity. Its values increased with the pollen addition (129.35–154.61), reaching maximum levels for 0.25 g/L, followed by 20 g/L and 5 g/L doses. Floral series reached maximum levels for doses between 0.25–1 g/L, while it began to decrease from the dose of 10 g/L. Spicy aromas were intensified an average 58% between 1–10 g/L of pollen dose. Fatty aroma series increased slightly in 5 g/L samples, increasing a 25% over $\Sigma OAV_T$. On 1 g/L, herbal odorant series showed a slight increase, mainly caused by C6 alcohols, α-terpineol, *n*-propyl alcohol.

Despite the increase in "negative" odorants series (fatty and grassy), their influence on the total was very low (Table 3). It should be noted that higher bee pollen doses increased chemical odorant series and 1 g/L, 5 g/L and 10 g/L dosages increased earthy notes as well. In all cases, the highest load was observed in fruity, floral and spicy series, with values greater than 90% and much higher in all pollen dosage wines (<94%). Despite varietal character, increases in fruity, floral and spicy series would suggest that bee pollen promoted an increase of compounds enhancing wine aromatic profile quality. Low and intermediate pollen doses were those with the maximum levels of ΣOAV (fruits, flowers, and spices). Besides, the increase in fatty and herbaceous aromas, produced during winemaking with high doses, dropped the ratio Σ (fruits, flowers, spices)/Σ (fatty, herbaceous). These results proved that low doses of bee pollen promote fruit and floral aromas, enhancing varietal sensory quality in red wines.

**Table 3.** Odor activity values summary (ΣOAV) grouped by odorant series.

| Odorant Series | Bee Pollen Doses in Tintilla de ROTA Red Wines | | | | | | |
|---|---|---|---|---|---|---|---|
| | Control | 0.1 g/L | 0.25 g/L | 1 g/L | 5 g/L | 10 g/L | 20 g/L |
| Fruity | 93.03 | 142.21 | 154.61 | 134.68 | 146.87 | 129.35 | 151.40 |
| Floral | 25.25 | 31.01 | 45.26 | 43.24 | 40.41 | 26.58 | 36.55 |
| Fatty | 8.50 | 8.67 | 8.15 | 8.46 | 10.89 | 10.47 | 10.69 |
| Grassy | 1.15 | 1.15 | 1.17 | 1.45 | 1.42 | 1.26 | 1.45 |
| Dry fruit | 0.05 | 0.05 | 0.06 | 0.05 | 0.03 | 0.03 | 0.04 |
| Earthy, mushrooms | 0.60 | 0.61 | 0.52 | 0.74 | 0.78 | 0.64 | 0.33 |
| Chemical | 0.10 | 0.22 | 0.13 | 0.15 | 0.35 | 0.29 | 0.39 |
| Spicy | 0.94 | 1.52 | 2.17 | 2.09 | 1.12 | 0.75 | 1.23 |
| Phenolic | 0.02 | 0.02 | 0.02 | 0.02 | 0.02 | 0.01 | 0.02 |
| $\Sigma OAV_T$ | 129.62 | 185.45 | 212.09 | 190.86 | 201.89 | 169.38 | 202.10 |

Odor Activity Values (OAV) sum of the main odorants found in Tintilla de Rota red wines produced (Control and different bee pollen doses addition (from 0.1 to 20 g/L)). ΣOAVT means the total sum of ΣOAV.

## 3.2. *Sensory Evaluation of the Resulting Wines.*

Generic and specific attributes average results with significant differences between the pollen wines and the control are shown in Figure 1a,b. Additional Table S1 show average numeric results of sensory analysis (generic and specific attributes). Each of the tasters was able to identify an average of 20 attributes. Tasters noted attributes related to fruit and floral aromas, and feelings of acidity and sweetness in mouth the most significant differences (ANOVA, $p < 0.05$) (Figure 1a). Low doses (0.1 and 0.25 g/L) were valued the best for fruit and floral attributes, being 0.25 g/L the best. Nevertheless, wines pungency increased with high bee pollen doses. An increase in spicy character would be related to a phenols and some esters increase, such as methyl vanilla, belonging to spicy odorant family.

For generic olfactory attributes, 0.25 g/L was the best evaluated followed by 0.10 g/L. Bitterness and astringency feelings, decreased from 5 g/L to 20 g/L compared to low doses and control, while sweetness feeling increased. For general taste attributes, 0.25 g/L and 1 g/L were the best scored wines. Thus, the 0.25 g/L bee pollen dose could be the best resulting dose in terms of general sensory (olfactory and gustatory) aspects. Wines with low doses of bee pollen improved their olfactory organoleptic qualities, increasing fruity and floral notes, highly appreciated by consumers in young red wines [46].

Compared with the control, all specific olfactory attributes (Figure 1b) showed significant differences excluding the white flower notes. Low doses were the best scored in red and black fruit attributes, being the best valued 0.25 g/L. Concerning citrus notes, a clear tendency was found with the pollen doses increase. Stone fruit and ripe or raisin fruit notes were significantly scored (ANOVA, $p < 0.05$) above low dose and control. This effect could be explained by oxidation notes produced from certain fatty acids [47], translated as ripe fruit or raisin notes by tasters.

Finally, it should be noted for red flower notes, low and intermediate pollen doses got the lowest values significantly compared to control. In contrast, the largest doses (10 and 20 g/L) had significantly lower values compared to the control. Also, all the large doses (5 g/L to 20 g/L) showed greater aromatic intensity in the vegetable notes. In addition, these doses showed a slight tendency to increase to spices, wood, toast, balsamic, minerals and animal's notes, compared to the low, intermediate doses and control (Figure 1b).

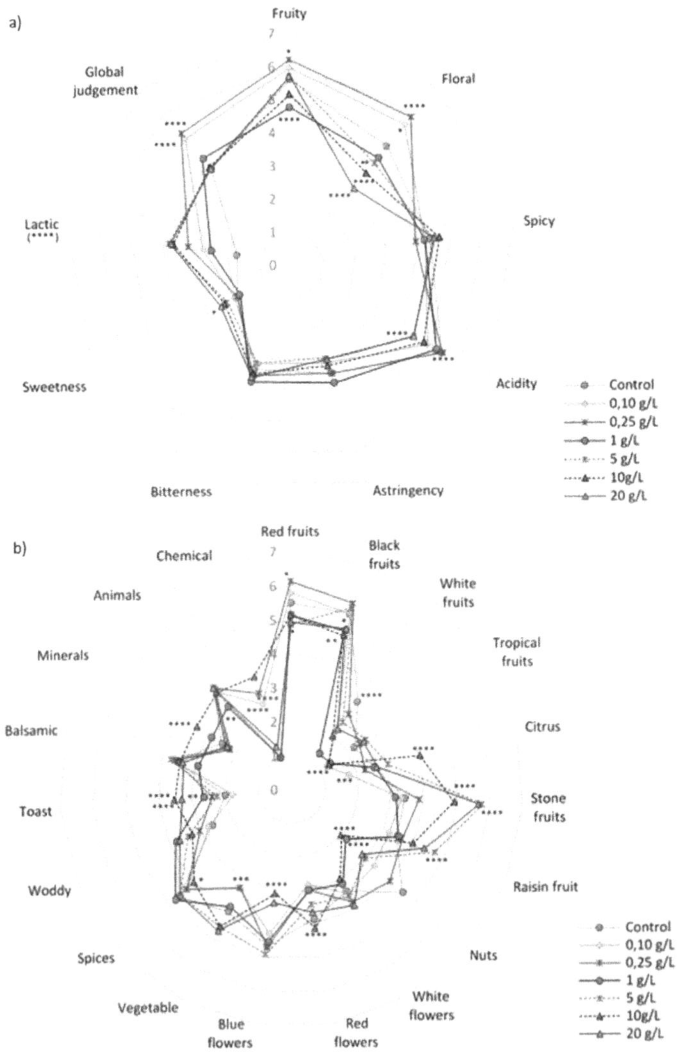


**Figure 1.** Tintilla de Rota red wines sensorial evaluation results of generic attributes (**a**), and specific olfactory attributes (**b**). * indicates level of significance for two-way ANOVA (BSD-test) (* $p < 0.05$, ** $p < 0.01$, *** $p < 0.001$ and **** $p < 0.0001$).

## 4. Conclusions

In conclusion, the contribution of multiflora bee pollen to Tintilla de Rota red grape musts increases the concentration of total volatile compounds of final wines, especially the families of higher alcohols, esters, terpenes, phenols, thiols, and norisoprenoids families. Lower pollen doses (0.1 and 0.25 g/L) increases the total levels of the OAV and the series of aromas associated with the fruity and floral character of red wines, whereas high bee pollen doses enhance the chemical, fatty, and grassy aromatic series mainly. In addition, high and intermediate dose (1 g/L) produced an increase in the earthy notes in the aromas. Descriptive sensory analysis determines that low doses of pollen (0.1 and 0.25 g/L) obtain the highest scores in the overall assessment and sensory attributes responsible for fruity and floral aromas in red wines, highlighting red and black fruit attributes. However, the high doses diverted the sensory profile towards fleshy stone fruit, fruit with raisins, and more typical aromas of red wines with some evolution or aging.

**Supplementary Materials:** The following are available online at http://www.mdpi.com/2304-8158/9/8/981/s1. Table S1. Generic and specific olfatory attributes results of Tintilla de Rota wines sensory analysis.

**Author Contributions:** Conceptualization, A.A.-A., P.S.-G., V.P. and A.J.-C.; data curation, A.A.-A. and V.P.; formal analysis, A.A.-A.; funding acquisition, A.A.-A. and V.P.; investigation, A.A.-A., P.S.-G., V.P. and A.J.-C.; methodology, A.A.-A., P.S.-G., V.P. and A.J.-C.; project administration, A.A.-A. and V.P.; supervision, V.P.; writing—original draft, A.A.-A., P.S.-G., A.J.-C. and V.P.; writing—review and editing, A.A.-A., P.S.-G., A.J.-C. and V.P. All authors have read and agreed to the published version of the manuscript.

**Funding:** This research was funded by the University of Cadiz for a Ph.D. student assistantship under the program 2013-023/PU/EPIF-FPI-CT/CP.

**Acknowledgments:** A.A.A. gratefully thanks the University of Cadiz for a Ph.D. student assistantship under the program 2013-023/PU/EPIF-FPI-CT/CP. The authors also thank the private winery Luis Pérez from Jerez de la Frontera, Cádiz, for grapes supporting.

**Conflicts of Interest:** The authors declare no conflict of interest.

## References

1. Mestre, M.V.; Maturano, Y.P.; Gallardo, C.; Combina, M.; Mercado, L.; Toro, M.E.; Carrau, F.; Vazquez, F.; Dellacassa, E. Impact on sensory and aromatic profile of low ethanol malbec wines fermented by sequential culture of Hanseniaspora uvarum and Saccharomyces cerevisiae native yeasts. *Fermentation* **2019**, *5*, 65. [CrossRef]
2. King, E.S.; Kievit, R.L.; Curtin, C.; Swiegers, J.H.; Pretorius, I.S.; Bastian, S.E.P.; Leigh Francis, I. The effect of multiple yeasts co-inoculations on Sauvignon Blanc wine aroma composition, sensory properties and consumer preference. *Food Chem.* **2010**, *122*, 618–626. [CrossRef]
3. Styger, G.; Prior, B.; Bauer, F.F. Wine flavor and aroma. *J. Ind. Microbiol. Biotechnol.* **2011**, *38*, 1145–1159. [CrossRef] [PubMed]
4. Polášková, P.; Herszage, J.; Ebeler, S.E. Wine flavor: Chemistry in a glass. *Chem. Soc. Rev.* **2008**, *37*, 2478–2489. [CrossRef] [PubMed]
5. Briche, E.; Beltrando, G.; Somot, S.; Quénol, H. Critical analysis of simulated daily temperature data from the ARPEGE-climate model: Application to climate change in the Champagne wine-producing region. *Clim. Chang.* **2014**, *123*, 241–254. [CrossRef]
6. Sacchelli, S.; Fabbrizzi, S.; Menghini, S. Climate change effects and adaptation strategies in the wine sector: A quantitative literature review. *Wine Econ. Policy* **2016**, *5*, 114–126. [CrossRef]
7. Zamora, F. Dealcoholised wines and low-alcohol wines. In *Wine Safety, Consumer Preference, and Human Health*; Springer: Cham, Switzerland, 2016; pp. 163–182.
8. De Orduña, R.M. Climate change associated effects on grape and wine quality and production Food Research International Climate change associated effects on grape and wine quality and production. *Food Res. Int.* **2017**, *43*, 1844–1855. [CrossRef]
9. Boulton, R.B.; Singleton, V.L.; Bisson, L.F.; Kunkee, R.E. Yeast and Biochemistry of Ethanol Fermentation. In *Principles and Practices of Winemaking*, 1st ed.; Boulton, R.B., Singleton, V.L., Bisson, L.F., Kunkee, R.E., Eds.; Springer: London, UK, 1996; pp. 141–143.
10. Webb, L.B.; Whetton, P.H.; Barlow, E.W.R. Modelled impact of future climate change on the phenology of winegrapes in Australia. *Aust. J. Grape Wine Res.* **2007**, *13*, 165–175. [CrossRef]
11. Hernández-Orte, P.; Ibarz, M.J.; Cacho, J.; Ferreira, V. Effect of the addition of ammonium and amino acids to musts of Airen variety on aromatic composition and sensory properties of the obtained wine. *Food Chem.* **2005**, *89*, 163–174. [CrossRef]
12. Lee, P.R.; Toh, M.; Yu, B.; Curran, P.; Liu, S.Q. Manipulation of volatile compound transformation in durian wine by nitrogen supplementation. *Int. J. Food Sci. Technol.* **2013**, *48*, 650–662. [CrossRef]
13. Trinh, T.T.T.; Woon, W.Y.; Yu, B.; Curran, P.; Liu, S.Q. Effect of L-isoleucine and L-phenylalanine addition on aroma compound formation during longan juice fermentation by a co-culture of Saccharomyces cerevisiae and Williopsis saturnus. *S. Afr. J. Enol. Vitic.* **2010**, *31*, 116–124. [CrossRef]
14. Barth, O.M.; Almeida-Muradian, L.B.; Pamplona, L.C.; Coimbra, S.; Barth, O.M. Chemical composition and botanical evaluation of dried bee pollen pellets. *J. Food Compos. Anal.* **2005**, *18*, 105–111. [CrossRef]

15. De-Melo, A.A.M.; Estevinho, L.M.; Moreira, M.M.; Delerue-Matos, C.; Freitas, A.D.S.D.; Barth, O.M.; de Almeida-Muradian, L.B.D.; De-Melo, A.A.M.; Estevinho, L.M.; Moreira, M.M.; et al. Phenolic profile by HPLC-MS, biological potential, and nutritional value of a promising food: Monofloral bee pollen. *J. Food Biochem.* **2018**, *42*, e12536. [CrossRef]
16. Human, H.; Nicolson, S.W. Nutritional content of fresh, bee-collected and stored pollen of Aloe greatheadii var. davyana (Asphodelaceae). *Phytochemistry* **2006**, *67*, 1486–1492. [CrossRef] [PubMed]
17. Paramás, A.M.G.G.; Bárez, J.A.G.; Marcos, C.C.; García-Villanova, R.J.; Sánchez, J.S.; Gonzalez-paramas, A.M.; Garcia-villanova, R.J.; Gonza, A.M.; Paramás, A.M.G.G.; Bárez, J.A.G.; et al. HPLC-fluorimetric method for analysis of amino acids in products of the hive (honey and bee-pollen). *Food Chem.* **2006**, *95*, 148–156. [CrossRef]
18. Dong, J.; Yang, Y.; Wang, X.; Zhang, H. Fatty acid profiles of 20 species of monofloral bee pollen from China. *J. Apic. Res.* **2017**, *54*, 503–511. [CrossRef]
19. Xu, X.; Sun, L.; Dong, J.; Zhang, H. Breaking the cells of rape bee pollen and consecutive extraction of functional oil with supercritical carbon dioxide. *Innov. Food Sci. Emerg. Technol.* **2009**, *10*, 42–46. [CrossRef]
20. Amores-Arrocha, A.; Sancho-Galán, P.; Jiménez-Cantizano, A.; Palacios, V. Bee Pollen as Oenological Tool to Carry out Red Winemaking in Warm Climate Conditions. *Agronomy* **2020**, *10*, 634. [CrossRef]
21. Amores-Arrocha, A.; Roldán, A.; Jiménez-Cantizano, A.; Caro, I.; Palacios, V. Effect on White Grape Must of Multiflora Bee Pollen Addition during the Alcoholic Fermentation Process. *Molecules* **2018**, *23*, 1321. [CrossRef]
22. Amores-Arrocha, A.; Roldán, A.; Jiménez-Cantizano, A.; Caro, I.; Palacios, V. Evaluation of the use of multiflora bee pollen on the volatile compounds and sensorial profile of Palomino fino and Riesling white young wines. *Food Res. Int.* **2018**, *105*, 197–209. [CrossRef]
23. Sancho-Galán, P.; Amores-Arrocha, A.; Jiménez-Cantizano, A.; Palacios, V. Use of multiflora bee pollen as a flor velum yeast growth activator in biological aging wines. *Molecules* **2019**, *24*, 1763. [CrossRef] [PubMed]
24. Tronchoni, J.; Curiel, J.A.; Sáenz-Navajas, M.P.; Morales, P.; De-la-Fuente-Blanco, A.; Fernández-Zurbano, P.; Ferreira, V.; Gonzalez, R. Aroma profiling of an aerated fermentation of natural grape must with selected yeast strains at pilot scale. *Food Microbiol.* **2018**, *70*, 214–223. [CrossRef] [PubMed]
25. Peinado, R.A.; Moreno, J.A.J.; Bueno, J.E.; Moreno, J.A.J.; Mauricio, J.C. Comparative study of aromatic compounds in two young white wines subjected to pre-fermentative cryomaceration. *Food Chem.* **2004**, *84*, 585–590. [CrossRef]
26. ISO NORM. *3591(1977) Sensory Analysis: Apparatus Wine Tasting Glass*; ISO: Geneva, Switzerland, 1977.
27. Jackson, R.S. *Wine Tasting: A Professional Handbook*; Academic Press: London, UK, 2009; p. 350.
28. Ribéreau-Gayon, P.; Dubourdieu, D.; Donéche, B.; Lonvaud, A. *Tratado de Enologia: Microbiología del Vino. Vinificaciones*; Mundi-Prensa: Buenos Aires, Argentina, 2003.
29. Togores, J.H. *Tratado de Enología I*; Mundi Prensa: Madrid, Spain, 2011.
30. Lonvaud-Funel, A.; Joyeux, A.; Desens, C. Inhibition of malolactic fermentation of wines by products of yeast metabolism. *J. Sci. Food Agric.* **1988**, *44*, 183–191. [CrossRef]
31. Zeeman, W.; Snyman, J.P.; van Wyk, C.J. The influence of yeast strain and malolactic fermentation on some volatile bouquet substances and on quality of table wines. In *Grape and Wine Centennial Symposium*; University of California: Davis, CA, USA, 1982.
32. Laurent, M.-H.; Henick-Kling, T.; Acree, T.E. Changes in he aroma and odor of Chardonnay wine due to malolactic fermentation. *Vitic. Enol. Sci.* **1994**, *49*, 3–10.
33. De Revel, G.; Martin, N.; Pripis-Nicolau, L.; Lonvaud-Funel, A.; Bertrand, A. Contribution to the knowledge of malolactic fermentation influence on wine aroma. *J. Agric. Food Chem.* **1999**, *47*, 4003–4008. [CrossRef] [PubMed]
34. Delaquis, P.; Cliff, M.; King, M.; Girard, B.; Hall, J.; Reynolds, A. Effect of two commercial malolactic cultures on the chemical and sensory properties of Chancellor wines vinified with different yeasts and fermentation temperatures. *Am. J. Enol. Vitic.* **2000**, *51*, 42–48.
35. Gambaro, A.; Boido, E.; Zlotejablko, A.; Medina, K.; Lloret, A.; Dellacassa, E.; Carrau, F. Effect of malolactic fermentation on the aroma properties of Tannat wine. *Aust. J. Grape Wine Res.* **2001**, *7*, 27–32. [CrossRef]
36. Swiegers, J.H.; Bartowsky, E.J.; Henschke, P.A.; Pretorius, I. Yeast and bacterial modulation of wine aroma and flavour. *Aust. J. Grape Wine Res.* **2005**, *11*, 139–173. [CrossRef]

37. Ferreira, V.; Lopez, R. The actual and potential aroma of winemaking grapes. *Biomolecules* **2019**, *9*, 818. [CrossRef]
38. Crouzet, J.; Flanzy, C. Les Enzymes En Oenologie. In *Œnologie, Fondements Scientifiques et Technologiques*; Tec & doc-Lavoisier: Paris, France, 1998; pp. 362–413.
39. Cheynier, V.; Schneider, R.; Salmon, J.M.; Fulcrand, H. Chemistry of Wine. In *Comprehensive Natural Products II*, 1st ed.; Hung-Wen, L., Mander, L., Eds.; Elsevier: Ámsterdam, The Netherlands, 2010; pp. 1119–1172.
40. Bayonove, C.L.; Baumes, R.L.; Crouzet, J.; Günata, Z. Aromas. In *Enología: Fundamentos científicos y tecnológicos*, 1st ed.; Flanzy, C., Ed.; Mundi-Prensa: Madrid, Spain, 2003; pp. 137–176.
41. Campos, M.G.R.; Bogdanov, S.; de Almeida-Muradian, L.B.; Szczesna, T.; Mancebo, Y.; Frigerio, C.; Ferreira, F. Pollen composition and standardisation of analytical methods. *J. Apic. Res.* **2008**, *47*, 154–161. [CrossRef]
42. da Silva, G.R.; da Natividade, T.B.; Camara, C.A.; da Silva, E.M.S.; dos Santos, F.D.A.R.; Silva, T.M.S. Identification of Sugar, Amino Acids and Minerals from the Pollen of Jandaíra Stingless Bees (Melipona subnitida). *Food Nutr. Sci.* **2014**, *05*, 1015–1021. [CrossRef]
43. Bogdanov, S. Pollen : Production, Nutrition and Health: A Review. 2017. Available online: https://www.bee-hexagon.net/pollen/ (accessed on 4 May 2018).
44. Palacios, V.; Roldán, A.; Jiménez-Cantizano, A.; Amores-Arrocha, A. Physicochemical and microbiological characterization of the sensory deviation responsible for the origin of the special sherry wines "palo cortado" type. *PLoS ONE* **2018**, *13*, e0208330. [CrossRef] [PubMed]
45. Jerkovic, I. Gas Chromatography and Mass Spectrometry in Honey Fingerprinting: The Occurrence of 3,4-dihydro-3-oxoedulan and (E)-4-(r-1′,t-2′,c-4′-trihdroxy-3′,6′,6′-trimethylcyclohexyl)-but-3-en-2-one. *Int. J. Chem. Mol. Eng.* **2018**, *12*, 81526.
46. Ferreira, V.; López, R.; Cacho, J.F. Quantitative determination of the odorants of young red wines from different grape varieties. *J. Sci. Food Agric.* **2000**, *80*, 1659–1667. [CrossRef]
47. Li, H.; Tao, Y.S.; Wang, H.; Zhang, L. Impact odorants of Chardonnay dry white wine from Changli County (China). *Eur. Food Res. Technol.* **2008**, *227*, 287–292. [CrossRef]

foods

MDPI

*Article*

# Influence of Two Different Cryoextraction Procedures on the Quality of Wine Produced from Muscat Grapes

**Ana Ruiz-Rodríguez, Enrique Durán-Guerrero, Ramón Natera †, Miguel Palma * and Carmelo G. Barroso †**

Department of Analytical Chemistry, Center for Agri-Food and Wine Research (IVAGRO), Faculty of Sciences, University of Cadiz, 11510 Puerto Real, Spain; ana.ruiz@uca.es (A.R.-R.); enrique.duranguerrero@uca.es (E.D.-G.); ramon.natera@uca.es (R.N.); carmelo.garcia@uca.es (C.G.B.)

* Correspondence: miguel.palma@uca.es

† Professor Natera and Professor Barroso have passed away.

Received: 28 August 2020; Accepted: 16 October 2020; Published: 24 October 2020

**Abstract:** Freezing grapes is a winemaking technique known as cryoextraction that intends to modify the composition of the final wines. The changes that take place in the frozen grapes facilitate the transfer of certain compounds from the grape skins into the musts because of the grape's unstructured tissues. For this study, the white grape variety Muscat of Alexandria was selected. Two different cryoextraction procedures have been analyzed as follows: (i) Ultrafast freezing, and (ii) liquid nitrogen freezing. The wines obtained using liquid nitrogen freezing exhibited higher levels of terpenoids, as well as higher levels of hydroxylic compounds and fatty acids than both the wines obtained through traditional methods and ultrafast freezing wines. In any case, both freezing techniques produced wines of a more intense aroma compared with those wines obtained by traditional methods. In fact, liquid nitrogen freezing produced the wines with the most intense aroma and were the best valued by the tasting panel.

**Keywords:** grapes; wines; cryoextraction; volatile compounds

## 1. Introduction

Numerous studies have been reported in the literature dealing with the phenomena involved in the extraction of aroma components from grape skins into grape juice during the preferential maceration of the juice and skin from white grapes [1–3]. In particular, and in the case of Muscat of Alexandria, the white grape discussed here, it has been found that the contact between grape juice and skin enhances the aromas of the variety and the resulting wines are notable for their mint and melissa aroma notes [4–6].

As a modification of the classical prefermentative maceration at low temperature, different techniques have been developed that involve freezing the grapes; a method known as cryoextraction [7]. Freezing the grapes does not necessarily affect their organoleptic qualities [8,9]. When the grapes are frozen, ice crystals are formed and these tear the pectocellulose walls, thus disorganizing their tissues and facilitating skin compound extraction processes. In fact, at a small-scale the freezing of grapes did not cause any adverse effects with regard to their vinification and facilitated a more rapid extraction of their aroma compounds [10]. In some cases, two different techniques were applied: Firstly, cryoextraction (freezing the grape prior to its pressing) and, secondly, supraextraction (freezing the grape followed by defrosting and pressing). It is clear that for both techniques the grapes need to be at least partially defrosted, although the difference lies with the time lapse before starting the pressing procedures, and not with the time required to freeze the grapes. Either of these two techniques gives place to fresher, more structured and pleasant wines with a long persistence [11]. Some interesting examples of wine production that involves grape freezing have been found, including the freezing of Syrah grapes by means of solid carbon dioxide that intends to provide wines with a more intense color [12], or the freezing of Nebbiolo grapes to produce more aromatic wines after their proper fermentation [13].

The freezing speed of the grapes is a particularly important aspect to be taken into account, since it is directly related to the degree of disorganization of the berrys' structure [14–16]. Casassa and Sari [17] studied the effect of external refrigeration by means of solid carbon dioxide on Malbec grapes and they discovered that more intensely colored wines were obtained when pellets of solid carbon dioxide were employed. Both solid carbon dioxide (−78.5 °C) and liquid nitrogen (−195.8 °C) induce a thermal shock that leads to a greater degradation of the berries, which in turn prolongs the contact between pulp and must, with the additional benefit of protecting the grapes from the effect of oxygen [18]. In contrast, the use of other external refrigeration methods—such as refrigeration chambers (between −18 °C and −28 °C)—even if less aggressive, does not isolate the macerating must from oxygen. On the other hand, chamber freezing allows a closer control of the process, so that the freezing of tissues can be avoided to reduce the level of bitter tannins transferred into the wine and thus favor the balance of the final product [19].

The effect of the different cryogenic techniques can be more easily evaluated in those grape varieties with specific compounds of interest in their skin. One such example is provided by the variety Muscat of Alexandria, where the level of the varietal characteristic aroma are particularly high in their berrys' skin. In fact, this grape variety has been frequently used to evaluate the effect of different cultivation conditions as well as winemaking techniques on the resulting wines [20]. Some such studies include the analysis of molecular aspects, while others have mainly focused on the effect of grape maturity and the development stages of its skin and pulp [21,22] on the final wine characteristic including the peppermint aroma [23]. In all of these studies, high levels of terpenes, such as linalool, geraniol, nerol and citronerol, were found to provide successful results.

The work described here intends to determine the influence that cryoextraction may have on the elaboration of white wine from Muscat of Alexandria grapes. Two freezing methods have been evaluated and these have been related to their specific freezing rates. The final objective is to determine the effect that each one of these two freezing procedures may have on the wines obtained and particularly with regard to their content in aroma related and phenolic components as well as on their actual sensory qualities.

## 2. Materials and Methods

### 2.1. Winemaking

Muscat of Alexandria grapes grown in the area of Jerez de la Frontera (Cádiz, Spain) were used to produce white wine. Three different vinification procedures were implemented (Figure 1).

Approximately 900 kg (100 kg × 3 replicates × 3 type of wines) of grapes from this variety were used to produce the wines. Three different tanks were used for each wine type. The first wine was the "reference" (R) sample, whose grapes had been crushed at room temperature (20 °C) and then pressed. The second wine was denoted as "ultra fast mechanical freezing" (UF). The grapes that were used to produce this second wine had been deep-frozen in a highly powerful freezing chamber that would bring the grapes down to −28 °C in 15 min. The third wine was labeled as "liquid nitrogen freezing" (LN) and the grapes used to produce this third wine type were frozen for less than 1 min by means of liquid nitrogen. The grapes frozen by either technique were stored in the same chamber at −18 °C for 2 h. The bunches were then allowed to thaw at room temperature for 3 h. The semi-thawed grapes were then crushed and pressed.

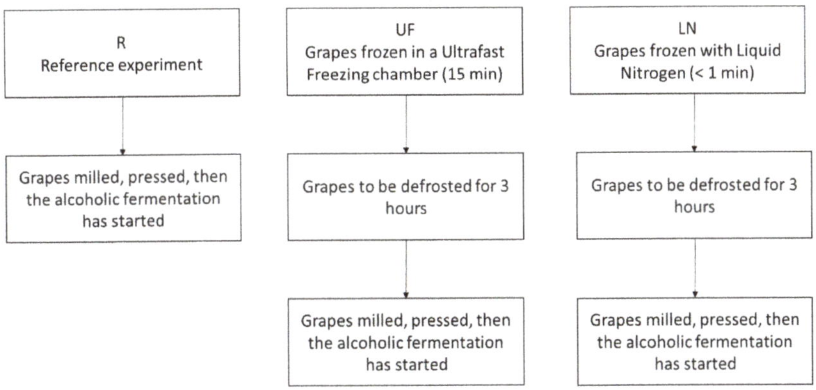


**Figure 1.** Winemaking processes for the three types of wine: R (reference), UF (ultra fast mechanical freezing) and LN (liquid nitrogen freezing).

The must for each winemaking process was corrected to a pH of 3.3 using tartaric acid and 40 mg $L^{-1}$ sulfur dioxide was also added as potassium metabisulfite. The yeast *Saccharomycescerevisiae* var. *bayanus*, commercially known as "viniferm PDM" commercialized by (Agrovin, Ciudad Real, Spain) was used for the fermentation process in 35 liter containers. The fermentation started 24 h after the inoculation of the yeast and carried on for 11 days, never exceeding 20 °C, until the end of the process when the reducing sugars reached below 5 g $L^{-1}$. The wine from each winemaking process was cold stabilized in a refrigerator at 4 °C. Sulfur dioxide was added to 80 mg $L^{-1}$ of total sulfur dioxide and it was filtered through 0.4–0.6 micron filtering plates.

### 2.2. Characterization of Musts and Wines

Both grape juices and wines were characterized by determining three routine parameters (sugar content, pH and total acidity) as well as total polyphenols. All the determinations were carried out in triplicate. The density measurements were performed by means of an Anton Paar DMA 4500M electronic densimeter (Graz, Austria) in order to determine the sugar levels in the musts based on their direct correlation with density. It was also used to determine ethanol level in 200 mL wine distillate samples. The acidity was calculated by acid-base titration using a Crison-Hach model pHmatic 23 (Dusseldorf, Germany) automatic titrator by adding NaOH 0.1 M until pH = 7.0. It was expressed as g $L^{-1}$ of tartaric acid.

The polyphenol content of the wines was determined according to the protocol by Mazza et al. after a minor modification [24]. Briefly, in order to measure the total polyphenols in the samples, their absorbance was registered at 280 nm using a Jasco V-530 UV/VIS spectrophotometer. Seven standard solutions (80–1000 mg $L^{-1}$ gallic acid) were used to express the result as gallic acid equivalents. For most wines and must, the results from this method correlated with the results from the Folin-Ciocalteau method [25].

### 2.3. Determination of Aroma Compounds

The stir bar sorptive extraction (SBSE) technique was used in conjunction with GC-MS to determine volatile components content of each wine. The extractions were carried out using 10 mm × 0.5 mm (length × film thickness) PDMS commercial stir bars, supplied by Gerstel (Mulheim a/d Ruhr, Germany). Each sample was run in duplicate and the average values of the three types of wine corresponding to each winemaking processes were used for later discussions. A volume of 25 mL of each wine type was pipetted and placed into a 100-mL Erlenmeyer flask containing 5.85 g of NaCl and 50 L of a solution formed by 4-methyl-2-pentanol (2.27 g $L^{-1}$ in Milli-Q water containing 5 g $L^{-1}$ of tartaric acid). The Erlenmeyer flask was placed on a 15-samples magnetic stirrer. The stir bar was stirred at 1250 rpm at 25 °C for 120 min. Then, the stir bar was removed from the wine sample and submerged for a few seconds into distilled water in order to remove any NaCl and subsequently gently wiped dry using a lint-free tissue. It was then transferred into a glass thermal desorption tube to be thermally desorbed by means of a commercial TDU thermal desorption unit (Gerstel) connected to a programmed-temperature vaporisation (PTV) injector CIS-4(Gerstel) through a heated transfer line. The PTV was installed on an Agilent 6890 GC-5973 MS system (Agilent Technologies, Palo Alto, CA, USA). An empty baffled liner was used with the PTV. The thermodesorption unit was equipped with a MultiPurpose Sampler (MPS) 2 L autosampler (Gerstel) with capacity for 98 coated stir bars. The desorption temperature was programmed from 40 to 300 °C (held for 10 min) at 60 °C $min^{-1}$ under a helium flow (75 mL $min^{-1}$) and the desorbed analytes were cryofocused using the PTV system with liquid nitrogen at −140 °C. Finally, the PTV system was programmed from −140 to 300 °C (held for 5 min) at 10 °C $s^{-1}$ for analysis by GC–MS. The capillary GC–MS analyses in the electron impact mode were performed on an Agilent 6890 GC-5973N MS system (Agilent, Little Falls, DE, USA), equipped with a DB-Wax capillary column (J&W Scientific, Folsom, CA, USA), 60 m × 0.25 mm I.D., with a 0.25 micron coating. Helium was used as the carrier gas at a flow rate of 1.1 mL $min^{-1}$. The detector temperature was 250 °C. The GC oven was programmed as follows: Held at 35 °C for 10 min, then ramped at 5 °C $min^{-1}$ up to 100 °C. Then it was raised to 210 °C at 3 °C $min^{-1}$ and held for 40 min. The peak identifications were carried out based on the Wiley library by analogy with mass spectra and this was confirmed by pattern retention indices, whenever possible, or according to the retention data reported in the literature. The quantitative data for the identified compounds were obtained by measuring the relative area of the molecular ion peak of each compound relative to that of 4-methyl-2-pentanol, the internal standard [26]. The relative standard deviation values using this method ranged from 3.27% (nerol) to 8.73% (4-vinylguaiacol) for terpenoids, from 2.19% (isoamyl alcohol) to 9.03% (1-hexanol) for hydroxylic compounds, from 2.54% (benzaldehyde) to 7.38 (nonanal) for aldehydes, from 1.36% (octanoic acid) to 11.06% (tetradecanoid acid) for fatty acids and from 3.64% (isoamyl acetate) to 7.40% (ethyl hexadecanoate) for ethyl esters.

### 2.4. Tasting Methodology

The sessions were carried out in a normalized tasting room (UNE-EN ISO 8589:2010) so that any influences from external stimuli on the judgments would be minimized. A panel formed by twelve members who were considered as experts took part in the tasting sessions. The panel members were either winemakers or equally experienced laboratory staff members selected for their consistent assessments.

The wine samples from the 3 replicates of each specific winemaking process were blended together before starting the sensory analysis. The sessions consisted of presenting the three types of wines to the judges for comparative purposes. Exactly 50 μL of wine was poured into a tasting glass (UNE-EN ISO 3591:1977) fitted with a lid to minimize aroma losses. Each glass was identified by a 3-digit code. The room temperature was set at 22 °C and the samples were presented randomly.

The sensory analysis of the wines was carried out in two sessions. In the first session, the judges were requested to perform an ordering test (UNE-EN ISO 8587:2010) to determine their preferences. This type of test is classified as discriminative and it is especially applicable when there are several samples to be compared, since it minimizes the consumption of the samples and the sensory fatigue of the judges. After presenting the wine samples before the judges, they were asked to evaluate and order them according to their preferences (from lowest to highest preference). The wine ID codes were noted down on the corresponding forms. Each judge was encouraged to include additional comments at his/her discretion, since such comments could contribute to a deeper interpretation of the quantitative data.

In the second session, the judges were asked to grant a score on the impact notes for each wine sample according to a 1 to 5 scale. The specific impact notes were selected from the conclusions of the first session.

### 2.5. Statistical Analysis

Wines were prepared in triplicate, i.e., three different fermentation tanks were used for each grape juice and analyses were done in triplicate. An analysis of variance (ANOVA) was performed on the continuous variables, i.e., sugar, total acidity, pH, total polyphenols and volatile compounds and the Xi-square test was applied to the discrete variable, i.e., the sensory panel results. The differences were considered as relevant when the *p*-value was less than 0.05. The statistical analyses were carried out by means of an SPSS V27 (IBM, Armonk, NY, USA).

## 3. Results and Discussion

### 3.1. Characteristics of the Resulting Musts

The grapes were frozen, thawed and then pressed. The resulting musts were analyzed in order to ascertain whether there were differences in their properties that could be attributed to the different treatments applied to the grapes. The results for total sugars, pH and total acidity in the musts are shown in Figure 2.

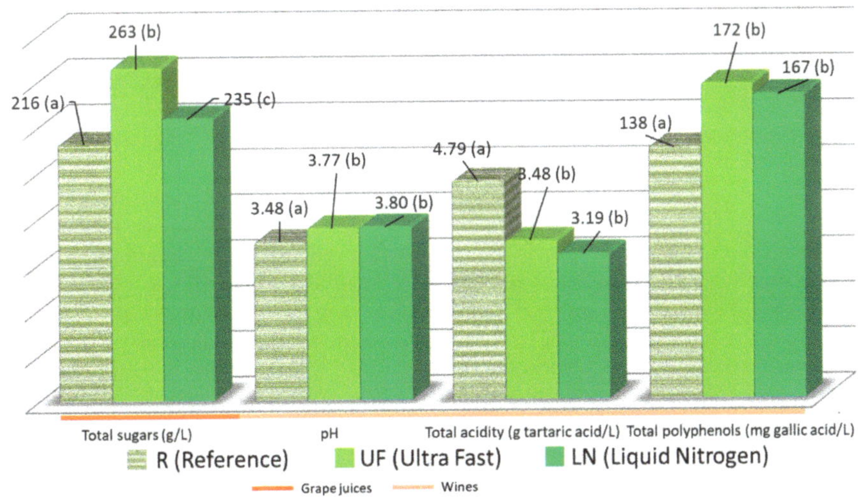


**Figure 2.** Average levels or the routine parameters in the different grape juices and wines (R, UF and LN). Different letters in brackets indicate a significant difference (*p*-value < 0.05) for each parameter.

Regarding the total acidity of the musts, it should be mentioned that the reference must, which was obtained without submitting its grape to any specific treatment, had the highest value (4.79 ± 0.43 vs. 3.48 ± 0.17 and 3.19 ± 0.21 g of tartaric acid $L^{-1}$). This result is a consequence of the precipitation of potassium salts during the freezing process and the lower solubility of the acidic salts when the grapes were defrosted—both effects had already been reported in the literature [27]. This process is also responsible for the different pH levels between the three musts. The highest pH values were related to the musts obtained after freezing/defrosting the grapes. No significant differences were found between the wines obtained using frozen grapes.

Regarding their sugar levels, the reference must (R) showed the lowest sugar content (216 ± 5) when compared to the musts obtained after freezing the grapes (263 ± 4 (UF) and 235 ± 4 (LN)) (Figure 2). One of the peculiarities of cryoextraction is that on pressing partially defrosted grapes, part of their juice defrosts faster because of a cryoscopic decrease of the melting point that takes place in the berry areas where the concentration of sugar is higher. Significant differences were found among the three grape musts.

Different grape freezing techniques have different effects on the final wines. When the grapes are frozen by means of liquid nitrogen (LN) the wines that are obtained have a greater total sugar content than those obtained when the grapes are frozen by ultra fast mechanical freezing (UF). The difference in sugar content might be a result of a faster defrosting process in the case of ultra fast mechanical freezing, while when grapes have been frozen by means of liquid nitrogen the effect of cryoscopic decrease would be more evident and, consequently, the musts obtained should exhibit higher sugar contents. In an effort to avoid the enzymatic oxidation of the grapes during the defrosting process, the grinding process was started just three hours after the defrosting of the grapes had begun. At that point, the thawing of the grapes was more advanced in those grapes that had undergone ultra fast mechanical freezing when compared to those which had been frozen by Liquid Nitrogen.

The alcoholic level of the R wine was 11.2 ± 0.21, 15.2 ± 0.36 for the LN wine, and 13.4 ± 0.41 for the UF wine. These values reflect the differences in must sugar content initial values and underline the first major effect caused by the two freezing techniques used.

### 3.2. Total Polyphenols in the Wines

Cryoextraction should produce wine with higher levels of polyphenols as they are easily extracted from the grape skins [12]. However, longer treatments before the alcoholic fermentation can promote the oxidation of polyphenols [28]. In order to identify the predominant effect, i.e., either greater extraction of polyphenols or greater oxidation of the total phenolic components, both effects were determined or the resulting values are presented in Figure 2.

It was confirmed that the wines from frozen grapes presented higher phenolic contents with no significant differences between them (138 ± 11 for the R wine, 172 ± 8 for the LN wine and 167 ± 9 for the UF wine expressed as mg $L^{-1}$ of gallic acid). No significant differences were found between the two wines obtained from frozen grapes (LN and UF). For this reason, it can be assumed that during the short freezing process, the freezing procedure does not matter, the grapes are degraded to a sufficient extent so that more polyphenols are extracted from their skin during the grinding and pressing process than in the case of berries that had not been subjected to freezing at all, even if both grinding and pressing are processes that are carried out at lower temperatures. This effect has been previously described in the literature although for longer freezing times [29,30]. If this is the case, a similar trend should be found with regard to those other compounds that originate in grape skin and that give the wine its particular aroma. Other aspects, such as solubility level may also be a relevant factor with regard to the prevalence of polyphenol extraction or phenolic oxidation.

### 3.3. Individual Aroma Components in the Wines

The aromas of Muscat of Alexandria wines were studied by analyzing the concentrations of terpenes, alcohols, aldehydes, fatty acids and ethyl esters and significant differences ($p$-value < 0.05) were found between the different winemaking techniques used in this study. Most of these compounds are present in grape skins, therefore, any differences between the wines from frozen grapes should be attributed to differences in the degradation level of the grapes skin over the freezing process. It has been previously described that freezing processes may affect some volatile compounds such as the thiols in Sauvignon blanc grapes [31], however, in those cases the effects were due to longer freezing periods that resulted in certain biochemical changes. In our study, a very short freezing period was applied in every case and, therefore, only mechanical changes in the grapes skin would explain the differences in the final composition of the wines.

Given the large number of the parameters analyzed related to wine aroma, out of convenience, only their average relative values were represented to be compared against the reference wine (R), which was given a value of 100%. The results are presented in Figures 3–7.

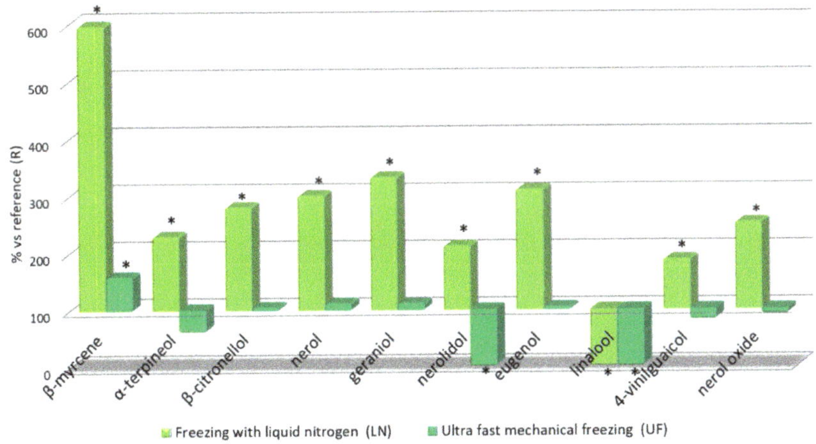


**Figure 3.** Relative levels of volatile terpenoids in the three wine types: Relative values for LN and UF vs R. * Significant difference vs the R wines ($p$-value < 0.05).

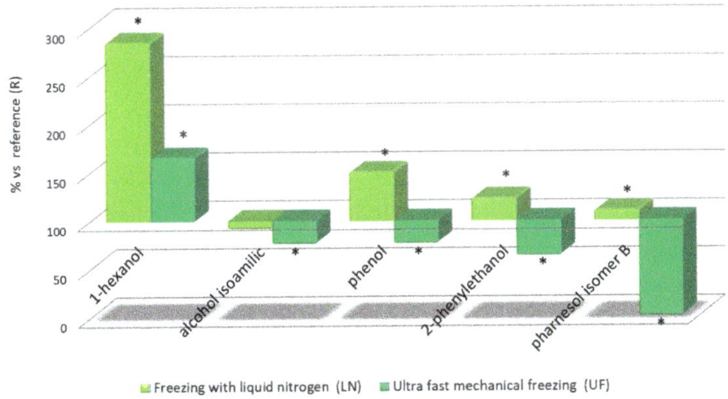


**Figure 4.** Relative levels of volatile alcohols in the three wine types: Relative values for LN and UF vs R. *Significant difference vs the R wines ($p$-value < 0.05).

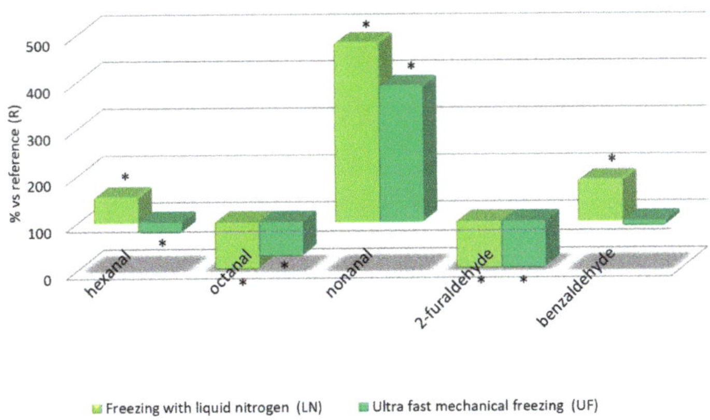


**Figure 5.** Relative levels of volatile aldehydes in the three wine types: Relative values for LN and UF vs. R. * Significant difference vs the R wines ($p$-value < 0.05).

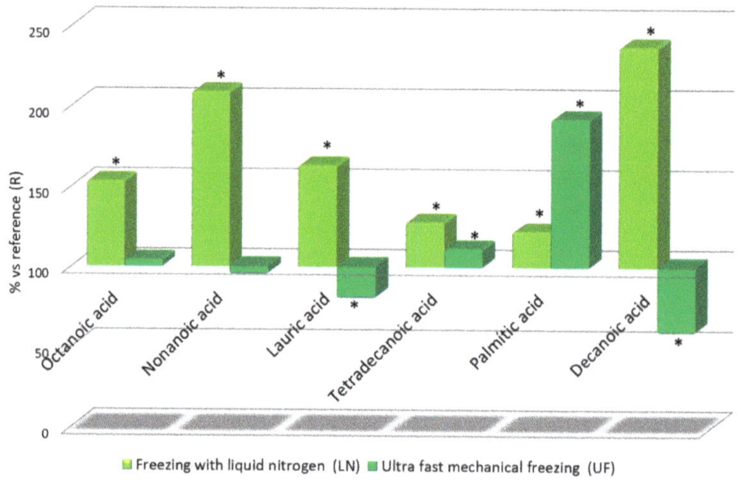


**Figure 6.** Relative levels of volatile fatty acids in the three wine types: Relative values for LN and UF vs. R. * Significant difference vs the R wines ($p$-value < 0.05).

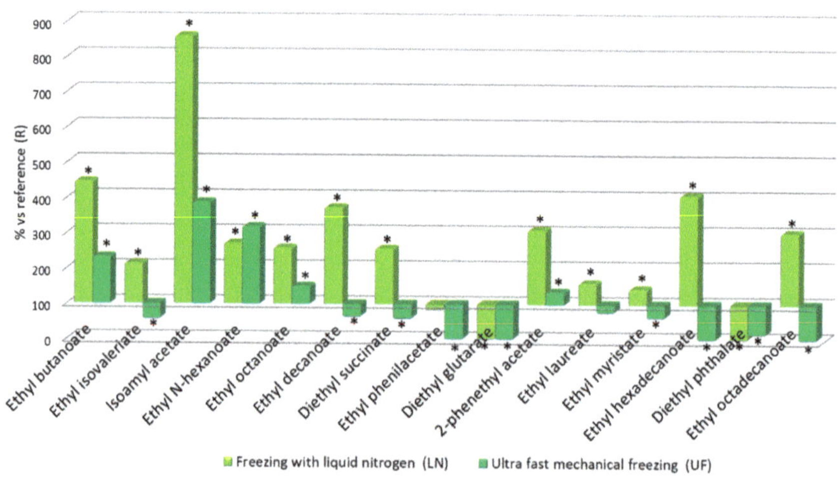


**Figure 7.** Relative levels of volatile ethyl esters in the three wine types: Relative values for LN and UF vs. R. * Significant difference vs the R wines ($p$-value < 0.05).

It can be seen in Figure 3 that several of the characteristic components of the varietal aroma that were analyzed exhibited higher values in the two wines from frozen grapes than those found in the reference wine (R) and, in many cases, such values were as much as three fold the ones in non-frozen grape wines. The effects of freezing techniques on terpenoids have been previously found also for other grape varieties [32]. These results are consistent with the previously mentioned skin degradation that occurs during freezing, which facilitates the transfer of skin origin components into the must. The above outlined results also seem to indicate that the net degradation of these components does not occur during the freezing and handling of the grapes in the manner previously described. The LN wine had a significantly higher concentration of terpenes than the reference wine, while the UF wine only exhibited slightly and hardly relevant higher values than the reference wine. The above observation is specified in percentage terms as follows: All the terpenes in the LN wine reached a value greater than 150% compared to wine R, while in the UF wine only β-myrcene exceeded that value (158%) with respect to wine R, only linalool showed a lower value in both LN and UF wines. These compounds are responsible for the aroma balsamic notes of wines [33], therefore freezing processes will increase the balsamic notes in Muscat wines.

β-myrcene content in the LN wine was almost 600% that in the reference wine (R), whereas nerol (301%), geraniol (332%) and eugenol (310%) exhibited three fold levels compared to the R wine. It is also worth highlighting other components that differed from those in the reference wine by a factor of two or more. For example, in the LN wine these components include terpineol (229%), although linalool was present at a lower level than in the R wine. Terpineol is related to mint aroma [34] and nerol oxide (238%) is related to flower aroma [33]. In the UF wine, only β-myrcene was present at higher levels than in the reference wine. Therefore, if we take into account the contribution of these compounds to the specific aroma of Muscat wines, grape freezing techniques seem to have a positive effect and such effect is clearly noticeable in the LN wines. However, it must be noted that no new compounds were detected on LN nor UF wines vs the R wines.

The data obtained from the determination of alcohol contents confirm that the LN wines always presented higher concentrations than the UF wines followed by the R wine. It is worth mentioning that, as can be seen in Figure 4, the difference in 1-hexanol content reached a significant 250% in the LN wines when compared to the reference wine. The UF wines also presented a marked difference with respect to the reference wine with a 50% increment. Similar results have been previously obtained in other investigations on the same and other grape varieties [35]. In this case, freezing would negatively affect the final wines in comparison to winemaking without freezing. 1-Hexanol provides herbaceous notes to wine aroma because of the enzymatic oxidation of the fatty acids in the grapes. This process is related to the breakage of the grain and the processes to which it is subjected from its harvesting until the beginning of the alcoholic fermentation. The LN wines had also higher concentrations of the other alcohols than both the reference and the UF wines. These differences were in most cases more evident when compared to the R wine, in any case, much lower than the differences found for 1-hexanol. Additionally, those alcohols are less important than 1-hexanol for the wines aroma properties. (Figure 4).

Regarding the analysis of aldehydes, their behavior is more heterogeneous with respect to the different techniques used in the winemaking process. Nonanal is the component with the most noticeable concentration difference with respect to the R wine: The LN wine reached a value of 482% and the UF wine went as high as 390%. Nonanal has been associated to citrus fruits aromas [36]. The LN wines also contained hexanal and benzaldehyde at 50% higher levels than the reference wine. These compounds are responsible for burnt sugar and almond related flavors in wines [33,36]. On the other hand, octanal and 2-furaldehyde had lower concentrations in the frozen grape wines than in the R wine but the differences were not so relevant (Figure 5). Although their final contribution to the aroma of wine is not well understood, aldehydes that have 8–10 carbon atoms are considered to be strong odorants. These compounds include (E)-2-nonenal, octanal, nonanal, decanal and (E,Z)-2,6-nonadienal [34]. Regarding the rest of the aldehydes, they have not been thoroughly understood and may give either pleasant or unpleasant aroma notes [37]. Therefore, because of the higher levels of compounds related to the citrus fruits aroma, the LN and UF wines will have more interesting aroma properties.

The study of fatty acids produced similar results to those corresponding to terpene-like components. The LN wines exhibited higher values for fatty acids than the UF wines. Their content in octanoic acid (153%), nonanoic acid (208%), lauric acid (163%) and decanoic acid (207%) were over that in the R wine by even more than 50%. A similar trend was detected in the UF wines for palmitic acid, with 190% its content in the reference wine, which is even higher than the levels found in the LN wines (Figure 6). It is again noteworthy that there were some cases where the content levels more than double those in the reference wine, such as nonanoic acid and decanoic acid in the LN wines. The UF wines, however, did not show such a marked increment with respect to the R wine.

Fermentation conditions can affect the composition of the fatty acids in a particular wine. Under anaerobic conditions, yeast produces medium-chain fatty acids and when fermentation is carried out in aerobic or semi-aerobic conditions more unsaturated fatty acids are produced [38]. However, the fermentation conditions for the three tests completed in our study were the same. Therefore, the only factor that could affect the composition of the initial fatty acids in the musts would be the treatment applied to the grapes (pressing, maceration or clarification), which in turn would affect the final composition of the wines.

Ethyl esters are formed because of the esterification of fatty acids during the alcoholic fermentation of must. The acetates from higher alcohols and the ethyl esters from fatty acids are associated to floral and fruity aromas in young wines [39]. Therefore, they are usually appreciated in young wines. Linear (C2–C4), medium (C6–C10), long (C6–C10) and branched (2-methyl propanoic, 2-methyl butanoic, etc.) volatile fatty acids are produced during fermentation, and it has been proven that as the length of their chains increase, their volatility decreases and wines' odor changes from acid to rancid [40]. The LN wines presented higher concentrations of these various components compared with the UF wines. The LN wines showed increments greater than 200% with respect to the reference wine for the following components: Ethyl butanoate (446%), ethyl isovalerate (213%), isoamyl acetate (858%), ethyl n-hexanoate (271%), ethyl octanoate (258%), ethyl decanoate (374%), diethyl succinate (257%), 2-phenylethyl acetate (311%), ethyl hexadecanoate (411%) and ethyl octadecanoate (305%), whereas their content levels in the UF wine were: Ethyl butanoate (232%), isoamyl acetate (389%), and ethyl n-caproate (319%) (Figure 7). The larger amounts of these compounds determined in the LN wine could be attributed to a greater presence of non-esterified organic acids.

### 3.4. Tasting Rating

On tasting the wines, all the members in the panel considered the reference wine to be the least aromatic of all the three wine types. The LN and UF wines were given significantly higher scores for fruit and flower notes with respect to the R wine that had be produced by traditional methods (Figure 8). This result is consistent with the analysis of the individual aroma components, such as ethyl butanoate, ethyl n-hexanoate, ethyl octaoate, ethyl decanoate, diethyl succinate and ethyl myristate, which were described as fruit flavors [33] and that had been found predominantly in the UF and LN wines. Likewise, nonanal, which provides citrus notes [36], was a major component in the UF and LN wines. While, 2-phenylethyl acetate and nerol oxide, usually associated to floral notes, was found in large concentrations in the LN wines [33].

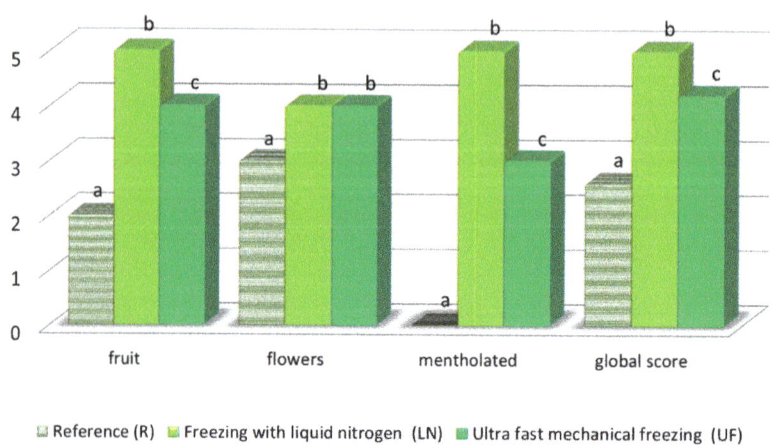


**Figure 8.** Panel tasting results for R, LN and UF wines. Different letters indicate a significant difference ($p$-value < 0.05) for each parameter.

The panelists were able to differentiate the LN from the UF wines by its remarkable menthol notes, which made the former wine unique and attractive. The compounds responsible for these notes seem to be monoterpenes and their derivatives [41,42]. β-Myrcene has been said to confer balsamic notes [33] while terpineol provides mint notes [34]; both of these compounds were determined at higher concentrations in the LN wines compared to the other two wines.

Regarding the overall rating (Figure 8), the LN wines reached the highest score, while the reference wine was the least valued one. Significant global score differences were noticeable between the reference wine against both LN and UF wines. However the differences between the two wines that had been produced from frozen grapes were not so considerable. This overall rating is in agreement with the above mentioned specific aroma and flavor characteristics exhibited by each wine type.

## 4. Conclusions

Freezing procedures reduced acidity, increased alcoholic strength and, above all, produced wines that were more aromatic, since they contain greater amounts of the components that are more closely associated to Muscat grape wine specific characteristics.

The wines produced using liquid nitrogen to freeze the grapes presented higher concentrations of terpenes in all the cases, whereas the wines produced using the ultrafast freezing chamber exhibited similar although just slightly higher values when compared to the reference wine. β-citronerol, nerol and geraniol contents were particularly high in the wines from frozen grapes using liquid nitrogen, with more than double their concentration with respect to the control wine. These results were consistent with the score granted by the tasting panel who judged them as characteristic wines of special interest.

With regard to the difference between wines produced by the two freezing techniques applied in this study, liquid nitrogen frozen grape wines, not only contained more of the compounds that provide Muscat wines with their characteristic fruity and mentholated notes, but they were also more appreciated by the judging panelist than those wines obtained from fast frozen grapes.

**Author Contributions:** Conceptualization, C.G.B. and M.P.; methodology, R.N., E.D.-G., and A.R.-R.; validation, A.R.-R. and E.D.-G.; investigation, A.R.-R. and R.N.; original draft writing, A.R.-R.; review and editing, A.R.-R., M.P. and E.D.-G.; supervision, C.G.B., M.P. and R.N.; funding acquisition, M.P. All authors have read and agreed to the published version of the manuscript.

**Funding:** This research was co-financed by the 2014-2020 ERDF Operational Programme and by the Department of Economy, Knowledge, Business and University of the Regional Government of Andalusia. Project reference: FEDER-UCA18-108366.

**Conflicts of Interest:** The authors declare no conflict of interest.

## References

1. Rice, S.; Tursumbayeva, M.; Clark, M.; Greenlee, D.; Dharmadhikari, M.; Fennell, A.; Koziel, J.A. Effects of harvest time on the aroma of white wines made from cold-hardy brianna and frontenac gris grapes using headspace solid-phase microextraction and gas chromatography-mass spectrometry-olfactometry. *Foods* **2019**, *8*, 29. [CrossRef] [PubMed]
2. Ayestarán, B.; Martínez-Lapuente, L.; Guadalupe, Z.; Canals, C.; Adell, E.; Vilanova, M. Effect of the winemaking process on the volatile composition and aromatic profile of Tempranillo Blanco wines. *Food Chem.* **2019**, *276*, 187–194. [CrossRef] [PubMed]
3. De Matos, A.D.; Longo, E.; Chiotti, D.; Pedri, U.; Eisenstecken, D.; Sanoll, C.; Robatscher, P.; Boselli, E. Pinot blanc: Impact of the winemaking variables on the evolution of the phenolic, volatile and sensory profiles. *Foods* **2020**, *9*, 499. [CrossRef] [PubMed]
4. Lukić, I.; Lotti, C.; Vrhovsek, U. Evolution of free and bound volatile aroma compounds and phenols during fermentation of Muscat blanc grape juice with and without skins. *Food Chem.* **2017**, *232*, 25–35. [CrossRef]
5. Jesus, D.; Campos, F.M.; Ferreira, M.; Couto, J.A. Characterization of the aroma and colour profiles of fortified Muscat wines: Comparison of Muscat Blanc "à petit grains" grape variety with Red Muscat. *Eur. Food Res. Technol.* **2017**, *243*, 1277–1285. [CrossRef]

6. Selli, S.; Canbas, A.; Cabaroglu, T.; Erten, H.; Günata, Z. Aroma components of cv. Muscat of Bornova wines and influence of skin contact treatment. *Food Chem.* **2006**, *94*, 319–326. [CrossRef]
7. Jackson, R. *Wine Science*; Academic Press: Cambridge, MA, USA, 2008; ISBN 9780123736468.
8. Olarte Mantilla, S.M.; Collins, C.; Iland, P.G.; Kidman, C.M.; Jordans, C.; Bastian, S.E.P. Comparison of sensory attributes of fresh and frozen wine grape berries using Berry Sensory Assessment. *Aust. J. Grape Wine Res.* **2013**, *19*, 349–357. [CrossRef]
9. Tomaz, I.; Šeparović, M.; Štambuk, P.; Preiner, D.; Maletić, E.; Karoglan Kontić, J. Effect of freezing and different thawing methods on the content of polyphenolic compounds of red grape skins. *J. Food Process. Preserv.* **2018**, *42*, e13550. [CrossRef]
10. Schmid, F.; Jiranek, V. Use of fresh versus frozen or blast-frozen grapes for small-scale fermentation. *Int. J. Wine Res.* **2011**, *3*, 25–30.
11. Bavčar, D.; Česnik, H.B.; Čuš, F.; Vanzo, A.; Gašperlin, L.; Košmerl, T. Impact of alternative skin contact procedures on the aroma composition of white wine. *South Afr. J. Enol. Vitic.* **2011**, *32*, 190–203. [CrossRef]
12. Gordillo, B.; López-Infante, M.I.; Ramírez-Pérez, P.; González-Miret, M.L.; Heredia, F.J. Influence of prefermentative cold maceration on the color and anthocyanic copigmentation of organic tempranillo wines elaborated in a warm climate. *J. Agric. Food Chem.* **2010**, *58*, 6797–6803. [CrossRef] [PubMed]
13. Petrozziello, M.; Guaita, M.; Motta, S.; Panero, L.; Bosso, A. Analytical and Sensory Characterization of the Aroma of "Langhe D.O.C. Nebbiolo" Wines: Influence of the Prefermentative Cold Maceration with Dry Ice. *J. Food Sci.* **2011**, *76*, 76. [CrossRef] [PubMed]
14. Correia, A.C.; Cosme, F.; Jordão, A.M. *The Impact of Modern Winemaking Technologies on Phenolic Composition of Red Wines*; Nova Science Publisher: New York, NY, USA, 2018; ISBN 9781536132885.
15. Lerno, L.A.; Panprivech, S.; Ponangi, R.; Brenneman, C.A.; Block, D.E.; Oberholster, A. Impact of cold soak duration on Cabernet Sauvignon fermentation and phenolic composition. *J. Sci. Food Agric.* **2019**, *99*, 805–815. [CrossRef] [PubMed]
16. Li, S.-Y.; Duan, C.-Q. Astringency, bitterness and color changes in dry red wines before and during oak barrel aging: An updated phenolic perspective review. *Crit. Rev. Food Sci. Nutr.* **2019**, *59*, 1840–1867. [CrossRef]
17. Casassa, L.F.; Sari, S.E. Sensory and chemical effects of two alternatives of prefermentative cold soak in Malbec wines during winemaking and bottle ageing. *Int. J. Food Sci. Technol.* **2015**, *50*, 1044–1055. [CrossRef]
18. Santesteban, L.G.; Miranda, C.; Royo, J.B. Influence of the freezing method on the changes that occur in grape samples after frozen storage. *J. Sci. Food Agric.* **2013**, *93*, 3010–3015. [CrossRef]
19. Tian, R.R.; Li, G.; Wan, S.B.; Pan, Q.H.; Zhan, J.C.; Li, J.M.; Zhang, Q.H.; Huang, W.D. Comparative study of 11 phenolic acids and five flavan-3-ols in cv. vidal: Impact of natural icewine making versus concentration technology. *Aust. J. Grape Wine Res.* **2009**, *15*, 216–222. [CrossRef]
20. Shin, H.W.; Kim, G.H.; Choi, C. Effects of plant growth regulators and floral cluster thinning on fruit quality of 'shine muscat' grape. *Hortic. Sci. Technol.* **2019**, *37*, 678–686.
21. Crespo, J.; Rigou, P.; Romero, V.; García, M.; Arroyo, T.; Cabellos, J.M. Effect of seasonal climate fluctuations on the evolution of glycoconjugates during the ripening period of grapevine cv. Muscat à petits grains blancs berries. *J. Sci. Food Agric.* **2018**, *98*, 1803–1812. [CrossRef]
22. Wu, Y.; Zhang, W.; Song, S.; Xu, W.; Zhang, C.; Ma, C.; Wang, L.; Wang, S. Evolution of volatile compounds during the development of Muscat grape 'Shine Muscat' (Vitis labrusca × V. vinifera). *Food Chem.* **2020**, *309*, 125778. [CrossRef]
23. Hjelmeland, A.K.; Ebeler, S.E. Glycosidically bound volatile aroma compounds in grapes and wine: A review. *Am. J. Enol. Vitic.* **2015**, *66*, 1–11. [CrossRef]
24. Mazza, G.; Fukumoto, L.; Delaquis, P.; Girard, B.; Ewert, B. Anthocyanins, phenolics, and color of Cabernet Franc, Merlot, and Pinot Noir wines from British Columbia. *J. Agric. Food Chem.* **1999**, *47*, 4009–4017. [CrossRef] [PubMed]
25. Aleixandre-Tudo, J.L.; Buica, A.; Nieuwoudt, H.; Aleixandre, J.L.; Du Toit, W. Spectrophotometric Analysis of Phenolic Compounds in Grapes and Wines. *J. Agric. Food Chem.* **2017**, *65*, 4009–4026. [CrossRef]
26. Guerrero, E.D.; Marín, R.N.; Mejías, R.C.; Barroso, C.G. Stir bar sorptive extraction of volatile compounds in vinegar: Validation study and comparison with solid phase microextraction. *J. Chromatogr. A* **2007**, *1167*, 18–26. [CrossRef] [PubMed]
27. Celotti, E.; Bornia, L.; Zoccolan, E. Evaluation of the electrical properties of some products used in the tartaric stabilization of wines. *Am. J. Enol. Vitic.* **1999**, *50*, 343–350.

28. Threlfall, R.; Main, G.; Morris, J. Effect of freezing grape berries and heating must samples on extraction of components and composition parameters of red wine grape varieties. *Aust. J. Grape Wine Res.* **2006**, *12*, 161–169. [CrossRef]
29. Sacchi, K.L.; Bisson, L.F.; Adams, D.O. A review of the effect of winemaking techniques on phenolic extraction in red wines. *Am. J. Enol. Vitic.* **2005**, *56*, 197–206.
30. García, S.; Santesteban, L.G.; Miranda, C.; Royo, J.B. Variety and storage time affect the compositional changes that occur in grape samples after frozen storage. *Aust. J. Grape Wine Res.* **2011**, *17*, 162–168. [CrossRef]
31. Chen, L.; Capone, D.L.; Nicholson, E.L.; Jeffery, D.W. Investigation of intraregional variation, grape amino acids, and pre-fermentation freezing on varietal thiols and their precursors for Vitis vinifera Sauvignon blanc. *Food Chem.* **2019**, *295*, 637–645. [CrossRef]
32. Ouellet, É.; Pedneault, K. Impact of frozen storage on the free volatile compound profile of grape berries. *Am. J. Enol. Vitic.* **2016**, *67*, 239–244. [CrossRef]
33. Zhang, S.; Petersen, M.A.; Liu, J.; Toldam-Andersen, T.B.; Ebeler, S.E.; Hopfer, H. Influence of pre-fermentation treatments on wine volatile and sensory profile of the new disease tolerant cultivar solaris. *Molecules* **2015**, *20*, 21609–21625. [CrossRef] [PubMed]
34. Eslahi, H.; Fahimi, N.; Sardarian, A.R. *Chemical Composition of Essential Oils*; Wiley Online Library: Hoboken, NJ, USA, 2017; ISBN 9781119149392.
35. Sánchez Palomo, E.; Pérez-Coello, M.S.; Díaz-Maroto, M.C.; González Viñas, M.A.; Cabezudo, M.D. Contribution of free and glycosidically-bound volatile compounds to the aroma of muscat "a petit grains" wines and effect of skin contact. *Food Chem.* **2006**, *95*, 279–289. [CrossRef]
36. Culleré, L.; Escudero, A.; Cacho, J.; Ferreira, V. Gas Chromatography-Olfactometry and Chemical Quantitative Study of the Aroma of Six Premium Quality Spanish Aged Red Wines. *J. Agric. Food Chem.* **2004**, *52*, 1653–1660. [CrossRef]
37. Ferreira, V.; Culleré, L.; López, R.; Cacho, J. Determination of important odor-active aldehydes of wine through gas chromatography-mass spectrometry of their O-(2,3,4,5,6-pentafluorobenzyl)oximes formed directly in the solid phase extraction cartridge used for selective isolation. *J. Chromatogr. A* **2004**, *1028*, 339–345. [CrossRef]
38. Restrepo, S.; Espinoza, L.; Ceballos, A.; Urtubia, A. Production of fatty acids during alcoholic wine fermentation under selected temperature and aeration conditions. *Am. J. Enol. Vitic.* **2019**, *70*, 169–176. [CrossRef]
39. Xiang, X.F.; Lan, Y.B.; Gao, X.T.; Xie, H.; An, Z.Y.; Lv, Z.H.; Lv, Z.H.; Duan, C.Q.; Wu, G.F. Characterization of odor-active compounds in the head, heart, and tail fractions of freshly distilled spirit from Spine grape (Vitis davidii Foex) wine by gas chromatography-olfactometry and gas chromatography-mass spectrometry. *Food Res. Int.* **2020**, *137*, 109388. [CrossRef]
40. Francis, I.L.; Newton, J.L. Determining wine aroma from compositional data. *Aust. J. Grape Wine Res.* **2005**, *11*, 114–126. [CrossRef]
41. Robinson, A.L.; Boss, P.K.; Solomon, P.S.; Trengove, R.D.; Heymann, H.; Ebeler, S.E. Origins of grape and wine aroma. Part 2. Chemical and sensory analysis. *Am. J. Enol. Vitic.* **2014**, *65*, 25–42. [CrossRef]
42. Zhang, P.; Fuentes, S.; Siebert, T.; Krstic, M.; Herderich, M.; Barlow, E.W.R.; Howell, K. Comparison data of common and abundant terpenes at different grape development stages in Shiraz wine grapes. *Data Brief* **2016**, *8*, 1127–1136. [CrossRef]

*foods* MDPI

*Article*

# Characterization of the Aromatic and Phenolic Profile of Five Different Wood Chips Used for Ageing Spirits and Wines

**María Guerrero-Chanivet [1,2], Manuel J. Valcárcel-Muñoz [2], M. Valme García-Moreno [1,*] and Dominico A. Guillén-Sánchez [1]**

1 Departamento de Química Analítica, Facultad de Ciencias, Instituto Investigación Vitivinícola y Agroalimentaria (IVAGRO), Campus Universitario de Puerto Real, Universidad de Cádiz, 11510 Cádiz, Spain; maria.guerreroch@uca.es (M.G.-C.); dominico.guillen@uca.es (D.A.G.-S.)

2 Bodegas Fundador S.L.U., Departamento de Investigación y Desarrollo, C/San Ildefonso, n° 3, 11403 Cádiz, Spain; m.valcarcel@bodegasfundador.com

* Correspondence: valme.garcia@uca.es; Tel.: +34-956016456

Received: 8 October 2020; Accepted: 4 November 2020; Published: 6 November 2020

**Abstract:** Wooden barrels and wood chips are usually used in the ageing of spirits and wines to improve their sensorial profile. Oak wood is the most popular material used in cooperage, but there are other interesting woods, such as cherry or chestnut, that could be considered for this purpose. In this study, a novel method for the determination of the aromatic profile of wood powder by Direct Thermal Desorption-Gas Chromatography-Mass Spectrometry (DTD-GC-MS) was optimized by experimental design. The volatile composition of five different types of wood chips was determined by direct analysis of wood powder by DTD-GC-MS method developed. Thirty-one compounds from wood were identified through this analysis, allowing the differentiation between woods. The aromatic and phenolic compound profile of the 50% hydroalcoholic extract of each type of wood studied was analyzed by Stir-bar Sorptive Extraction-Gas Chromatography-Mass Spectrometry (SBSE-GC-MS) and Ultra-High-Performance Liquid Chromatography (UHPLC) to determine which wood compounds are transferred to spirits and wine after ageing. Different phenolic profiles were found by UHPLC in each wood extract, allowing their differentiation. However, results obtained by SBSE-GC-MS did not allow distinguishing between wood extracts. The analysis of wood in solid state, without any type of previous treatment except grinding, by DTD-GC-MS does not imply any loss of information of the aromatic compounds present in wood as other techniques. This is a potential method to identify aromas in wood that, in addition, allows different types of wood to be differentiated.

**Keywords:** oak; cherry; chestnut; wood chips; phenolic compounds; aroma; ageing

## 1. Introduction

Ageing spirits and wines in wooden barrels or the use of wood chips are industrial common practices that change the sensorial profile of the product. The structural characteristics and chemical composition of the wood are responsible for many of the processes that take place during the maturation period, affecting the composition of the spirits and wines, modulating their sensorial quality and complexity, such as aroma, structure or astringency and contributing to their stability. Wood characteristics, such as the geographical origin and botanical species [1–6], volume of the barrel [7] or chip size [8,9] and toasting level [3,10,11], affected the sensorial profile of the final product. There are many spirits such as armagnac, cognac, brandy, whisky, rum, tequila or grappa as well as wines, that are aged in barrels or use wood chips in their ageing processes in order to obtain a special aroma profile.

The barrel or the chips are key elements during the ageing process. They are active contributors to the sensorial properties of the distillates that are in touch with them. There are several physical-chemical phenomena where components of the spirit (made from wine, cane sugar, malt, agave, etc.) or the wine compounds from the wood are involved [10,12,13]. Most of them are extraction processes, but other chemical reactions take place, such as oxidation, esterification, hydrolysis, ethanolysis, Maillard reactions, polymerization, and polycondensation reactions. There are also physical phenomena, as evaporation or the perspiration of water molecules to the outside through the wood, that take place too during ageing process. All of them depend on many variables, as the composition of the wood, the atmospheric conditions or the type of distillate and its alcoholic strength.

Wood is composed, mostly, of holocelluloses (cellulose and hemicellulose) and lignin. They represent around 90% of the total of wood. There are other compounds, as phenolic compounds (polyphenols or simple phenols), fatty acids, alcohols or inorganic substances, that represent 10% of the wood composition [13]. Wood can play a significant role in contributing flavor to alcoholic beverages. Most of these compounds are responsible for the sensorial profile of the final product. Lignin is a polymer that can suffer thermal degradation during the manufacturing of barrels or by ethanolysis and hydrolysis during spirit and wine maturation powered by their acid character [14]. Compounds from hemicellulose as furfural and derivatives [15,16] and compounds from lignin as guaiacyl-type aldehydes (vanillin and coniferylaldehyde), syringyl-type aldehydes (syringaldehyde and sinapaldehyde), and cinnamic and benzoic acids [10,13] are the most significant components extracted from wood during maturation. Other compounds, as hydrolysable tannins, as gallotannins and ellagitannins, are highly soluble in ethanol-water solutions and their transformation into gallic acid or ellagic acid by hydrolysis is very common [13].

The geographical origin and botanical species affect the composition of the wood. Oak is the main material used in cooperage to make barrels as well as wood chips destined to aged spirits and wines, but also chestnut and cherry are used for this purpose. Traditionally, American and French oak wood (*Quercus alba* and *Quercus petraea* and *robur*) are the most employed type of wood in cooperage companies to make barrels and wood chips. Chestnut wood is characterized by a higher porosity than oak, and high quantities of polyphenols may be transferred to the distillate. Cherry wood is characterized by a high porosity too and oxygen permeation, and is usually used for short ageing times [17]. Chestnut wood has proved to be as sustainable for cooperage and has interesting properties for the ageing of brandies [18,19] and could be interesting to age other spirits and wines. As regards cherry wood, it has also been considered as a possible source of wood for the production of wines or spirits [20–23].

References related to the direct analysis of wood in order to characterize its chemical and aromatic profile were not found in the bibliography. It is interesting to know which compounds are present in wood and if they could be transfer to the alcoholic beverage during its ageing. The main goal of this work is to know the aromatic and phenolic profile of five different wood chips (*Quercus alba*, *Quercus petraea*, *Quercus pyrenaica*, *Castanea sativa*, *Prunus avium*) used in ageing processes and find out which compounds could contribute to spirits and wines during its maturation in order to optimize a methodology that could be applied for the analysis of not only the wood chips but also the staves of the wooden barrels used in cooperage. In order to characterize the aromatic profile of each type of wood, the volatile composition was studied by DTD-GC-MS (Direct Thermal Desorption-Gas Chromatography-Mass Spectrometry). Moreover, grounded wood chips were extracted by 50% hydroalcoholic solution in order to determine which compounds could be released into the spirit or the wine. The extracts were characterized by GC-MS (Gas Chromatography-Mass Spectrometry), SBSE-GC-MS (Stir-bar Sorptive Extraction–Gas Chromatography-Mass Spectrometry), UHPLC (Ultra-High-Performance Liquid Chromatography) and TPI (Total Phenolic Index).

## 2. Materials and Methods

### 2.1. Wood Samples

Five different kinds of wood were studied: American oak (*Quercus alba*), French oak (*Quercus petraea*) and Spanish oak (*Quercus pyrenaica*) with a medium toasting level; and Cherry (*Prunus avium*) and Chestnut (*Castanea sativa*) wood without toasting.

Samples were in the form of wood chips (5–15 mm length × 5–10 mm width × 2 mm thickness) and 100 g of each wood were ground to a 0.25 μm grain size powder with an ultra-centrifugal mill ZM 200 (Retsch GmbH, Haan, Germany) before analysis and extraction. Chips were obtained from Roble Enológico, S.L. (Cantabria, Spain). For the optimization of the DTD-GC-MS methodology a mixture of equal parts of the five studied wood chips was used.

### 2.2. Reagents

The rectified wine distillate at 96% vol. used in this study was supplied by Bodegas Fundador, S.L.U. (Jerez de la Frontera, Spain). For the wood extraction experiments, it was diluted with ultrapure water from EMD Millipore (Bedford, MA, USA) until it reached 50% vol. of alcoholic strength.

UHPLC grade acetonitrile from Panreac (Barcelona, Spain) and acetic acid from Merck (Darmstadt Germany) were used to prepare the UHPLC phases. Standards for calibration were purchased from Sigma-Aldrich (Saint Louis, MO, USA). Ultrapure water from EMD Millipore (Bedford, MA, USA) was used to prepare the chromatography phases, reagents and standards for calibration.

4-methyl-2-pentanol (Sigma-Aldrich, Steinheim, Germany) was employed as an internal standard in SBSE-GC-MS and DTD-GC-MS analysis.

### 2.3. Analysis of Volatile Compounds of Wood Powder by DTD-GC-MS

This study was based on two factorial experiments: a factorial design of $2^4$ was chosen to determine the most influential parameters of the direct thermal desorption process and a $3^2$ experiment was carried out to establish the optimum values of these parameters. The final conditions considered to be optimal were the following: heating 10 mg of the sample at 250 °C during 7 min, desorbing the sample at 250 °C during 6 min and transferring the desorbed compounds to the line at 1:10 ratio split. The Statgraphics Statistical Computer Package "Statgraphics Centurion 18.0" was used for data treatment.

After the method optimization, the analysis of volatile compounds of wood powder were carried out by DTD-GC-MS. An amount of 10 mg of ground wood was placed together with 5 μL of a solution of 4-methyl-2-pentanol (303 mg $L^{-1}$ in an ethanol-water solution at 40% of alcohol) in the desorption tube, plugged at both ends with silanized glass wool. The desorption tube was heated to 250 °C for 7 min. The volatile compounds were desorbed in a stream of helium and collected into a cold trap (−15 °C). The desorption was carried out at 250 °C during 6 min and the volatile compounds were transferred (split 1:10) to the chromatographic column through a line heated to 225 °C. The experiments were carried out in a GCMS-TQ8040 Shimadzu gas chromatograph with mass detection (Shimadzu, Kyoto, Japan) equipped with a DB-Wax capillary column (J&W Scientific, Folsom, CA, USA), 60 m × 0.25 mm I.D., with a 0.25 μm coating. The chromatographic conditions were the same as the ones used previously. Samples were analyzed in duplicate. The relative area of each compound was obtained by measuring the area of the chromatographic signal produced by largest mass fragment (base peak) with respect to that of the internal standard, 4-methyl-2-pentanol. The results were expressed in relative area values.

### 2.4. Ultrasound-Assisted Extraction from Wood Powder

A total 1.1 g of wood powder was extracted with 200 mL of rectified wine distillate at 96% vol./water (1:1) hydroalcoholic solution at 40 °C during 2.5 h using an ultrasonic bath system (JP Selecta, S.A., Abrera, Spain) with 38.5 W $L^{-1}$ power as accelerating energy of the extraction process. The powder was washed with 50 mL of the same mixture used before. The extract was centrifuged at 5000 rpm during 5 min and transferred to a 250 mL volumetric flask. The extracts were used for the SBSE-GC-MS, GC-MS and UHPLC determinations. Each wood extraction was carried out in duplicate. The analysis of each extraction was also carried out in duplicate. Samples were stored in darkness under refrigeration.

### 2.5. *Analysis of Volatile Compounds of Wood Extracts by SBSE-GC-MS and GC-MS*

Volatile compounds of wood extracts were analyzed by SBSE-GC-MS and GC-MS techniques. PDMS commercial stir bars (10 mm length × 0.5 mm film thickness) provided by Gerstel (Mülheim a/d Ruhr, Germany) were used for the extractions. The procedure stablished in previous investigations of our research group was followed [24]: a volume of 35 mL of sample was placed in an Erlenmeyer flask and was diluted 1:1 (*v*/*v*) with ultrapure water. Then, 140 µL of a solution of 4-methyl-2-pentanol (2.3056 g $L^{-1}$ in an ethanol-water solution at 50% of alcohol) was added as an internal standard. Once the stir bar was added, the flask was placed on a 15-position magnetic stirrer (Mülheim a/d Ruhr, Germany) under agitation during 100 min at 1100 rpm at 25 °C. Finally, the stir bar was removed and washed and transferred into a thermal desorption glass where the thermal desorption was carried out. A commercial TDU (thermal desorption unit, Gerstel) with a programmed temperature vaporisation injector CIS-4 (cooled injection system, Gerstel) was used to carried out the thermal desorption of the coated stir bars. The desorption temperature was programmed from 40 to 300 °C (held for 10 min) at 60 °C $min^{-1}$ under a helium flow (75 mL $min^{-1}$) and the desorbed compounds were cryofocused in the CIS-4 system with liquid nitrogen at −140 °C. Finally, the CIS-4 was programmed from −140 °C to 300 °C (held for 5 min) at 10 °C/s for analysis by GC-MS. An Agilent 6890 GC-5973N MS system (Agilent, Little Falls, DE, USA), equipped with a DB-Wax capillary column (J&W Scientific, Folsom, CA, USA), 60 m × 0.25 mm I.D., with a 0.25 µm coating, was used to carried out the capillary GC-MS analyses in the electron impact mode. Helium was used as the carrier gas at a flow rate of 1.0 mL $min^{-1}$. The GC oven was programmed as follows: held at 35 °C for 10 min, then ramped at 5 °C $min^{-1}$ to 100 °C. Then it was raised to 210 °C at 3 °C $min^{-1}$ and held for 40 min. The mass detector operated in EI+ mode at 70 eV in a range from 30 to 400 amu. The identification of the compounds was carried out by analogy with mass spectra held in the Wiley Library (*Wiley Registry of Mass Spectral Data*, 7th Edition, 2000) and confirmed by retention indices of standards. The relative area of each compound was obtained by measuring the area of the chromatographic signal produced by largest mass fragment (base peak) with respect to that of the internal standard, 4-methyl-2-pentanol. Seventeen compounds were identified: ethyl butyrate, isoamyl acetate, limonene, ethyl caprylate, ethyl caprate, isopentyl octanoate, ethyl 2-phenylacetate, ethyl laureate, caprylic acid, ethyl myristate, capric acid, ethyl palmitate, ethyl 9-hexadecenoate, ethyl stearate, lauric acid, myristic acid and pentadecanoic acid. The results were expressed in relative area values.

Regarding the GC-MS analysis, the followed temperature program was the same as in the SBSE-GC-MS. The experiments were carried out in a GCMS-TQ8040 Shimadzu gas chromatograph with mass detection (Shimadzu, Kyoto, Japan) equipped with a DB-Wax capillary column (J&W Scientific, Folsom, CA, USA), 60 m × 0.25 mm I.D., with a 0.25 µm coating (the same column as the SBSE-GC-MS equipment).

### 2.6. *Analysis of Phenolic Compounds and Furfurals in the Wood Extracts*

Nine phenolic compounds (gallic acid, ellagic acid, caffeic acid, vanillic acid, vanillin, syringic acid, syringaldehyde, sinapaldehyde and coniferylaldehyde) and two furanic aldehydes (furfural and 5-hydroxymethylfurfural) were identified and quantified by UHPLC. Two eluents were used: a phase that consisted of 3% acetonitrile, 2% acetic acid and 95% ultrapure water, and B phase that consisted of 85% acetonitrile, 2% acetic acid, and 13% ultrapure water The method stablished in previous investigations of our research group was followed [25] for these analysis: 0 min, 100% A; 3 min, 90% A; 4 min, 90% A; 6.5 min, 25% A with a flow rate of 0.7 mL $min^{-1}$ and a column temperature of 47 °C. The injection volume was 2.5 µL. The column was washed with 100% B for 3 min and equilibrated with 100% A for 3 min. Then, 0.22 μm nylon membranes were used to filtered samples and standards. The detection by UV absorption was conducted by scanning between 250 and 400 nm, with a resolution of 1.2 nm. The comparison of retention times and UV-Vis spectra of the peaks in samples with those previously obtained by the injection of standards allows the identification of each compounds. Samples and standards were injected in duplicate. The results were expressed in mg of compound per liter of sample.

### 2.7. Total Polyphenol Index in the Wood Extracts

Total Polyphenol Index (TPI) of the wood extracts was determined by the measure of the absorbance at 280 nm. Samples were measured directly or diluted with ultrapure water where necessary. A Lambda 25 spectrophotometer (Perkin Elmer, Boston, MA, USA) was used for the analysis. The calibration curve was prepared with gallic acid solutions ranging from 0 to 50 mg $L^{-1}$. A glass cell with a 10 mm optical path was used. Sample measurements were carried out in duplicate. The results are expressed in mg of gallic acid equivalent (GAE) per litre of sample.

### 2.8. Statistical Analysis

The Statgraphics 18 software package (Statgraphics Technologies, Inc., The Plains, VA, USA) was employed for factorial design experiments and ANOVA. Microsoft Excel 2016 (Microsoft Corp., Redmond, WA, USA) was employed for other statistical parameters.

## 3. Results and Discussions

### 3.1. DTD-GC-MS Condition Optimization

Heating temperature, heating time, desorption temperature and desorption time of the direct thermal desorption process were evaluated to achieve the best overall analytical conditions. No references related with wood and the volatile compounds that it could bring to spirits and wines determined by this method were found in bibliography and, therefore, it had to be optimized.

To optimize the direct thermal desorption conditions, we chose a sequential exploration of the response, which was carried out in two stages. In the first stage, a factorial design of $2^4$ was chosen to analyze the influence of heating temperature, heating time, desorption temperature and desorption time using a mixture of all the wood types studied, as described in Section 2.1, in order to consider all the compounds that could be present in the wood samples. In the second stage, a factorial design of $3^2$ was chosen to optimize the heating temperature and heating time.

#### 3.1.1. Screening by a $2^4$ Factorial Design

The values corresponding to the low (−) and high (+) levels for each factor are shown in Table 1. The design involved sixteen experiments in duplicate. Total area values and chromatographic peak number of each experiment evaluated in the $2^4$ factorial design are shown in Table 2. The data obtained for the heating temperature, heating time, desorption temperature and desorption time were evaluated by ANOVA at the 0.05 significance level (Table 3).

**Table 1.** Levels of the $2^4$ factorial design.

| Factor | Low (−) | High (+) |
|---|---|---|
| Heating temperature (°C) | 180.0 | 220.0 |
| Heating time (min) | 1.0 | 15.0 |
| Desorption temperature (°C) | 180.0 | 250.0 |
| Desorption time (min) | 3.0 | 10.0 |

**Table 2.** Conditions, total area and number of chromatographic peaks obtained in each experiment of the $2^4$ factorial design.

| Experiment | Heating Temperature (°C) | Heating Time (min) | Desorption Temperature (°C) | Desorption Time (min) | Total Area | Number of Peaks |
|---|---|---|---|---|---|---|
| 1.1 | 180 | 15 | 180 | 10 | 45,266,253 | 57 |
| 1.2 | 180 | 15 | 180 | 3 | 45,342,152 | 40 |
| 1.3 | 180 | 15 | 250 | 10 | 52,191,184 | 63 |
| 1.4 | 180 | 1 | 250 | 10 | 20,703,659 | 16 |
| 1.5 | 220 | 1 | 180 | 10 | 24,144,870 | 6 |
| 1.6 | 220 | 15 | 250 | 3 | 97,177,735 | 110 |
| 1.7 | 180 | 1 | 180 | 10 | 41,301,225 | 21 |
| 1.8 | 180 | 1 | 180 | 3 | 19,116,371 | 22 |
| 1.9 | 220 | 1 | 180 | 3 | 11,964,472 | 14 |
| 1.10 | 220 | 15 | 250 | 10 | 112,801,779 | 100 |
| 1.11 | 180 | 1 | 250 | 3 | 61,846,897 | 22 |
| 1.12 | 220 | 1 | 250 | 3 | 23,463,052 | 5 |
| 1.13 | 220 | 15 | 180 | 3 | 80,695,112 | 91 |
| 1.14 | 220 | 1 | 250 | 10 | 29,182,299 | 11 |
| 1.15 | 180 | 15 | 250 | 3 | 62,341,956 | 53 |
| 1.16 | 220 | 15 | 180 | 10 | 80,828,868 | 98 |

**Table 3.** Main effects and interactions in the $2^4$ factorial design for the number of chromatographic peaks and total area.

| Effect | No. of Chromatographic Peaks | | Total Area | |
|---|---|---|---|---|
| | F Ratio | *p* Value | F Ratio | *p* Value |
| A: Heating temperature | 2.41 | 0.1643 | 1.39 | 0.2763 |
| B: Heating time | 29.73 | 0.0010 | 13.19 | 0.0084 |
| C: Desorption temperature | 0.12 | 0.7428 | 1.37 | 0.2806 |
| D: Desorption time | 0.03 | 0.8734 | 0.00 | 0.9638 |
| AC | 0.00 | 0.9746 | 0.04 | 0.8477 |
| AD | 0.08 | 0.7910 | 0.44 | 0.5294 |
| BC | 0.29 | 0.6061 | 0.13 | 0.7331 |
| CD | 0.03 | 0.8734 | 0.46 | 0.5197 |

Parameters related with the heating process had a significant positive influence on the total area and the number of chromatographic peaks, appearing statistically as the most influential effect (Figure 1). The effect of the parameters heating time and heating temperature is positive for the two variables considered, that is, high temperature levels and high heating time produce the extraction of larger amounts of volatile compounds (Figure 2), being the heating time the only parameter that presents a significant effect ($p$ value $< 0.05$, Table 3), both for the number of peaks and for the total area. The heating temperature is the next parameter that most affects the variables considered, although its effect is not significant ($p$ value $> 0.05$, Table 3). As the heating time and the heating temperature increase (15 min and 220 °C, respectively), the response obtained in both variables is greater (Figure 2).

The parameters related with the desorption process, temperature desorption and time desorption do not show a significant influence on the total area or the number of chromatographic peaks ($p$ value $> 0.05$, Table 3). However, the desorption at 250 °C showed better results than the desorption at 180 °C, being selected as the optimal value for the following analysis (Figure 2). No differences between the high and low level of the desorption time were found, so an average value (6 min) was selected.

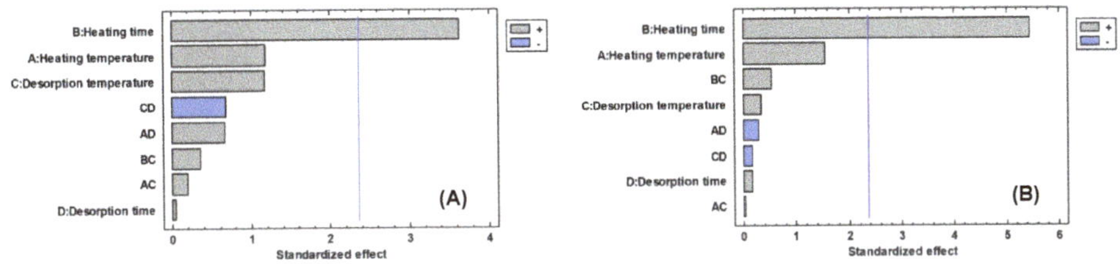


**Figure 1.** Pareto chart of main effects in the $2^4$ factorial design for number of chromatographic peaks (**A**) and total area (**B**).

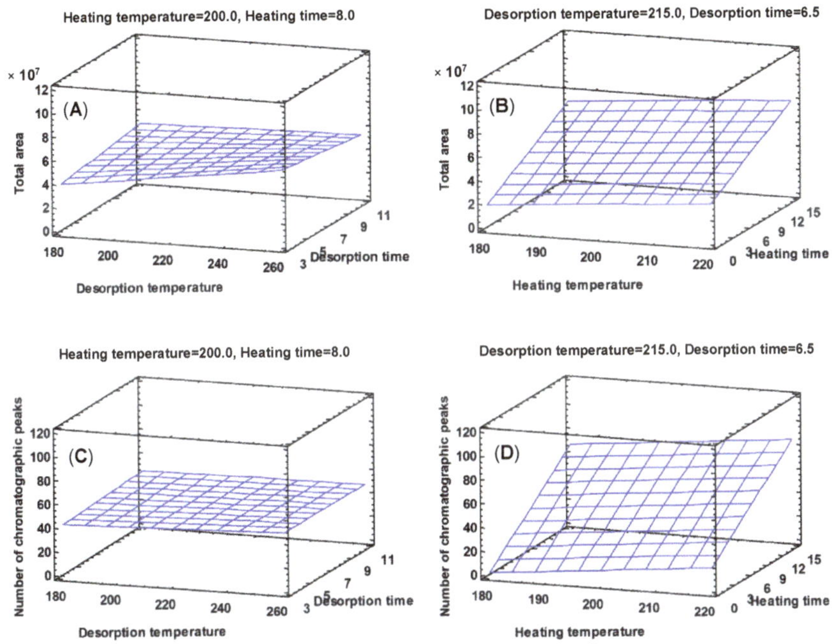


**Figure 2.** Interaction heating time–heating temperature. Estimated response surface for total chromatographic area (**A**) and for the number of chromatographic peaks (**C**). Interaction desorption time–desorption temperature. Estimated response surface for total chromatographic area (**B**) and for the number of chromatographic peaks (**D**).

Heating parameters turned out to be the most influential ones in the direct thermal desorption process. For these analysis, 20 mg of a mixture of the five woods was used. Some chromatographic peaks were saturated and so, for the optimized method experiments, the sample amount employed was lower than before in the following factorial design.

In summary, the best conditions obtained in this first optimization study were the following: heating time, 15 min; heating temperature, 220 °C; desorption time, 6 min; and desorption temperature, 250 °C.

### 3.1.2. Optimization by a $3^2$ Factorial Design

In order to optimize the parameters of the direct thermal desorption method, the most influent variables resulting from the first factorial design were studied. Three levels of heating temperature and heating time were established. The design involved nine experiments in duplicate. The values corresponding to the low (−) and high (+) levels for each factor are shown in Table 4.

**Table 4.** Levels of the $3^2$ factorial design.

| Factor | Low (−) | High (+) |
|---|---|---|
| Heating temperature (°C) | 200.0 | 250.0 |
| Heating time (min) | 5.0 | 15.0 |

After the results obtained in the $2^4$ factorial design experiments, the desorption conditions established for the analysis were the following: 6 min and 250 °C. A total of 10 mg of the mixture sample was used in the study. Total area values and chromatographic peak number of each experiment evaluated in the $3^2$ factorial design were shown in Table 5. The data obtained for the heating temperature and the heating time were evaluated by ANOVA at the 5% significance level (Table 6).

**Table 5.** Conditions, total area and number of chromatographic peaks obtained in each experiment of the $3^2$ factorial design.

| Experiment | Heating Temperature (°C) | Heating Time (min) | Total Area | Number of Peaks |
|---|---|---|---|---|
| 2.1 | 220 | 10 | 83,670,835 | 91 |
| 2.2 | 200 | 15 | 56,175,969 | 76 |
| 2.3 | 250 | 10 | 96,680,416 | 110 |
| 2.4 | 250 | 5 | 103,356,194 | 102 |
| 2.5 | 220 | 15 | 54,743,149 | 70 |
| 2.6 | 220 | 5 | 84,269,330 | 74 |
| 2.7 | 200 | 5 | 54,598,381 | 52 |
| 2.8 | 250 | 15 | 82,280,240 | 102 |
| 2.9 | 200 | 10 | 65,838,463 | 55 |

**Table 6.** Main effects and interactions in the $3^2$ factorial design for number of chromatographic peaks and total area.

| Effect | No. of Chromatographic Peaks | | Total Area | |
|---|---|---|---|---|
| | F Ratio | *p* Value | F Ratio | *p* Value |
| A: Heating temperature | 24.47 | 0.0159 | 33.52 | 0.0103 |
| B: Heating time | 0.57 | 0.5050 | 7.21 | 0.0747 |
| AA | 0.35 | 0.5975 | 0.18 | 0.6969 |
| AB | 1.23 | 0.3480 | 2.31 | 0.2259 |
| BB | 0.62 | 0.4898 | 3.24 | 0.1695 |

Heating temperature had a significant positive influence on the total area and the number of chromatographic peaks, appearing as the statistically main effect (Figure 3). The effect of heating time and heating temperature is positive for the two variables considered, that is, high temperature levels produce the extraction of larger amounts of volatile compounds as the heating time value is between 5 and 10 min (Figure 4). The heating temperature is the only parameter that presents a significant effect ($p$ value < 0.05, Table 6), both for the number of peaks and for the total area. The heating temperature does not significantly affect the total area or the number of chromatographic peaks ($p$ value > 0.05, Table 6). However, the heating time range from 5 to 10 min showed the best results, and so, an average value (7 min) was selected as the optimum value.

Taking into account all the results obtained, the final direct thermal desorption conditions considered to be optimal were as follows: heating 10 mg of the sample at 250 °C during 7 min and desorbing the sample at 250 °C during 6 min.

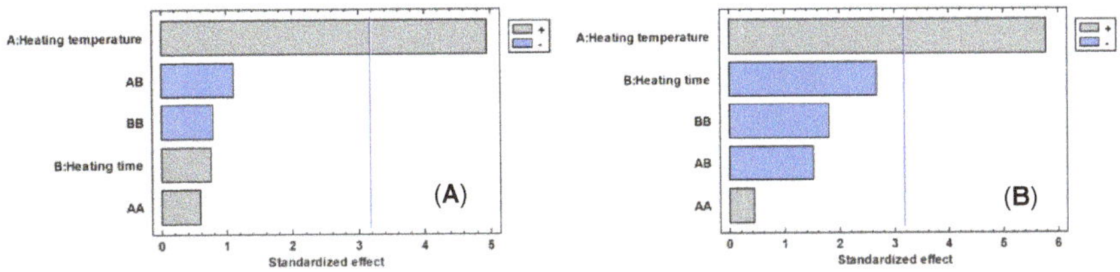


**Figure 3.** Pareto chart of main effects in the $3^2$ factorial design for the number of chromatographic peaks (**A**) and total area (**B**).

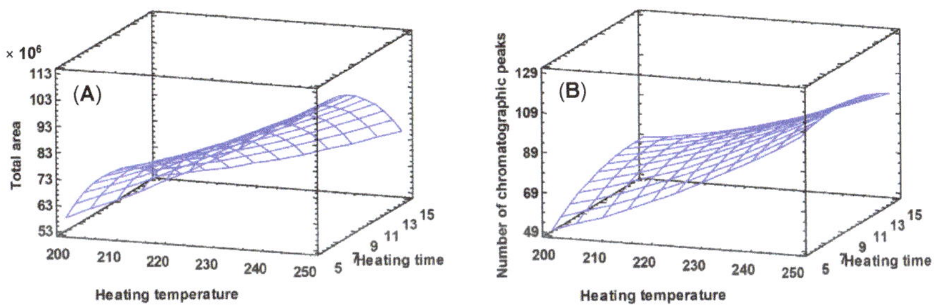


**Figure 4.** Interaction heating time–heating temperature. Estimated response surface for total chromatographic area (**A**) and for the number of chromatographic peaks (**B**).

### *3.2. Analysis of Volatile Compounds of Wood Powder by DTD-GC-MS*

Five different types of wood (American oak, Spanish oak, French oak, Chestnut and Cherry) were analyzed (in duplicate) employing the DTD-GC-MS method optimized. For this analysis, the wood chips were grounded to a 0.25 µm grain size. All the factorial design experiments were carried out in a splitless mode. High peak densities were obtained in all of them. In order to avoid detector saturation during the analysis of the real samples, they were injected in split mode. Different split ratios were tested 1:30, 1:20, 1:10 and 1:5. The split ratio 1:10 showed the best results.

The amount of the volatile compounds detected in each type of wood has been obtained by means of the relative integration with respect to the internal standard, 4-methyl-2-pentanol (Table 7). The results were evaluated by ANOVA at the 5% significance level (Table 7).

As expected, many similarities were found regarding the volatile composition of three oak woods studied. As can be seen in Figure 5, their aroma profile is very similar. However, there are some differences, as the levels of formic acid, acrylic acid and furanone that are significantly different, at 5% of significance level, in American oak with respect to Spanish and French oaks (Table 7). There are significant differences between the amount of vanillin and syringaldehyde between American and French oak, but there are none as compared to Spanish oak (Table 7). These compounds could be affected by the toasting level of the chip wood. All the studied oaks have a medium toasting level. Acetic acid, furfural, formic acid, 5-methylfurfural, vanillin and syringaldehyde are the most abundant components [20,23,26,27]. Many of them are generated during the heat treatment processes during the toasting of the chips. There are compounds as whiskey lactones that are only identified in American and French oaks. According to the literature, these compounds are commonly present in oak wood, being more abundant in the *Quercus alba* species [20,22]. 2-phenylethanol is only detected in Spanish oak and chestnut. 4-cyclopentene-1,3-dione is only detected in American oak. Myristic acid is detected in French and Spanish oak [5] but it is not detected in American oak.

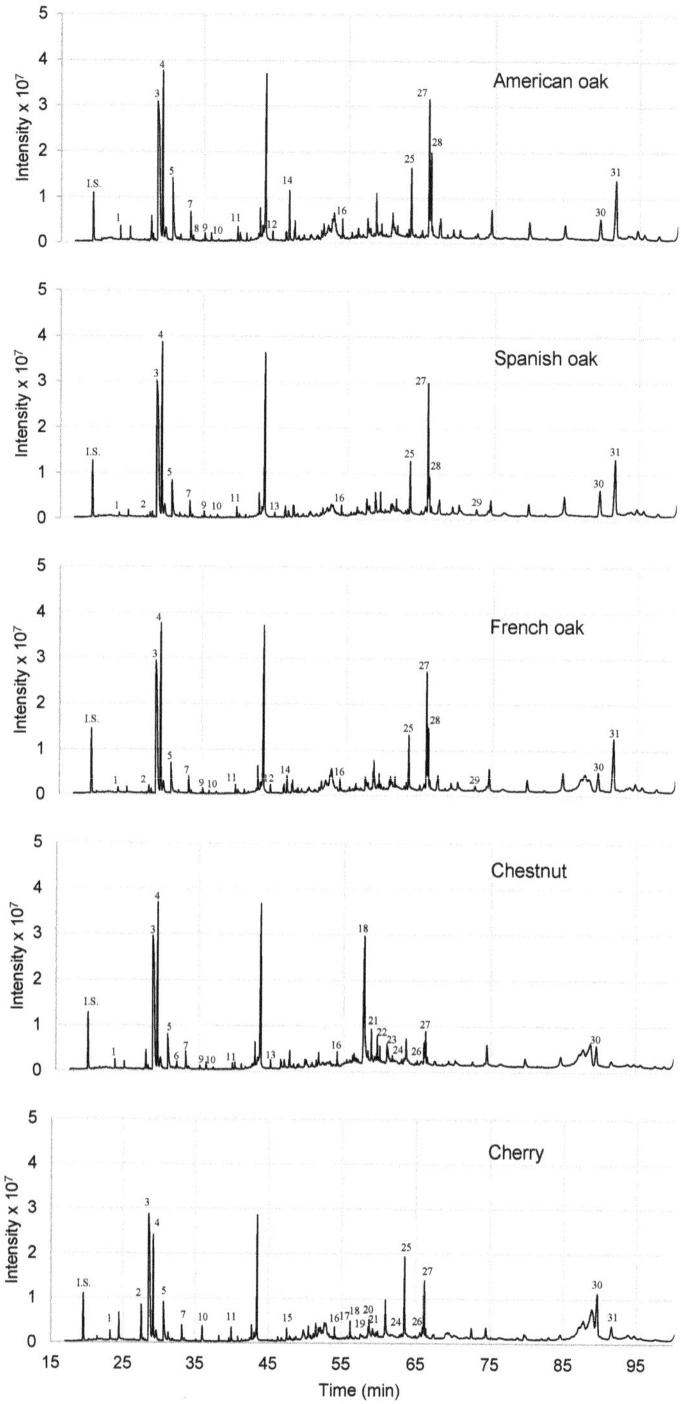


**Figure 5.** DTD-GC-MS chromatogram of the hydroalcoholic wood extracts. IS: Internal Standard (4-methyl-2-pentanol). The key for the compounds is in Table 7.

**Table 7.** Relative area values of volatile compounds determined by DTD-GC-MS of wood powder.

| Compound | American Oak | Spanish Oak | French Oak | Chestnut | Cherry |
|---|---|---|---|---|---|
| Acetol **(1)** | 0.112 ± 0.025 $^{a}$ | 0.045 ± 0.003 $^{a}$ | 0.046 ± 0.008 $^{a}$ | 0.054 ± 0.011 $^{a}$ | 0.501 ± 0.104 $^{b}$ |
| 2-*O*-cyclobutyl 1-*O*-octadecyl oxalate **(2)** | n.d. | 0.020 ± 0.028 | 0.089 ± 0.014 | n.d. | 0.372 ± 0.526 |
| Acetic acid **(3)** | 7.158 ± 1.123 | 5.906 ± 0.453 | 5.079 ± 0.639 | 4.779 ± 0.799 | 4.673 ± 2.868 |
| Furfural **(4)** | 2.813 ± 0.425 | 2.574 ± 0.245 | 2.497 ± 0.364 | 2.213 ± 0.327 | 1.976 ± 0.241 |
| Formic acid **(5)** | 1.653 ± 0.276 $^{a}$ | 0.825 ± 0.043 $^{b}$ | 0.687 ± 0.112 $^{b}$ | 0.714 ± 0.118 $^{b}$ | 1.191 ± 0.520 $^{a,b}$ |
| 2,3-butanediol **(6)** | n.d. | n.d. | n.d. | 0.082 ± 0.013 | n.d. |
| 5-methylfurfural **(7)** | 0.466 ± 0.118 | 0.231 ± 0.027 | 0.277 ± 0.041 | 0.242 ± 0.021 | 0.237 ± 0.294 |
| 4-cyclopentene-1,3-dione **(8)** | 0.054 ± 0.016 | n.d. | n.d. | n.d. | n.d. |
| Acrylic acid **(9)** | 0.103 ± 0.020 $^{a}$ | 0.069 ± 0.006 $^{b}$ | 0.059 ± 0.008 $^{b,c}$ | 0.040 ± 0.004 $^{c}$ | n.d. |
| Furfuryl alcohol **(10)** | 0.118 ± 0.020 $^{a}$ | 0.028 ± 0.007 $^{a}$ | 0.038 ± 0.008 $^{a}$ | 0.073 ± 0.012 $^{a}$ | 0.390 ± 0.087 $^{b}$ |
| 2(5H)-Furanone **(11)** | 0.159 ± 0.036 $^{a}$ | 0.108 ± 0.008 $^{b}$ | 0.094 ± 0.008 $^{b,c}$ | 0.058 ± 0.004 $^{b,c}$ | 0.048 ± 0.022 $^{c}$ |
| *Trans*-whiskey lactone **(12)** | 0.115 ± 0.025 | n.d. | 0.101 ± 0.013 | n.d. | n.d. |
| 2-phenylethanol **(13)** | n.d. | 0.052 ± 0.003 $^{a}$ | n.d. | 0.087 ± 0.012 $^{b}$ | n.d. |
| *Cis*-whiskey lactone **(14)** | 0.707 ± 0.142 $^{a}$ | n.d. | 0.251 ± 0.041 $^{b}$ | n.d. | n.d. |
| Cyclopropylcarbinol **(15)** | n.d. | n.d. | n.d. | n.d. | 0.192 ± 0.196 |
| 4-hydroxy-2-methylacetophenone **(16)** | 0.194 ± 0.049 | 0.091 ± 0.014 | 0.126 ± 0.015 | 0.121 ± 0.030 | 0.125 ± 0.078 |
| Pyranone **(17)** | n.d. | n.d. | n.d. | n.d. | 0.212 ± 0.217 |
| Glycerin **(18)** | n.d. | n.d. | n.d. | 3.554 ± 0.587 | n.d. |
| Levulinic acid **(19)** | n.d. | n.d. | n.d. | n.d. | 0.137 ± 0.151 |
| p-acetylacetophenone **(20)** | n.d. | n.d. | n.d. | n.d. | 0.268 ± 0.119 |
| *Trans*-isoeugenol **(21)** | n.d. | n.d. | n.d. | 0.289 ± 0.044 $^{a}$ | 0.199 ± 0.026 $^{b}$ |
| Ethyl hydrogen succinate **(22)** | n.d. | n.d. | n.d. | 0.211 ± 0.040 | n.d. |
| 2,3-dihydrobenzofuran **(23)** | n.d. | n.d. | n.d. | 0.137 ± 0.021 | 0.175 ± 0.119 |
| Benzoic acid **(24)** | n.d. | n.d. | n.d. | 0.337 ± 0.008 | 0.471 ± 0.496 |
| 5-HMF **(25)** | 1.206 ± 0.231 | 0.794 ± 0.040 | 0.834 ± 0.098 | n.d. | 1.347 ± 1.024 |
| Methoxyeugenol **(26)** | n.d. | n.d. | n.d. | 0.317 ± 0.043 | 0.845 ± 1.140 |
| Vanillin **(27)** | 2.491 ± 0.372 $^{a}$ | 1.990 ± 0.164 $^{a,b}$ | 1.815 ± 0.176 $^{b}$ | 0.483 ± 0.040 $^{c}$ | 0.138 ± 0.130 $^{c}$ |
| Ethriol **(28)** | 1.609 ± 0.326 | 0.511 ± 0.061 | 0.939 ± 0.082 | n.d. | 0.865 ± 0.956 |
| Myristic acid **(29)** | n.d. | 0.172 ± 0.058 | 0.122 ± 0.021 | n.d. | 0.700 ± 0.618 |
| Palmitic acid **(30)** | 1.095 ± 0.079 | 1.154 ± 0.029 | 0.768 ± 0.201 | 0.692 ± 0.040 | 1.759 ± 1.241 |
| Syringaldehyde **(31)** | 2.731 ± 0.390 $^{a}$ | 2.235 ± 0.205 $^{a,b}$ | 2.057 ± 0.130 $^{b}$ | n.d. | 0.453 ± 0.270 $^{c}$ |

Data are mean value ± standard deviation; values in the same row with different letters are significantly different ($p < 0.05$); n.d.: Not detected.

In Figure 5, chestnut and cherry show a slightly different aromatic profile both between them and with respect to the oak types studied. Most of the detected compounds were in significant different amounts with respect the other woods studied (Table 7). However, they contain many compounds present in oak wood too, like acetic acid, furfural, formic acid, 5-methylfurfural and palmitic acid [20,22,26], and their content is not significantly different among woods at 5% significance level (Table 7), except for formic acid content. It should be noted that the cherry and chestnut wood studied in both cases is an untoasted wood; however, they are also rich in compounds that come from the toasting of the wood, such as furfurals. This is due to the prior heating of the sample as part of the analysis method. The peak profile during the first 40 min of the analysis is very similar for all the wood chips studied. Cherry is the wood that has the most different volatile composition profile. 2,3-butanediol, glycerin and ethyl hydrogen succinate are only present in chestnut. This wood is similar to the studied oak woods, but also shares some similarities with cherry wood. Compounds as 2-phenylethanol or acrylic acid are detected in oak and chestnut wood, but are not present in cherry wood. On the other hand, compounds as trans-isoeugenol, 2,3-dihydrobenzofuran, benzoic acid and methoxyeugenol are present in chestnut and cherry wood but not in oak wood. According to the literature, methoxyeugenol is present in significant levels in cherry and chestnut wood, it could be present in oak wood too but, in this case it was not detected [5,28]. Cherry wood has the most particular aroma profile. Cyclopropyl carbinol, pyranone [20], levulinic acid and p-acetylacetophenone are only detected in this wood. There are some compounds as 5-HMF, ethriol and syringaldehyde that are detected also in oak wood, but they are not in chestnut wood. Therefore, there are compounds or profiles that are characteristics of each type of wood studied, thus being able to be targets to identify each species in DTD-GC-MS analysis.

All the mentioned studies analyzed the volatile compounds through a previous hydroalcoholic extraction of the wood; the novelty of this work is the direct analysis of wood aromas by DTD-GC-MS. Although there are studies in oak wood that work with DTD-GC-MS (but using a solid support to trap the volatile compounds) [29], in the rest of cited references, the woods have not been studied in this way. The previous direct thermal desorption stage could alter the sample, since heating could increase the toasting level of wood sample studied and increase the concentration of those compounds that are related to this process. However, there is not a loss of information, as could happen in extraction processes, since the aromatic profile is measured directly.

### 3.3. *Analysis of Volatile Compounds of Wood Extracts by SBSE-GC-MS and GC-MS*

An analysis on hydroalcoholic wood extracts was carried out to compare the compounds that are present in the wood detected by the analysis by DTD-GC-MS with those volatile compounds that could be transferred to the spirit or the wine during their ageing through the wood chips. In order to determine if they could be detected in aged alcoholic beverages, an hydroalcoholic ultrasound assisted extraction was carried out.

As regards the volatile composition of wood extracts, a low amount of the compounds was found in the samples. Relative area values of volatile compounds determined by SBSE-GC-MS of wood extracts are shown in Table 8. The aromatic profile is different in all the studied wood but the amount of each detected compound is very low. In order to complement this information, the samples were analyzed by GC-MS. However, any compounds were not detected with this technique.

**Table 8.** Relative area values of volatile compounds determined by Stir-bar Sorptive Extraction–Gas Chromatography–Mass Spectrometry (SBSE-GC-MS) of wood extracts.

| Compound | Extractant | American Oak | Spanish Oak | French Oak | Chestnut | Cherry |
|---|---|---|---|---|---|---|
| Ethyl butyrate | n.d. | d. | d. | 0.010 ± 0.007 | 0.010 ± 0.002 | 0.020 ± 0.014 |
| Isoamyl acetate | n.d. | d. | n.d. | 0.265 ± 0.243 | 0.392 ± 0.170 | 0.258 ± 0.058 |
| Limonene | n.d. | d. | 0.031 ± 0.002 | d. | d. | d. |
| Ethyl caprylate | 0.693 $^{a}$ | 0.181 ± 0.024 $^{b}$ | 0.159 ± 0.003 $^{b}$ | 0.294 ± 0.157 $^{b}$ | 0.737 ± 0.253 $^{b}$ | 0.810 ± 0.359 $^{b}$ |
| Ethyl caprate | 1.523 $^{a}$ | 0.422 ± 0.039 $^{b,c,d}$ | 0.326 ± 0.017 $^{b,c}$ | 0.574 ± 0.414 $^{b,d}$ | 1.027 ± 0.640 $^{e,f}$ | 0.721 ± 0.336 $^{e,f}$ |
| Isopentyl octanoate | n.d. | n.d. | n.d. | n.d. | n.d. | d. |
| Ethyl 2-phenyl acetate | n.d. | n.d. | n.d. | d. | 0.057 ± 0.041 | n.d. |
| Ethyl laureate | 0.140 | 0.292 ± 0.029 | 0.228 ± 0.025 | 0.422 ± 0.342 | 0.395 ± 0.267 | 0.295 ± 0.212 |
| Caprylic acid | 0.191 | 0.163 ± 0.089 | 0.088 ± 0.018 $^{a}$ | 0.147 ± 0.062 | 0.184 ± 0.059 $^{b}$ | 0.202 ± 0.068 $^{b}$ |
| Ethyl myristate | 0.078 | 0.095 ± 0.036 | d. | d. | 0.266 ± 0.255 | 0.202 ± 0.181 |
| Capric acid | 0.214 | 0.215 ± 0.163 | d. | d. | 0.090 ± 0.040 | 0.091 ± 0.032 |
| Ethyl palmitate | 0.233 | 0.261 ± 0.207 | d. | d. | 0.152 ± 0.077 | 0.119 ± 0.052 |
| Ethyl 9-hexadecenoate | 0.035 | d. | d. | d. | d. | d. |
| Ethyl stearate | n.d. | d. | n.d. | n.d. | n.d. | n.d. |
| Lauric acid | 0.731 $^{a}$ | 0.131 ± 0.051 $^{b}$ | 0.054 ± 0.038 $^{c}$ | 0.090 ± 0.028 $^{b,c}$ | 0.085 ± 0.014 $^{b,c}$ | 0.040 ± 0.028 $^{c}$ |
| Myristic acid | 0.057 $^{a}$ | 0.263 ± 0.085 $^{b}$ | 0.180 ± 0.016 $^{a,b,c}$ | 0.200 ± 0.081 $^{b,c}$ | 0.170 ± 0.051 $^{a,c}$ | 0.138 ± 0.023 $^{a,c}$ |
| Pentadecanoic acid | n.d. | d. | d. | 0.097 ± 0.053 | 0.069 ± 0.019 | 0.079 ± 0.014 |

Data are mean value ± standard deviation; values in the same row with different letters are significantly different ($p < 0.05$); n.d.: Not detected; d.: Detected.

Attending to the results, shown in Table 8, wood related compounds were not detected. As regards the relative area values of the compounds present in the extractant (1:1 hydroalcoholic mixture of rectified wine distillate at 96% vol. and water solution), similarities with the relative area values of wood extracts were found. The relative area values obtained were evaluated by ANOVA at the 5% significance level (Table 8). As it can be seen, most of them are not significantly different, which means that the compounds detected are not influenced by wood. It seems that it only contributes to trace levels of them. According to the literature, there are fatty acids as caprylic acid, myristic acid or palmitic acid present in wood composition [30–32]. The contribution of these compounds to the extracts and their respective esters could be due to the extraction procedure, in which the wood powder was extracted with an hydroalcoholic solution under 40 °C. The esterification of the fatty acids in the presence of ethanol at this temperature resulting in the corresponding esters as ethyl caprylate, ethyl myristate, ethyl palmitate or ethyl laureate could take place during the extraction process. This fact could explain the increase of the ethyl laureate and myristic acid in all the wood extracts studied, ethyl caprylate in chestnut and cherry wood extracts and ethyl palmitate in the American oak extract. The only compound detected in both analysis (DTD-GC-MS and SBSE-GC-MS) was myristic acid.

In the direct analysis of the wood powder by DTD-GC-MS, numerous volatile compounds were identified, which make it a very interesting technique. These compounds are transferred from the wood to the spirit or the wine during their ageing, modifying its sensorial profile. During ultrasound-assisted extractions, these compounds were extracted by the hydroalcoholic mixture used. However, once the hydroalcoholic extracts were analyzed, no volatile compounds were detected by GC-MS and very few compounds and at very low levels were detected by SBSE-GC-MS. Therefore, there is a loss of information regarding the analysis of volatile compounds once the ultrasound-assisted extraction is performed. However, the hydroalcoholic extracts were useful to characterize the phenolic compounds that wood could contribute to spirits and wines and to complete the aromatic profile of the woods studied. The majority of the compounds identified in SBSE-GC-MS analysis (Table 8) are characteristic of wines, wine spirits or brandies, so their presence in the wood extracts studied could be also due to the origin of the extractant used, that has a part of a grape derived alcoholic beverage.

### 3.4. Phenolic Composition of the Wood Extracts and Total Polyphenol Index

The TPI data of the studied samples, expressed in mg of equivalent gallic acid (GAE) per litre, are shown in Table 9. Of all the woods studied, Spanish oak released the highest amount of phenolic compounds into the alcoholic beverage. The lowest TPI values of oak were found in American oak wood. The TPI values for cherry wood (without toasting) are between medium toasted French and American wood. Chestnut (without toasting) has the lowest composition in phenolic compounds of all those studied. The results of the one-way analysis of the variance (ANOVA) proved that all the wood extracts are statistically different, with a probability of 95%.

**Table 9.** Phenolic compounds contents (mg $L^{-1}$) and total polyphenol index (TPI) (mg gallic acid equivalent (GAE) $L^{-1}$) of wood extracts.

| Compound | American Oak | Spanish oak | French oak | Chestnut | Cherry |
|---|---|---|---|---|---|
| Gallic acid **(1)** | 0.98 ± 0.14 $^a$ | 3.35 ± 0.17 $^b$ | 1.69 ± 0.15 $^c$ | 0.23 ± 0.06 $^d$ | 0.17 ± 0.32 $^e$ |
| Hydroxymethylfurfural **(2)** | 0.28 ± 0.06 $^a$ | 0.44 ± 0.09 $^b$ | 0.40 ± 0.05 $^b$ | 0.08 ± 0.04 $^c$ | n.d. |
| Furfural **(3)** | 0.10 ± 0.01 $^a$ | 0.41 ± 0.03 $^b$ | 0.33 ± 0.04 $^c$ | n.d. | n.d. |
| Caffeic acid **(4)** | n.d. | n.d. | n.d. | 0.28 ± 0.02 | n.d. |
| Vanillic acid **(5)** | 1.10 ± 0.19 $^a$ | 0.59 ± 0.05 $^b$ | 1.48 ± 0.019 $^c$ | 0.55 ± 0.11 $^b$ | n.d. |
| Syringic acid **(6)** | 0.32 ± 0.09 $^a$ | 0.46 ± 0.04 $^b$ | 0.36 ± 0.05 $^a$ | 0.14 ± 0.01 $^c$ | 0.08 ± 0.16 $^d$ |
| Vanillin **(7)** | 0.87 ± 0.04 $^a$ | 2.00 ± 0.08 $^b$ | 1.46 ± 0.04 $^c$ | 0.39 ± 0.03 $^d$ | n.d. |
| Syringaldehyde **(8)** | 0.90 ± 0.03 $^a$ | 2.18 ± 0.07 $^b$ | 1.37 ± 0.07 $^c$ | 0.25 ± 0.05 $^d$ | n.d. |
| Ellagic acid **(9)** | 4.65 ± 0.10 $^a$ | 8.82 ± 0.59 $^b$ | 8.03 ± 0.50 $^c$ | 3.36 ± 0.15 $^d$ | n.d. |
| Coniferylaldehyde **(10)** | 1.21 ± 0.04 $^a$ | 2.51 ± 0.07 $^b$ | 1.83 ± 0.06 $^c$ | 0.19 ± 0.05 $^d$ | n.d. |
| Sinapaldehyde **(11)** | 1.53 ± 0.02 $^a$ | 4.59 ± 0.08 $^b$ | 2.93 ± 0.10 $^c$ | 0.25 ± 0.05 $^d$ | n.d. |
| Total Phenolic Index | 184.61 ± 1.68 $^a$ | 355.26 ± 3.51 $^b$ | 285.15 ± 10.52 $^c$ | 120.28 ± 1.20 $^d$ | 252.34 ± 0.72 $^e$ |

Data are mean value ± standard deviation; values in the same row with different letters are significantly different ($p < 0.05$); n.d.: Not detected.

The content in low molecular weight phenolic compounds determined by means of UHPLC in the wood extracts, expressed in mg $L^{-1}$, is also shown in Table 9. As regards the phenolic acids studied, gallic acid, vanillic acid, caffeic acid, syringic acid and ellagic acid were found. Gallic and ellagic acids come from the hydrolysis of gallotannins and ellagitannins under an acidic environment [28]. The oxidation and hydrolysis of the compounds derived from the degradation of lignin is the origin of vanillic and syringic acids [16,33]. A significant amount of phenolic aldehydes (p-hydroxybenzaldehyde, vanillin, syringaldehyde, coniferylaldehyde and sinapaldehyde) was found in some of the samples studied (Table 9). Their origin is in the thermal degradation of lignin [14,16,33], a process that takes place during the manufacturing of the barrel due to the toasting of the wood and its thermal treatments [10]. 5-hydroxymethylfurfural and furfural have been detected in significant amounts in some of the samples studied (Table 9). The presence of furfural is due to the heating of the pentoses, while 5-hydroxymethylfurfural has its origin in the thermal degradation of the glucose and cellulose. Their presence depends on the toasting of the wood [10].

Phenolic composition is very similar in American, Spanish and French oak (Figure 6). The same compounds were found in the hydroalcoholic extracts of all of them, however their proportion was not the same. Gallic acid, ellagic acid, vanillic acid, vanillin, syringaldehyde, coniferylaldehyde and sinapaldehyde are the most abundant compounds detected in the oak wood extracts. Spanish oak has the highest amount of phenolic compounds, while American oak has the lowest quantity. The amount of phenolic compounds present in each wood is influenced by the origin and the heat treatment of the wood during the manufacturing of the barrel or chips [10]. During the ageing period, wood characteristics as porosity affect the extraction of phenolic and volatile compounds. Spanish and French oak are more porous than American oak, and this has a positive influence during the extraction process.

All the compounds detected in oak wood, except furfural, were also found in Chestnut extracts. This wood was untreated, without toasting treatment, so it explains the absence of this compound. However, due to its porosity, a great amount of phenolic compounds was found in the hydroalcoholic extracts studied. Chestnut wood has high levels of gallic acid [34], in this case the wood studied was not toasted, so the level of this compound is lower than expected.

As regards the cherry wood chromatogram, a high level of phenolic compounds was detected. There are many signals at the end of the chromatogram indicating that, according to their retention time, these are low-polar compounds. According to the literature, they could be flavonoid-type compounds [23] as (+)-catechin [35], taxifolin [36], naringenin [27], aromadendrin [37] or kaempferol [37], that are very common in cherry wood. Flavonoids were only detected in cherry wood (Figure 6), what makes its aromatic profile very interesting. It would be interesting to be able to identify these compounds in the future, because they make cherry wood an alternative material for the ageing of spirits and wines and to obtain different sensorial profiles from oak or chestnut. There are other compounds, as vanillin, vanillic acid, syringaldehyde, sinapaldehyde or coniferylaldehyde that come from lignin degradation and are also present in cherry wood [23], but their presence is higher when the wood is toasted. In this sample, these compounds are at trace level, below the limit of quantification, and thus, they could not be quantified.

In summary, a similar profile has been observed in the three types of oak studied. Oak wood is rich in ellagitannins and low molecular weight acids and aldehydes. No flavonoids have been detected in any of them [26]. Besides being from the same family, the three oak wood chips analyzed were toasted. During the toasting process, wood increases its concentration of compounds derived from lignin, and a different reduction between phenolic profiles was observed for the different woods [26]. On the other hand, untoasted chestnut chip wood was studied. This wood is slightly similar to oak wood; it is also rich in ellagitannins and low molecular weight acids and aldehydes. In this wood, there is also an absence of flavonoids [26]. However, as this wood was untoasted, it has a low concentration in compounds derived from lignin. As regards cherry wood, a phenolic profile totally different from the rest of the woods was observed. This wood is rich in flavonoid-type compounds and has a certain deficiency in ellagitannins and derivatives. It was also untoasted, so a low concentration of compounds derived from lignin was found. Cherry wood has a specific profile of low molecular weight compounds [18,23,26,38].

Furfural, hydroxymethylfurfural, vanillin and syringaldehyde were also detected in DTD-GC-MS analysis of wood powder. Furfural and hydroxymethylfurfural levels are higher in DTD-GC-MS experiments than in UHPLC analysis. In DTD-GC-MS, during the prior direct thermal desorption process there is a heating of the sample, reaching high temperatures that affect it, and these compounds are related with the toasting level of the wood. Vanillin and syringaldehyde are also found in a higher amount in DTD-GC-MS, except for chestnut wood. They are compounds that come from lignin thermal degradation, being also affected by high temperatures. During direct thermal desorption, wood is toasted, which produces an alteration of the initial sample that generates differences between the level of the compounds determined by both analyses.

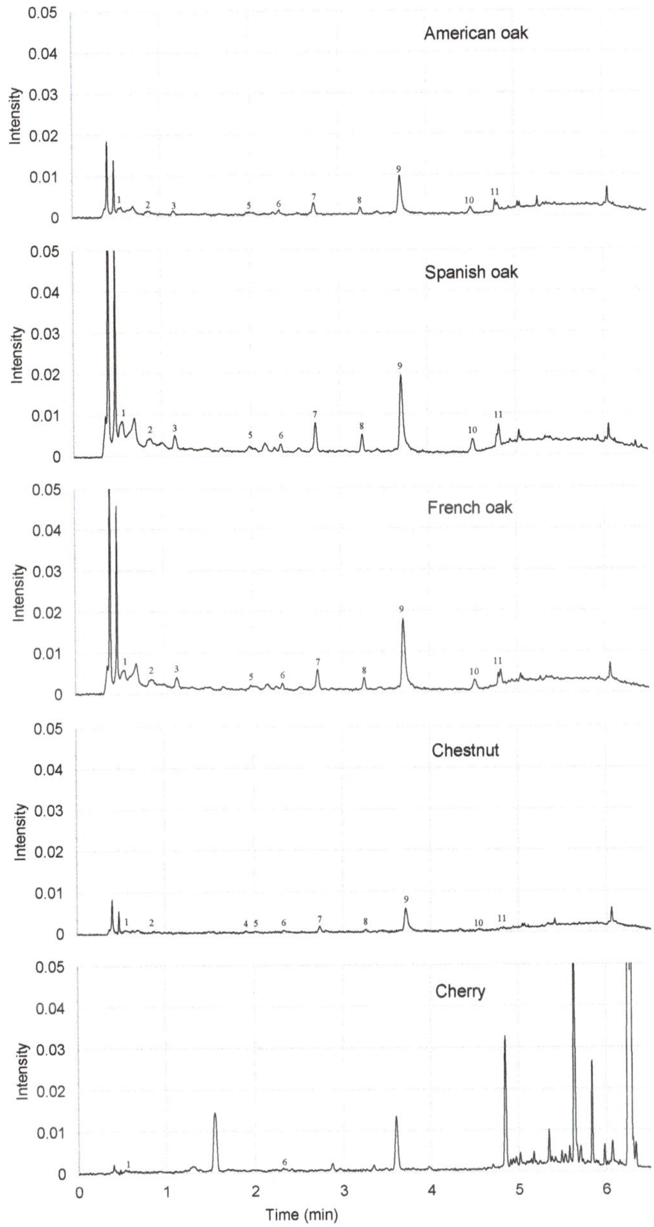


**Figure 6.** Ultra-High-Performance Liquid Chromatography (UHPLC) chromatogram comparison of the hydroalcoholic wood extracts at 280 nm. The key for the compounds is in Table 9.

The phenolic compound content determined by UHPLC was evaluated by ANOVA at the 5% significance level (Table 9). As can be seen, most of them are significantly different, which means that the phenolic profile is characteristic of each wood. Although there are similarities in the presence of some compounds in the three oak woods studied or with chestnut wood, the proportion of them in each wood is different. Only a few similarities were found regarding syringic acid (no significant differences between American and French oak at the 5% significance level were found), vanillic acid (no significant differences between Spanish oak and chestnut at the 5% significance level were found) and hydroxymethylfurfural (no significant differences between Spanish and French oak at the 5% significance level were found).

## 4. Conclusions

The conditions for the DTD-GC-MS of wood samples were optimized and a method for the direct characterization of wood chips studied was established. The optimal direct thermal desorption conditions determined were the following: heating 10 mg of the sample at 250 °C during 7 min, desorbing the sample at 250 °C during 6 min and transferring the desorbed compounds to the line at 1:10 ratio split.

In this study, five different wood chips were characterized. The characterization of their aroma profile by DTD-GC-MS was carried out and different profiles of each wood were determined. Compounds as acetic acid, furfural, 5-methyfurfural and palmitic acid were found in all types of wood studied. However, compounds as whiskey lactones (American and French oak), 4-cyclopentene-1,3-dione (American oak), 2,3-butanediol (Chestnut), glycerin (Chestnut), ethyl hydrogen succinate (Chestnut), cyclopropyl carbinol (Cherry), pyranone (Cherry), levulinic acid (Cherry) or p-acetylacetophenone (Cherry) are only present in certain types of wood, which make them interesting target compounds to identify these woods. Similar volatile compounds were detected in all the woods studied during the first 40 min of the analysis. However, the aromatic profile is totally different from one wood to another at the end of the chromatogram. These differences allow them to be distinguished. This method is a potential technique to identify aromas in wood that, in addition, allows to differentiate between different types of wood.

To analyze the aromatic profile of the hydroalcoholic wood extracts, SBSE-GC-MS was employed. The differentiation of the woods was not possible by this technique due to the similarity of all the chromatograms obtained between them and the extractant used. However, the phenolic profile of wood extracts, determined by UHPLC, allowed this differentiation. There were differences between the phenolic profile of oak woods, and chestnut and cherry wood extracts. Cherry wood extract has the most particular phenolic profile of the wood extracts studied.

Direct determination of wood aromas is possible due to the direct thermal desorption. This technique allows the analysis of wood in solid state, without any type of previous treatment, except grinding. DTD-GC-MS analysis allows the determination of the aromatic profile, without any loss of information. There are other techniques that involve a previous treatment of the sample, such as extraction, where some aromatic compounds may be lost. Furthermore, the information obtained by DTD-GC-MS was characteristic of each type of wood, allowing its differentiation.

This work established and optimized a novel method for the characterization of wood chips in a direct way, and for the characterization of their extractable compounds, allowing its application to other types of samples as barrel staves. This is an interesting strategy that could be applied, not only for the analysis of wood chips, but also for wooden barrels used during the ageing process of spirits and wines.

**Author Contributions:** Conceptualization, methodology, formal analysis, D.A.G.-S., M.G.-C., M.J.V.-M. and M.V.G.-M.; investigation, M.G.-C.; writing—original draft preparation, M.G.-C.; writing—review and editing, D.A.G.-S., M.J.V.-M. and M.V.G.-M.; supervision, D.A.G.-S., M.J.V.-M. and M.V.G.-M.; project administration, D.A.G.-S. and M.V.G.-M.; funding acquisition, D.A.G.-S. and M.V.G.-M. All authors have read and agreed to the published version of the manuscript.

**Funding:** This research was supported by the University of Cádiz and Bodegas Fundador, S.L.U. (ref.: OT2019/108). The authors wish to thank the University of Cádiz and Bodegas Fundador, S.L.U. for the industrial predoctoral contract TDI-8-18, granted to the author María Guerrero Chanivet.

**Conflicts of Interest:** The authors declare no conflict of interest.

## References


1. Caldeira, I.; Mateus, A.M.; Belchior, A.P. Flavour and odour profile modifications during the first five years of Lourinhã brandy maturation on different wooden barrels. *Anal. Chim. Acta* **2006**, *563*, 264–273. [CrossRef]
2. Jordao, A.M.; Ricardo-da-Silva, J.M.; Laureano, O. Comparison of volatile composition of cooperage oak wood of different origins (Quercus pyrenaica vs. Quercus alba and Quercus petraea). *Mitt. Klosterneubg.* **2005**, *55*, 22–31.
3. Martínez-Gil, A.; del Alamo-Sanza, M.; Sánchez-Gómez, R.; Nevares, I. Different Woods in Cooperage for Oenology: A Review. *Beverages* **2018**, *4*, 94. [CrossRef]
4. Canas, S.; Silva, V.; Belchior, A.P. Wood related chemical markers of aged wine brandies. *Cienc. Tec. Vitivinic.* **2008**, *23*, 45–52.
5. De Rosso, M.; Cancian, D.; Panighel, A.; Dalla Vedova, A.; Flamini, R. Chemical compounds released from five different woods used to make barrels for aging wines and spirits: Volatile compounds and polyphenols. *Wood Sci. Technol.* **2009**, *43*, 375–385. [CrossRef]
6. Prida, A.; Puech, J.L. Influence of geographical origin and botanical species on the content of extractives in American, French, and East European oak woods. *J. Agric. Food Chem.* **2006**, *54*, 8115–8126. [CrossRef]
7. Pérez-Prieto, L.J.; López-Roca, J.M.; Martínez-Cutillas, A.; Pardo Mínguez, F.; Gómez-Plaza, E. Maturing wines in oak barrels. Effects of origin, volume, and age of the barrel on the wine volatile composition. *J. Agric. Food Chem.* **2002**, *50*, 3272–3276. [CrossRef]
8. Bautista-Ortín, A.B.; Lencina, A.G.; Cano-López, M.; Pardo-Mínguez, F.; López-Roca, J.M.; Gómez-Plaza, E. The use of oak chips during the ageing of a red wine in stainless steel tanks or used barrels: Effect of the contact time and size of the oak chips on aroma compounds. *Aust. J. Grape Wine Res.* **2008**, *14*, 63–70. [CrossRef]
9. Fernández De Simón, B.; Muiño, I.; Cadahía, E. Characterization of volatile constituents in commercial oak wood chips. *J. Agric. Food Chem.* **2010**, *58*, 9587–9596. [CrossRef]
10. Canas, S. Phenolic Composition and Related Properties of Aged Wine Spirits: Influence of Barrel Characteristics. A Review. *Beverages* **2017**, *3*, 55. [CrossRef]
11. Soares, B.; Garcia, R.; Freitas, A.M.C.; Cabrita, M.J. Phenolic compounds released from oak, cherry, chestnut and robinia chips into a syntethic wine: Influence of toasting level. *Cienc. Tec. Vitivinic.* **2012**, *27*, 17–26.
12. Nishimura, K.; Matsuyama, R.R. Maturation and Maturation Chemistry. In *The Science and Technology of Whiskies*; Piggott, J., Sharp, R., Duncan, R.E.B., Eds.; Longman Scientific and Technical, London: London, UK, 1989; pp. 253–263.
13. Mosedale, J.; Puech, J.-L. Wood maturation of distilled beverages. *Trends Food Sci. Technol.* **1998**, *9*, 95–101. [CrossRef]
14. Conner, J.M.; Paterson, A.; Piggott, J.R. Analysis of lignin from oak casks used for the maturation of Scotch whisky. *J. Sci. Food Agric.* **1992**, *60*, 349–353. [CrossRef]
15. Sarni, F.; Moutounet, M.; Puech, J.L.; Rabier, P. Effect of heat treatment of oak wood extractable compounds. *Holzforschung* **1990**, *44*, 461–466. [CrossRef]
16. Le Floch, A.; Jourdes, M.; Teissedre, P.-L. Polysaccharides and lignin from oak wood used in cooperage: Composition, interest, assays: A review. *Carbohydr. Res.* **2015**, *417*, 94–102. [CrossRef]
17. ISO 11664-4 Colorimetry. Part 4: CIE 1976 L*a*b* Colour Space. Available online: https://www.iso.org/standard/52497.html (accessed on 5 November 2020).
18. Canas, S.; Leandro, M.C.; Spranger, M.I.; Belchior, A.P. Low molecular weight organic compounds of chestnut wood (*Castanea sativa* L.) and corresponding aged brandies. *J. Agric. Food Chem.* **1999**, *47*, 5023–5030. [CrossRef]
19. Canas, S.; Belchior, A.P.; Falcao, A.; Gonçalves, J.; Spranger, M.; Bruno de Sousa, R. Effect of heat treatment on the thermal and chemical modifications of oak and chestnut wood useed in brandy ageing. *Ciênc. Tec. Vitivinic.* **2007**, *22*, 5–14.
20. Fernández de Simón, B.; Esteruelas, E.; Muñoz, A.M.; Cadahía, E.; Sanz, M. Volatile Compounds in Acacia, Chestnut, Cherry, Ash, and Oak Woods, with a View to Their Use in Cooperage. *J. Agric. Food Chem.* **2009**, *57*, 3217–3227. [CrossRef]
21. Fernández De Simón, B.; Martínez, J.; Sanz, M.; Cadahía, E.; Esteruelas, E.; Muñoz, A.M. Volatile compounds and sensorial characterisation of red wine aged in cherry, chestnut, false acacia, ash and oak wood barrels. *Food Chem.* **2014**, *147*, 346–356. [CrossRef]

22. Setzer, W.N. Volatile components of oak and cherry wood chips used in aging of beer, wine, and sprits. *Am. J. Essent. Oils Nat. Prod.* **2016**, *4*, 37–40.
23. Sanz, M.; Cadahía, E.; Esteruelas, E.; Muñoz, Á.M.; Fernández De Simón, B.; Hernández, T.; Estrella, I. Phenolic compounds in cherry (*Prunus avium*) heartwood with a view to their use in cooperage. *J. Agric. Food Chem.* **2010**, *58*, 4907–4914. [CrossRef]
24. Delgado, R.; Durán, E.; Castro, R.; Natera, R.; Barroso, C.G. Development of a stir bar sorptive extraction method coupled to gas chromatography-mass spectrometry for the analysis of volatile compounds in Sherry brandy. *Anal. Chim. Acta* **2010**, *672*, 130–136. [CrossRef]
25. Schwarz, M.; Rodríguez, M.C.; Guillén, D.A.; Barroso, C.G. Development and validation of UPLC for the determination of phenolic compounds and furanic derivatives in Brandy de Jerez. *J. Sep. Sci.* **2009**, *32*, 1782–1790. [CrossRef] [PubMed]
26. Sanz, M.; Fernández, B.; Simón, D.; Cadahía, E.; Esteruelas, E.; Teresa, M.; Estrella, I.; Ángel, M. Analytica Chimica Acta Polyphenolic profile as a useful tool to identify the wood used in wine aging. *Anal. Chim. Acta* **2012**, *732*, 33–45. [CrossRef]
27. Cordenonsi, L.M.; Sponchiado, R.M.; Campanharo, S.C.; Garcia, C.V.; Raffin, R.P.; Schapoval, E.E.S. Study of flavonoids present in pomelo (*citrus maxima*) by dsc, uv-vis, ir, 1h and 13c nmr and ms. *Drug Anal. Res.* **2017**, *1*, 31–37. [CrossRef]
28. Álvarez Batista, M.Á. *Influencia del "Envinado" de las Botas de Roble Americano en la Evolución de los Aguardientes de vino Durante su Envejecimiento*; Universidad de Sevilla: Sevilla, Spain, 1997.
29. Pérez-Coello, M.S.; Sanz, J.; Cabezudo, M.D. Analysis of volatile components of oak wood by solvent extraction and direct thermal desorption-gas chromatography-mass spectrometry. *J. Chromatogr. A* **1997**, *778*, 427–434. [CrossRef]
30. Salem, M.Z.M.; Ali, H.M.; Mansour, M.M. Fatty acid methyl esters from air-dried wood, bark, and leaves of Brachychiton diversifolius R. Br: Antibacterial, antifungal, and antioxidant activities. *BioResources* **2014**, *9*, 3835–3845. [CrossRef]
31. Pearce, R.B.; Holloway, P.J. Suberin in the sapwood of oak (*Quercus robur* L.): Its composition from a compartmentalization barrier and its occurrence in tyloses in undecayed wood. *Physiol. Plant Pathol.* **1984**, *24*, 71–81. [CrossRef]
32. Dodd, R.S.; Rafiit, Z.A.; Zavarin, E. Chemosystematic Variation in Acorn Fatty Acid of Californian Live Oaks. *Biochem. Syst. Ecol.* **1993**, *21*, 279–285. [CrossRef]
33. Cernîşev, S. Analysis of lignin-derived phenolic compounds and their transformations in aged wine distillates. *Food Control* **2017**, *73*, 281–290. [CrossRef]
34. Bonerz, D.P.M.; Bloomfield, D.G.; Dykes, S.I.; Rawel, H.M.; Rohn, S.; Kroll, J.; Creasy, G.L.; Pour Nikfardjam, M.S. A new gentle fining agent for Pinot Noir. *Mitt. Klosterneuburg.* **2004**, *54*, 86–95.
35. Li, X.; Xie, H.; Jiang, Q.; Wei, G.; Lin, L.; Li, C.; Ou, X.; Yang, L.; Xie, Y.; Fu, Z.; et al. The mechanism of (+) taxifolin's protective antioxidant effect for ·OH-treated bone marrow-derived mesenchymal stem cells. *Cell. Mol. Biol. Lett.* **2017**, *22*, 1–11. [CrossRef]
36. Sasaya, T.; Demachi, S.; Terazawa, M. Studies on the Extractives of Larch Report 11: Determination of Flavonoids in Larix leptolepis. *Res. Bull. Coll. Exp. For.* **1997**, *27*, 429–443.
37. Kim, B.G. Biosynthesis of bioactive isokaemferide from naringenin in Escherichia coli. *J. Appl. Biol. Chem.* **2019**, *62*, 1–6. [CrossRef]
38. Sanz, M.; De Simón, B.F.; Cadahía, E.; Esteruelas, E.; Muñoz, A.M.; Hernández, T.; Estrella, I.; Pinto, E. LC-DAD/ESI-MS/MS study of phenolic compounds in ash (*Fraxinus excelsior* L. and *F. americana* L.) heartwood. Effect of toasting intensity at cooperage. *J. Mass Spectrom.* **2012**, *47*, 905–918. [CrossRef]

foods

MDPI

Article

# Evolution of the Aroma of Treixadura Wines during Bottle Aging

Iván Vázquez-Pateiro [1], Uxía Arias-González [1], José Manuel Mirás-Avalos [2,*] and Elena Falqué [1,3,*]

1 Depto. Química Analítica, Facultad de Ciencias, Universidade de Vigo, As Lagoas s/n, 32004 Ourense, Spain; ivpateiro@yahoo.es (I.V.-P.); uxiaariasgonzalez@gmail.com (U.A.-G.)
2 Unidad de Suelos y Riegos (Asociada a EEAD-CSIC), Centro de Investigación y Tecnología Agroalimentaria de Aragón (CITA), Avda. Montañana 930, 50059 Zaragoza, Spain
3 Clúster de Investigación y Transferencia Agroalimentaria del Campus Auga (CITACA), As Lagoas s/n, 32004 Ourense, Spain
* Correspondence: jmmiras@cita-aragon.es (J.M.M.-A.); efalque@uvigo.es (E.F.)

Received: 7 September 2020; Accepted: 3 October 2020; Published: 8 October 2020

**Abstract:** Aroma is a crucial attribute for wine quality, particularly in white wines. Traditionally, the consumption of young white wines is recommended over the year following grape harvest due to potential aroma losses that would worsen wine quality. This study aimed to investigate the evolution of volatile compounds, odor activity value-based aroma notes, and sensory perception in Treixadura (*Vitis vinifera* L.) dry white wines during a 24-month bottle-aging period. Volatile composition was determined by gas chromatography, and wine sensory evaluation was performed by experts. Wine samples had similar volatile compositions at the time of bottling. The volatile contents of the wines were respectively 322.9, 302.7, 323.0, and 280.9 mg $L^{-1}$ after 6, 12, 18, and 24 months of bottle storage. Most of the volatiles tended to maintain constant concentrations, or with slight increases in all families of volatiles except for acetates and carbonyl compounds, until two years after harvest (18 months of bottle storage) and, then, concentrations reduced sharply. After 24 months of storage in the bottle, the concentrations of terpenes, C6 compounds, higher alcohols, ethyl esters, fatty acids, acetates, carbonyl compounds, and volatile phenols were reduced by 32%, 47%, 11%, 39%, 50%, 74%, 41%, and 54%, respectively. The 18-month bottle-aged wines showed the highest concentrations of volatiles, as well as the best performance in the sensory evaluation, suggesting that a good balance of the aroma attributes was achieved on this date. In conclusion, the current study suggests that Treixadura wines expressed their maximum aroma potential two years after grape harvest.

**Keywords:** bottle aging; flavor profile; sensory evaluation; volatile composition; white wine

## 1. Introduction

Wine aroma is produced by the interactions of hundreds of chemical compounds derived from multiple sources [1]. According to their origin, wine aroma compounds can be grape-derived such as monoterpenes and norisoprenoids [2,3]; microbially-derived secondary metabolites formed from sugar and amino acid metabolism during the fermentation [1,4]; and those compounds formed during wine storage, either in oak barrels [5,6] or in bottles [7,8]. The major groups of aroma compounds are monoterpenes, norisoprenoids, aliphatics, higher alcohols, esters, phenylpropanoids, methoxypyrazines, and volatile sulfur [2,9]. However, identifying one single compound that defines the character of a given grapevine variety has seldom been accomplished [1]. Therefore, the varietal character depends on the overall profile of odor-active compounds present in the grape and corresponding wine [1]. This character is extremely important for wine typicity and commercial success, as most wineries rely on this concept for marketing campaigns.

Wine aroma slightly evolves during bottle aging because the amounts of oxygen that penetrate through the closures are low [7,10,11]. Oxygen penetrates through the stoppers at a rate between 0.005 and 5 mg $L^{-1}$ $year^{-1}$ [12], depending on the type of closure used [13,14]. Small doses of oxygen may have a favorable effect on wine aroma, such as the decomposition of sulfur compounds responsible for negative flavors; however, an excess of oxygen can have adverse effects on wine aroma [14], leading to the question of how long can a given wine type be stored or aged in the bottle. In this context, the redox status can affect the release of certain varietal aromas from amino acid metabolism [15], but also can lead to the appearance of reductive aromas from sulfur compounds such as dimethyl sulfide [16].

In the northwest of the Iberian Peninsula (the regions of Galicia in Spain and Tras-dos-Montes in Portugal), white grapevine varieties are predominantly grown. Among the traditional cultivars from these regions, Treixadura is one of the most important because it is used to obtain balanced wines with a high aromatic potential [17], especially those monovarietal wines from the Ribeiro Designation of Origin (DO) in Galicia. Similar to other white wines, higher alcohols are the most important volatiles from the quantitative point of view in Treixadura wines, whereas ethyl esters, acetates, and fatty acids are qualitatively relevant for the aroma of these wines [18]. In fact, nine volatiles have a significant relevance on the aroma of Treixadura wines, including higher alcohols (2-methyl-1-butanol, 3-methyl-1-butanol, and 2-phenylethanol), acetates (isoamyl acetate and ethyl acetate), and esters (ethyl butyrate, ethyl hexanoate, ethyl octanoate, and ethyl decanoate) [18]. The concentrations of higher alcohols in Treixadura wines can be explained by the high contents in amino acids observed in grapes from this variety [19]. However, the volatile compounds that have the most relevant role in the aroma of Treixadura wines are ethyl octanoate, ethyl hexanoate, and isoamyl acetate, which provide fruity nuances [17,20]. In contrast, the contents of monoterpenes and norisoprenoids in Treixadura wines are very low and, consequently, they do not play a relevant role in the aromatic profile of the wines from this variety [17,21]. Despite being present in Treixadura wines, linalool, citronellol, and geraniol appear at low concentrations that usually do not surpass their respective odor thresholds [22]. In addition, bounded terpenes do not appear in high concentrations [23]. Finally, Treixadura wines have low concentrations of sulfur compounds, although 3-methyl-propyl acetate and 4-methyl-1-butanol may provide onion, garlic, and fungal nuances [24]. Therefore, the volatile composition of wines from this variety has been previously described under several situations [17,20,25,26]; however, no information is available about the evolution of the aroma of Treixadura wines over their storage in bottles and this leads to a debate on when the optimum time for consumption is. Furthermore, investigations monitoring the evolution of dry white wines' flavor profiles during bottle aging are limited [7].

In this context, the aim of the current study was to assess, on a six-month basis, the alterations in the volatile composition and sensory properties of Treixadura wines from the Ribeiro DO produced over a 24-month period of bottle aging. Finally, the optimal period for consumption of Treixadura wines was determined based on the evolution of the volatile and aroma properties, providing useful information to winemakers for managing their wine stock and developing marketing campaigns.

## 2. Materials and Methods

### 2.1. Wine Samples

Treixadura wines used in the current study, corresponding to the 2013 vintage, were made at industrial scale by several wineries from the Ribeiro DO employing their standard winemaking protocols. Bottling was performed on May 2014 at the packaging line of each winery to ensure a 750 mL volume of each bottle. Wine bottles coming from the same fermentation tank were stored in a cool place under dark conditions until analysis. All bottles had the same type of closure in order to avoid different oxygen penetration rates into the bottles. Wines were analyzed on a 6-month basis: November 2014 (M6), May 2015 (M12), November 2015 (M18), and May 2016 (M24).

### 2.2. Determination of Volatile Compounds

Methanol and higher alcohols were determined in triplicate by direct injection of 2 μL, from 5 mL of wine to which 1 mL of 4-methyl-2-pentanol (1 g $L^{-1}$) was added as internal standard, into a Hewlett Packard 5890 gas chromatograph using an HP-Innowax capillary column (60 m × 0.25 mm i.d.; film thickness 0.25 μm) as described by Bertrand and Ribéreau-Gayon [27].

The extraction of the rest of volatile compounds was performed according to Armada et al. [28]. Briefly, a wine sample of 100 mL containing 2 mL of 3-octanol (20 mg $L^{-1}$) and 2 mL of 3,4-dimethyl-phenol (100 mg $L^{-1}$) as internal standards was extracted three times (10, 5, and 5 mL) with dichloromethane. Then, the organic extract was dried and concentrated to 0.5 mL under nitrogen, and 3 μL were injected in triplicate in splitless mode (purge time, 30 s; purge rate, 70) in a Hewlett Packard HP 5890-I gas chromatograph coupled to a Hewlett Packard HP 5970 mass spectrometer. Spectra were recorded in the electron impact mode (ionization energy, 70 eV; source temperature, 250 °C), using an HP-Innowax column (60 m × 0.25 mm i.d.; film thickness 0.25 μm). The carrier gas was helium (18 psi). The temperature program was isothermal at 45 °C for 1 min, then 3 °C $min^{-1}$ to 230 °C with a final isotherm of 25 min. The acquisition was made in scanning mode (mass range, 30–300 amu; 1.9 spectra $s^{-1}$).

The identification of the volatile compounds was confirmed by comparing their mass spectra (MS Chemstation Wiley 7N library) and their retention times with those of the pure compounds. For obtaining the calibration curves, five known amounts of the analytes were subjected to the same liquid–liquid extraction as that for the wine samples, and the quantification was carried out by the interpolation of relative peak areas with respect to the response of internal standards. Those substances for which pure compounds were not available were referred as a function of the normalized area respect to the internal standard (3-octanol). Each wine sample was analyzed in triplicate.

### 2.3. Aromatic Index

In order to estimate the influence of each volatile on the Treixadura wine aroma, odor activity values (OAV) were computed as the ratio between the concentration of a given compound and its corresponding perception threshold [29]. Theoretically, OAV should be greater than the unity [29]; however, due to synergic effects among different substances, those compounds with values greater than 0.2 can be considered as active aromas [30]. The odor thresholds for the compounds considered in this study, along with their corresponding aromatic descriptors, are shown in Table 1.

### 2.4. Sensory Evaluation

Four wine sensory assessments were carried out over the study period, each one approximately 15 days after the performance of the gas chromatography determinations. The panel consisted of 6 to 10 professional enologists (25–50 years of age, 25% females and 75% males), most of them from the wineries that supplied the wine samples. All wines were tasted in the same session, but the sessions were not replicated due to the availability of the tasting panel. The wines were served in standard tasting glasses coded with random numbers and covered with a watch-glass to minimize the loss of volatile compounds. Testing temperature was 10 °C and room temperature was 20–22 °C. A card of 7 aromatic attributes (floral, fruity, grass, spicy, woody, sulfurous, and caramel) accompanied by a scale from 0 to 10 to rate the intensity of each nuance in each wine sample, where 0 indicated that the descriptor was not perceived and 10 indicated the highest intensity. In addition, panellists must score the global quality of the wine sample both at the aroma (olfactory) and taste (mouthfeel) levels, as well as provide a global mark for the wine overall quality.

**Table 1.** Odor thresholds, matrix in which they were obtained, and descriptors for several volatile compounds. References for the thresholds are included.

| Family | Compound | Odor Threshold (mg L$^{-1}$) | Matrix | Descriptor | Reference |
|---|---|---|---|---|---|
| Terpenes | linalool | 0.050 | Wine | Rose | [31] |
| | α-terpineol | 0.400 | Wine | Flowers, linden | |
| C6 Compounds | 1-hexanol | 4 | Ethanol (11%) | Herbaceous | [32] |
| | *cis*-3-hexen-1-ol | 1 | Ethanol (10%) | Green, bitter | [33] |
| | *trans*-3-hexen-1-ol | 13 | Beer | | [34] |
| Higher Alcohols | 1-propanol | 30 | Not specified | Ripe fruit | [35] |
| | 1-butanol | 11 | Not specified | Medicine | |
| | isobutanol | 75 | Ethanol (10%) | Clove | [33] |
| | isoamyl alcohol | 40 | Ethanol (10%) | Fusel | [36] |
| | 2-phenylethanol | 14 | Ethanol (10%) | Rose, honey | [37] |
| Alcohols | methanol | 2000 | Not specified | Alcohol | [35] |
| | benzyl alcohol | 900 | Beer | Blackberry | [34] |
| Carbonyl Compounds | benzaldehyde | 2 | Ethanol (10%) | Almond | [33] |
| | furfural | 150 | Beer | Toasted | [34] |
| | acetoine | 150 | Ethanol (12%) | - | [38] |
| Ethyl Esters | ethyl butyrate | 0.4 | Ethanol (10%) | Blueberry | [33] |
| | ethyl hexanoate | 0.08 | | Green apple | |
| | ethyl octanoate | 0.58 | | Sweet, flower | |
| | ethyl decanoate | 0.5 | | Brandy, grape | |
| | ethyl lactate | 150 | | Butter | |
| | diethyl succinate | 1.2 | | Melon | |
| Acetates of Higher Alcohols | isoamyl acetate | 0.16 | Ethanol (10%) | Banana | [33] |
| | hexyl acetate | 0.67 | | Pear, apple, cherry | |
| | 2-phenylethyl acetate | 1.8 | | Rose, flower | |
| Volatile Fatty Acids | butyric acid | 4 | Ethanol (9.5%) | Butter, cheese | [39] |
| | isobutyric acid | 2.3 | Ethanol (11%) | - | [37] |
| | isovaleric acid | 0.03 | | - | |
| | hexanoic acid | 3 | Beer | Cheese, fatty | [34] |
| | octanoic acid | 10 | | Fatty, rancid | |
| | decanoic acid | 6 | | | |
| Volatile Phenols | 4-vinyl-guaiacol | 0.440 | Ethanol (12%) | Paint, watercolor | [40] |
| | 4-vinyl-phenol | 0.375 | | Pharmacy, clove | |

### 2.5. Statistical Analysis

Significant differences among times after bottling for the concentrations of each volatile compound were assessed with a one-way analysis of variance (ANOVA). Post-hoc comparison of means was performed using the Fischer's Least Significant Difference (LSD) test. Similarly, ANOVA was used to determine the influence of time after bottling on the OAV of each compound. Principal Component Analysis (PCA) was applied to discriminate among the means of families of volatile compounds in the samples according to the time after bottling. Statistical analysis was carried out using R environment v.3.6.2 [41].

## 3. Results

### 3.1. Evolution of the Concentrations of Volatile Compounds Over Storage Time in the Bottle

Eight monovarietal Treixadura wines made at industrial scale were analyzed and the volatile composition of each of them is shown in the Supplementary Tables S1 to S8. A total of 44 volatiles were detected in the Treixadura wine samples studied, including terpenes, C6 compounds, higher alcohols, esters, volatile fatty acids, acetates, carbonyl compounds, volatile phenols, and other compounds, and the average value at each sampling date is displayed in Table 2. Terpenes appeared at low concentrations and the most relevant volatile within this family was linalool (Table 2). Among C6 compounds, 1-hexanol was the most quantitatively important volatile in Treixadura wines (Table 2). Isoamyl acetate and methanol were the most relevant higher alcohols detected in the samples studied (Table 2). The most relevant ester was ethyl octanoate, whereas octanoic acid was the most quantitatively important fatty acid in the Treixadura wines studied (Table 2). Finally, the most relevant volatiles among acetates, carbonyl compounds, and volatile phenols were, respectively, isoamyl acetate, acetoine, and 4-vinyl-phenol (Table 2).

**Table 2.** Average concentrations of volatile compounds (mean ± standard error) in Treixadura wines from the Ribeiro Designation of Origin (DO) at different times of bottle aging. M6, M12, M18, and M24 indicate 6, 12, 18, and 24 months after bottling.

| Family | Compound [1] | M6 | M12 | M18 | M24 | *P*-Value [1] |
|---|---|---|---|---|---|---|
| Terpenes | linalool * | 29.6 ± 2.4 a | 29.7 ± 1.0 a | 32.6 ± 1.6 a | 18.4 ± 0.7 b | 0.003 |
| | α-terpineol * | 21.9 ± 5.2 | 22.0 ± 6.8 | 34.9 ± 5.4 | 16.7 ± 2.2 | 0.597 |
| C6 Compounds | 1-hexanol | 1.3 ± 0.1 a | 1.1 ± 0.1 ab | 1.4 ± 0.1 a | 0.7 ± 0.0 b | 0.011 |
| | *cis*-3-hexen-1-ol | 0.22 ± 0.05 | 0.18 ± 0.04 | 0.19 ± 0.04 | 0.11 ± 0.02 | 0.736 |
| | *trans*-3-hexen-1-ol | 0.20 ± 0.08 | 0.26 ± 0.11 | 0.28 ± 0.12 | 0.11 ± 0.04 | 0.861 |
| Alcohols | methanol | 63.3 ± 3.5 | 58.4 ± 2.3 | 63.4 ± 2.6 | 63.1 ± 2.9 | 0.710 |
| | 1-propanol | 15.1 ± 1.3 | 12.8 ± 1.0 | 13.3 ± 1.4 | 16.1 ± 1.4 | 0.494 |
| | isobutanol | 20.2 ± 1.2 | 19.2 ± 1.1 | 22.3 ± 1.6 | 17.9 ± 1.2 | 0.297 |
| | 1-butanol | 1.3 ± 0.1 | 1.3 ± 0.1 | 1.3 ± 0.1 | 0.9 ± 0.1 | 0.134 |
| | isoamyl alcohol | 164.3 ± 6.9 | 158.0 ± 7.2 | 159.5 ± 8.0 | 152.3 ± 6.4 | 0.824 |
| | benzyl alcohol | 0.08 ± 0.01 | 0.29 ± 0.08 | 0.21 ± 0.02 | 0.09 ± 0.01 | 0.129 |
| | 2-phenylethanol | 8.5 ± 0.5 a | 7.8 ± 0.3 a | 9.1 ± 0.5 a | 3.7 ± 0.3 a | <0.001 |
| | 3-methyl-1-pentanol | 0.07 ± 0.01 ab | 0.07 ± 0.00 ab | 0.07 ± 0.00 a | 0.04 ± 0.00 b | 0.031 |
| | 3-ethoxy-1-propanol # | 38.3 ± 7.9 | 38.9 ± 7.0 | 35.1 ± 7.2 | 19.9 ± 3.9 | 0.456 |
| | 1,2-propanodiol # | 8.6 ± 0.7 | 8.0 ± 0.3 | 7.4 ± 0.7 | 4.8 ± 0.2 | 0.221 |
| | 1,3-butanediol # | 355.8 ± 35.1 a | 258.9 ± 15.1 ab | 247.1 ± 19.8 ab | 132.8 ± 15.8 b | <0.001 |
| | 2,3-butanediol # | 78.9 ± 7.2 a | 66.0 ± 5.9 a | 59.5 ± 4.1 ab | 32.9 ± 3.7 b | 0.003 |
| Ethyl Esters | ethyl butyrate | 0.44 ± 0.04 | 0.39 ± 0.03 | 0.44 ± 0.03 | 0.31 ± 0.02 | 0.167 |
| | ethyl hexanoate | 0.65 ± 0.03 a | 0.54 ± 0.05 a | 0.63 ± 0.02 a | 0.35 ± 0.02 b | <0.001 |
| | ethyl octanoate | 1.41 ± 0.13 a | 1.48 ± 0.06 a | 1.21 ± 0.05 a | 0.63 ± 0.02 b | <0.001 |
| | ethyl decanoate | 0.58 ± 0.02 a | 0.63 ± 0.04 a | 0.63 ± 0.02 a | 0.30 ± 0.01 b | <0.001 |
| | ethyl-3-hydroxybutyrate | 0.14 ± 0.01 a | 0.14 ± 0.01 a | 0.16 ± 0.01 a | 0.08 ± 0.01 b | 0.006 |
| | ethyl-4-hydroxybutyrate # | 92.7 ± 13.6 a | 62.7 ± 6.3 ab | 56.4 ± 6.6 ab | 22.3 ± 2.2 b | 0.003 |
| | ethyl lactate | 13.8 ± 1.7 | 9.7 ± 0.5 | 15.1 ± 2.1 | 8.4 ± 1.3 | 0.170 |
| | monoethyl succinate # | 50.3 ± 3.0 b | 54.2 ± 2.5 b | 79.3 ± 3.9 a | 27.7 ± 1.4 c | <0.001 |
| | diethyl succinate | 0.78 ± 0.06 b | 1.16 ± 0.11 b | 1.80 ± 0.11 a | 0.85 ± 0.08 b | <0.001 |
| Volatile Fatty Acids | isobutyric acid | 1.01 ± 0.08 a | 0.93 ± 0.07 a | 0.93 ± 0.06 a | 0.56 ± 0.03 b | 0.002 |
| | butyric acid | 2.88 ± 0.13 a | 2.75 ± 0.11 a | 2.90 ± 0.13 a | 1.62 ± 0.14 b | <0.001 |
| | isovaleric acid | 0.75 ± 0.04 a | 0.73 ± 0.02 a | 0.74 ± 0.03 a | 0.40 ± 0.02 b | <0.001 |
| | hexanoic acid | 4.7 ± 0.1 a | 4.2 ± 0.1 a | 4.7 ± 0.1 a | 2.2 ± 0.1 b | <0.001 |
| | octanoic acid | 6.4 ± 0.2 a | 6.1 ± 0.2 a | 6.9 ± 0.1 a | 3.2 ± 0.1 b | <0.001 |
| | decanoic acid | 2.1 ± 0.1 a | 2.0 ± 0.1 a | 2.1 ± 0.0 a | 0.9 ± 0.0 b | <0.001 |
| | lauric acid | 0.16 ± 0.02 | 0.16 ± 0.02 | 0.15 ± 0.02 | 0.18 ± 0.07 | 0.987 |
| | *trans*-2-hexenoic acid # | 16.6 ± 1.9 | 16.7 ± 2.2 | 19.0 ± 1.9 | 11.2 ± 0.9 | 0.262 |
| Acetates of Higher Alcohols | isoamyl acetate | 2.1 ± 0.2 a | 1.4 ± 0.1 a | 1.0 ± 0.1 a | 0.5 ± 0.1 b | <0.001 |
| | hexyl acetate | 0.14 ± 0.02 | 0.17 ± 0.05 | 0.15 ± 0.04 | 0.08 ± 0.02 | 0.738 |
| | 2-phenylethyl acetate | 0.10 ± 0.01 a | 0.09 ± 0.01 a | 0.07 ± 0.01 ab | 0.02 ± 0.00 b | 0.003 |
| Carbonyl Compounds | furfural | 0.03 ± 0.00 c | 0.04 ± 0.00 bc | 0.07 ± 0.01 a | 0.05 ± 0.00 b | <0.001 |
| | benzaldehyde | 0.02 ± 0.00 | 0.03 ± 0.00 | 0.03 ± 0.0 | 0.01 ± 0.00 | 0.126 |
| | acetoine | 3.1 ± 0.5 | 2.1 ± 0.2 | 2.6 ± 0.4 | 1.8 ± 0.3 | 0.427 |
| Volatile Phenols | 4-vinyl-phenol | 3.8 ± 0.3 b | 5.5 ± 0.4 ab | 5.7 ± 0.4 a | 1.9 ± 0.1 c | <0.001 |
| | 4-vinyl-guaiacol | 1.4 ± 0.1 b | 1.6 ± 0.1 ab | 2.0 ± 0.1 a | 0.5 ± 0.0 c | <0.001 |
| Others | γ-butyrolactone | 1.8 ± 0.2 ab | 1.4 ± 0.1 ab | 1.9 ± 0.2 a | 1.0 ± 0.1 b | 0.016 |
| | methionol # | 34.2 ± 2.4 a | 34.1 ± 2.0 a | 37.0 ± 2.1 a | 16.4 ± 1.4 b | <0.001 |

[1] Concentrations in mg $L^{-1}$; except for those compounds marked with * in µg $L^{-1}$ and # as normalized area. Different letters in the row indicate significant differences among times after bottling for a given compound.

Bottle storage time significantly affected the concentrations of 26 of these volatiles (Table 2). In general, concentrations were lower at the final measurement date, while no significant differences were detected among the rest of measurement dates (Table 2). Despite this lack of differences, concentrations tended to decline with storage time, except for the monoethyl and diethyl succinates and furfural, which appeared at higher concentrations on the third measurement date (Table 2).

It must be noted that not all the compounds listed in Table 2 were detected in all the Treixadura wine samples (Supplementary Tables S1 to S8). However, the main findings regarding the effect of storage time on the concentrations of volatile compounds were observed for each sample, although some exceptions to these general observations existed. In wine sample 1, the concentration of ethyl butyrate increased over time until the third date of measurements, leading to a greater content of esters on that date (Supplementary Table S1). In wine sample 2, α-terpineol was not detected and the concentrations of 2-phenylethanol and benzyl alcohol increased up to the third measurement date

(Supplementary Table S2). In wine sample 3, the concentrations of volatile fatty acids were rather low but increased on the third date of measurements (Supplementary Table S3). In wine sample 4, no terpenes were detected and the concentrations of higher alcohols and acetates were lower than in the rest of the samples (Supplementary Table S4). In wine sample 5, terpenes appeared at greater concentrations on the third measurement date; the contents of higher alcohols were the greatest compared to the rest of the samples, whereas the carbonyl compounds were detected at the lowest concentrations (Supplementary Table S5). In wine sample 6, C6 compounds were detected at the lowest concentrations when compared with the rest of the samples studied; moreover, their concentrations were significantly higher on the third date of measurements (Supplementary Table S6). Wine sample 7 had the highest and lowest concentrations of terpenes and isoamyl acetate, respectively, when compared with the rest of the samples; while methanol concentration increased with storage time (Supplementary Table S7). In wine sample 8, C6 compounds appeared at low concentrations while volatile fatty acids were detected at high concentrations when compared with the rest of the samples (Supplementary Table S8).

The PCA applied to the average concentrations of the different families of volatiles (Figure 1) explained 97.3% of the variability within the wine samples. The first component (PC1) explained 85.9% of this variability and depended on the concentrations of all families of compounds, whereas PC2 explained 11.4% of the variability and depended on the concentrations of phenols, terpenes, higher alcohols, and acetates (Figure 1). In the bi-plot, M6 was located on the positive side of PC1 and the negative side of PC2, due to the high concentrations of acetates in this sample. Wines from M12 were located in the center of the bi-plot, indicating that these samples did not have outstanding concentrations of any of the families of compounds. Wines from M18 were located on the positive sides of both PC due to their high concentrations of phenols and terpenes. Finally, samples from M24 were located on the negative sides of both PC, indicating that their concentrations on all the families of compounds were lower than those from the rest of the samples (Figure 1).

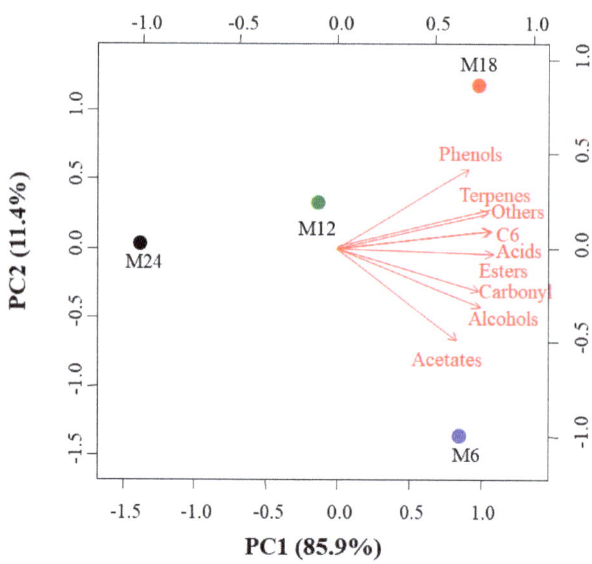


**Figure 1.** Principal component analysis (PCA) of Treixadura wines after several months of aging in the bottle: Bi-plot of the first two components (PC) for families of volatile compounds related to wine aroma. M6, M12, M18, and M24 indicate 6, 12, 18, and 24 months after bottling.

### 3.2. *Effect of Bottle Storage Time on Odor Activity Values*

Table 3 shows the OAV for the 32 volatile compounds for which odor thresholds were available. The volatiles with the highest OAV were isovaleric acid, isoamyl acetate, 4-vinyl-phenol, and ethyl hexanoate. From Table 3, a total of 21 compounds showed OAV greater than 0.2, except for the last measurement date in which 19 compounds showed OAV over this threshold. Moreover, 10 compounds had OAV greater than 1 in the first two measurement dates, 11 compounds on the third date, and 7 compounds on the last date of measurements.

**Table 3.** Odor activity values of volatile compounds in Treixadura wines from the Ribeiro DO at different times of bottle aging. M6, M12, M18, and M24 indicate 6, 12, 18, and 24 months after bottling.

| Family | Compound | M6 | M12 | M18 | M24 | *p*-Value |
|---|---|---|---|---|---|---|
| Terpenes | linalool [1] | 0.6 a | 0.6 a | 0.7 a | 0.4 b | 0.006 |
| | α-terpineol | 0.0 | 0.0 | 0.1 | 0.0 | 0.301 |
| C6 Compounds | 1-hexanol | 0.3 ab | 0.3 ab | 0.4 a | 0.2 b | 0.008 |
| | *cis*-3-hexen-1-ol | 0.2 | 0.2 | 0.2 | 0.2 | 0.994 |
| | *trans*-3-hexen-1-ol | 0.0 | 0.0 | 0.0 | 0.0 | 0.819 |
| Alcohols | methanol | 0.0 | 0.0 | 0.0 | 0.0 | 0.999 |
| | 1-propanol | 0.5 | 0.4 | 0.5 | 0.5 | 0.572 |
| | isobutanol | 0.3 | 0.3 | 0.3 | 0.2 | 0.332 |
| | 1-butanol | 0.1 | 0.1 | 0.1 | 0.1 | 0.240 |
| | isoamyl alcohol | 4.1 | 4.0 | 4.0 | 3.8 | 0.781 |
| | benzyl alcohol | 0.0 | 0.0 | 0.0 | 0.0 | 0.999 |
| | 2-phenylethanol | 0.6 a | 0.5 a | 0.6 a | 0.3 b | <0.001 |
| Ethyl Esters | ethyl butyrate | 1.1 | 1.0 | 1.1 | 0.8 | 0.306 |
| | ethyl hexanoate | 8.2 a | 6.7 ab | 7.8 a | 4.8 b | 0.003 |
| | ethyl octanoate | 2.5 a | 2.5 a | 2.1 a | 1.2 b | <0.001 |
| | ethyl decanoate | 1.2 a | 1.3 a | 1.3 a | 0.7 b | 0.001 |
| | ethyl lactate | 0.1 a | 0.1 a | 0.1 a | 0.0 b | 0.003 |
| | diethyl succinate | 0.7 b | 1.0 a | 1.5 a | 0.7 b | <0.001 |
| Volatile Fatty Acids | isobutyric acid | 0.5 a | 0.4 a | 0.4 a | 0.2 b | 0.002 |
| | butyric acid | 0.7 a | 0.7 a | 0.7 a | 0.4 b | <0.001 |
| | isovaleric acid | 25.1 a | 24.4 a | 24.7 a | 13.3 b | <0.001 |
| | hexanoic acid | 1.6 a | 1.4 a | 1.6 a | 0.7 b | <0.001 |
| | octanoic acid | 0.6 a | 0.6 a | 0.7 a | 0.3 b | <0.001 |
| | decanoic acid | 0.4 | 1.1 | 1.1 | 0.1 | 0.471 |
| Acetates of Higher Alcohols | isoamyl acetate | 13.3 a | 8.7 ab | 6.1 bc | 2.8 c | <0.001 |
| | hexyl acetate | 0.2 | 0.2 | 0.2 | 0.1 | 0.877 |
| | 2-phenylethyl acetate | 0.1 | 0.1 | 0.0 | 0.0 | 0.059 |
| Carbonyl Compounds | furfural | 0.0 | 0.0 | 0.0 | 0.0 | 0.999 |
| | benzaldehyde | 0.0 | 0.0 | 0.0 | 0.0 | 0.999 |
| | acetoine | 0.0 | 0.0 | 0.0 | 0.0 | 0.451 |
| Volatile Phenols | 4-vinyl-phenol | 10.2 b | 14.7 ab | 15.2 a | 5.2 c | <0.001 |
| | 4-vinyl-guaiacol | 3.2 b | 3.6 b | 4.6 a | 1.1 c | <0.001 |

[1] Different letters in the row indicate significant differences among times after bottling for a given compound.

The OAV of 16 compounds were significantly affected by storage time in the bottle (Table 3). In general, OAV were lower on the last date of measurements except for diethyl succinate, 4-vinyl-phenol, and 4-vinyl-guaiacol for which OAV on the last date did not significantly differ from those of the first measurement date (Table 3). In general, all wine samples showed the same profile with 20–22 volatiles with OAV greater than 0.2; from these substances, 10–12 volatiles had OAV greater than 1 (Supplementary Tables S9 and S10). Despite the fact that significant and marked reductions in the concentrations of the volatiles were detected at the end of the period of bottle storage, the reductions in OAV were less relevant. In this sense, some substances passed from having OAV greater than 1 at M6 to OAV in the range of 0.2–1 at M24; however, they can still contribute to wine aroma.

### 3.3. *Evolution of the Sensory Profile of Treixadura Wines over Bottle Storage*

The panellists gave the highest marks to the fruity, floral and grass descriptors (Figure 2), whereas the rest of aroma descriptors did not reach more than two points in the sensory evaluations. Four descriptors (floral, fruity, grass, and caramel) showed significantly different marks depending on the storage time. In the case of floral, wines from M18 had higher marks than those from M12. In the case of fruity, wines from M18 had higher marks than those from M12 and M24. In the case of grass, wines from M24 had lower marks than those from M12 and M18. Finally, wines from M12 received lower marks for caramel than those from M18 and M24.

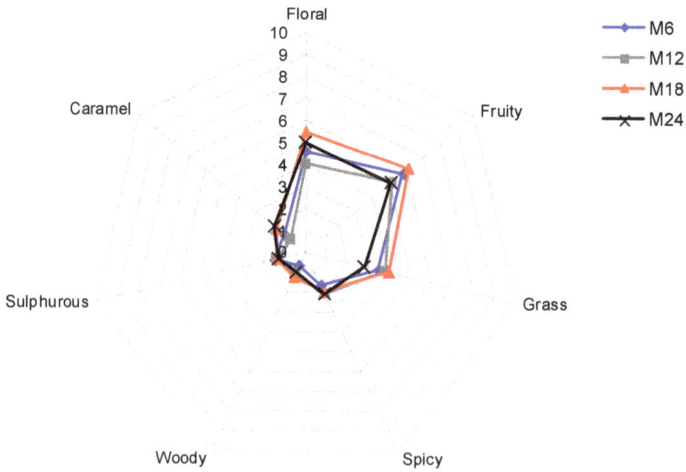


**Figure 2.** Aroma profile of Treixadura wines as affected by storage time in the bottle. Data are averages for the 8 wine samples considered in the current study. M6, M12, M18, and M24 indicate 6, 12, 18, and 24 months after bottling.

No significant differences among storage time in the bottle were detected for the olfactory, mouthfeel, and global marks given to the Treixadura wines, although a trend to higher marks was observed for M6 and M18 (Figure 3). Despite of a certain variability among samples, the highest global quality marks were given to samples after 18 months of bottling (M18). Some of the samples maintained these high marks six months later, but most of them suffered from a decline in this global quality mark by the end of this experiment (M24).

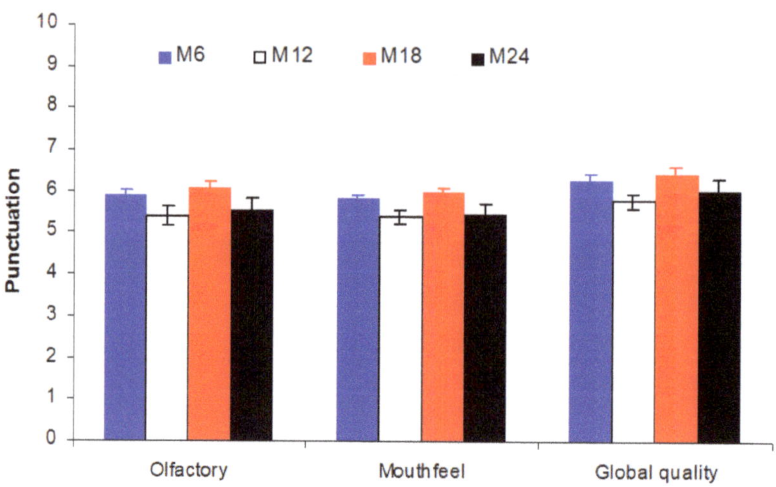


**Figure 3.** Punctuations for the olfactory and mouth phases of the sensory evaluation, as well as for the global quality of the Treixadura wines at different storage times in the bottle. Error bars represent standard errors. M6, M12, M18, and M24 indicate 6, 12, 18, and 24 months after bottling.

## 4. Discussion

This study confirmed that Treixadura wines do not have a terpenic aroma profile and, consequently, Treixadura cannot be considered rich in varietal compounds [17,20,25]. In contrast, the studied wines had high contents in ethyl esters and isoamyl acetate, which provide fruity nuances [33], and vinyl-phenols that provide aroma to paint, watercolor, and clove [40]. In the current study, Treixadura wines had a similar volatile composition at the time of bottling, despite coming from different wineries that, likely, used different protocols for winemaking. However, over the process of bottle aging, several reactions occurred and altered the volatile composition of the wines. Previous research reported that reactions such as oxidation, hydrolysis, and reactions caused by charge transfer and formation of covalent bonds influenced the evolution of wine flavor during bottle aging [42,43]. In the case of white wines, scarce research efforts have been devoted to elucidate the mechanisms that produce changes in aromatic composition during bottle aging [7]. Nevertheless, it has been suggested that an oxidation of alcohols into aldehydes is produced, as well as an increase followed by a diminishing of terpenes, acetates, and ethyl esters, while there is a formation or an increase in the concentrations of norisoprenoids, thiols, and sulfur compounds of low molecular weight [14].

In the current work, the volatile compounds detected in Treixadura wines followed one of three patterns during their evolution over bottle aging. First, the volatile compounds detected in the wines from the current study maintained their concentrations up to the third measurement date (2 years after grape harvest, 18 months in the bottle) and declined sharply in concentration on the last measurement date (30 months after grape harvest, 24 months in the bottle). Compounds relevant to wine aroma, including linalool, 2-phenylethanol, ethyl hexanoate, ethyl octanoate, isovaleric acid, and isoamyl acetate, followed this pattern over bottle aging. A second pattern was observed for the concentrations of other volatiles, which decreased steadily over the period of bottle aging, including 1,3-butanediol, isobutyric acid, isoamyl acetate, and acetoine. Finally, the concentrations of 19 volatiles did not significantly vary over the period of bottle aging (third pattern). A previous study on Cabernet Sauvignon wines pointed out similar patterns of evolution [11], although the specific pattern for a given compound differed from that observed in the current study, likely to differences in the variety and experimental setup used. In contrast, research on a white variety, Chardonnay [7], provided similar results as those presented here. In this sense, alcohols (1-hexanol, *cis*-3-hexen-1-ol, isobutanol) tended to remain stable over bottle aging, whereas ethyl esters (ethyl hexanoate, octanoate, and decanoate) and fatty acids (hexanoic, octanoic, and decanoic) tended to appear at low concentrations by the end of the bottle aging period. This diminishment of critical aroma compounds such as ethyl esters, terpenes, and norisoprenoids could reduce the perception of fruity and floral nuances at the sensory level [7,11,44]. Overall, bottle aging within 18 months enhanced the accumulation of volatile compounds and wine maturation in this study.

The changes in concentrations discussed above modified the relevance of the volatiles on wine aroma. In the current study, the compounds that had the highest OAV and, consequently, contributed significantly to Treixadura wine aroma were isovaleric acid, 4-vinyl-phenol, isoamyl acetate, and ethyl hexanoate; with OAV ranging from 2.8 to 25, depending on the compound and the date after bottling. These compounds coincide with those reported by Vilanova et al. [20] for wines of the same variety. In addition, Cortés and Blanco [18] indicated that ethyl octanoate, ethyl butyrate, isoamyl alcohol, and 2-phenylethanol also had a relevant contribution to the aroma of Treixadura wines, although their concentrations depended on the yeast strain used for fermentation. In the current study, these compounds were present and their OAV were from 0.5 (2-phenylethanol) to 4 (isoamyl alcohol), thus they contributed to wine aroma.

In this sense, the 'fruity' descriptor received the highest marks in all wines. In fact, several authors reported that Treixadura wines have a characteristic flavor related to fruits and pointed out several descriptors including 'banana', 'apple', 'citrus', and 'pear' [18]; 'stone fruit' and 'ripen fruit' [20]; and 'fresh fruit' [26]. In the current study, the 'fruity' descriptor received lower marks on the second date (12 months after bottling) but these marks increased six months later to decrease again on the last date of sensory evaluations. This pattern is similar to that of compounds such as ethyl esters, isoamyl acetate, and diethyl succinate, which appeared at higher concentrations on the third date of determinations and showed high OAV, which might have caused Treixadura wines to have this 'fruity' character. A similar behavior over bottle aging, namely, maximum values after two years from harvesting, was observed for the 'floral' descriptor, which might have been produced by compounds such as 2-phenylethanol and linalool. Despite the fact that these compounds appeared at OAV between 0.2 and 1, synergistic effects could have caused their detection and contributed to the Treixadura wine aroma, as previously reported [18,20,45].

The 'grass' descriptor received marks around 3–4 units and these were higher on the second and third dates of assessment. This nuance could be produced by isoamyl alcohols, 1-hexanol, and *cis*-3-hexenol [32,33,36], which appeared at significant concentrations in the Treixadura wines studied. These compounds are synthesized from the branched-chain amino acids [9], which are abundant in Treixadura when compared to other grape varieties [19]. The intensity of the remaining aroma descriptors was low, although some of them ('caramel', 'spicy') have been previously encountered in Treixadura wines [20,25]. Marks for 'caramel' increased with bottle aging, as previously reported for Riesling [46] and Semillon [47] wines, being explained by an increase of the concentration of furfural [48]; however, this was not observed in the current study. Nevertheless, the marks for this descriptor were low, up to 2 in the case of the M18 and M24 samples, and the observed increase of these marks with bottle aging can be caused by several factors such as, for instance, spontaneous malolactic fermentation or the oxidation of wine [49]; moreover, the M18 sample had a higher concentration of $\gamma$-butyrolactone, a compound that provides caramel and sweet nuances [33].

Finally, it must be noted that the nonvolatile matrix exerts a powerful impact on wine aroma perception, which has been reported similar to that of the volatile composition [50,51]. In the current study, the mouthfeel quality of the wines tended to a greater quality during the first 18 months of the bottle-aging period and this could have positively impacted the assessment of fragrance attributes, as previously reported for Cabernet Sauvignon wines [11].

## 5. Conclusions

In conclusion, by assessing the wine volatile composition, OAV, and sensory perception, a comprehensive understanding of the evolution of flavor profiles of Treixadura wines was established. Most volatile compounds in the studied wines showed stable, or even increased, concentrations up to two years after harvest (18 months of bottle aging). Then, their contents sharply decreased. The concentrations of acetates, mainly of isoamyl acetate, progressively decreased during bottle aging, being reduced up to four to five times when compared to the initial concentration in the wines. Sensory evaluation showed that the most-valued aromatic descriptors ('fruity' and 'floral') received the highest marks in the samples from 18 months of bottle aging (two years after harvest); these samples also reached the highest marks both for the olfactory and mouthfeel levels, as well as for the global quality of the wine. From the results obtained, and against the common belief that Galician white wines must be consumed within the year following their production, it would be advised that Treixadura wines were consumed two years after harvest (18 months in the bottle). Therefore, the current study has extended the research into the evolution of aroma compounds in white wines; however, further attention should be given to wine flavor chemistry and quality during bottle aging.

**Supplementary Materials:** The following are available online at http://www.mdpi.com/2304-8158/9/10/1419/s1, Tables S1 to S8: Concentrations of volatile compounds (mean ± standard deviation SD) in Treixadura wine samples 1 to 8 from the Ribeiro DO at different times of bottle aging, Table S9: Bottling storage effects on the odor activity values of wines from Treixadura in Ribeiro Designation of Origin (Samples 1–4), Table S10: Bottling storage effects on the odor activity values of wines from Treixadura in Ribeiro Designation of Origin (samples 5–8).

**Author Contributions:** Conceptualization, I.V.-P. and E.F.; methodology, U.A.-G., I.V.-P., and E.F.; validation, U.A.-G., I.V.-P., J.M.M.-A., and E.F.; formal analysis, U.A.-G. and J.M.M.-A.; investigation, U.A.-G., I.V.-P., and E.F.; resources, I.V.-P. and E.F.; data curation, E.F.; writing—original draft preparation, J.M.M.-A. and E.F.;

writing—review and editing, J.M.M.-A. and E.F.; project administration, E.F. All authors have read and agreed to the published version of the manuscript.

**Funding:** This research received no external funding and the APC was funded by CITACA Strategic Partnership ED431E 2018/07 (Xunta de Galicia, Spain).

**Acknowledgments:** The authors express their gratitude to the eight wineries from the Ribeiro Designation of Origin that kindly provided their wines for this study (Aurea Lux, Casal de Armán, Coto de Gomariz, Finca Teira, Pazo Vieite, Quinta do Avelino, Ramón do Casar, and Viña do Campo). We also thank the enologists that participated on the sensory evaluation of the wines for the unvaluable help.

**Conflicts of Interest:** The authors declare no conflict of interest. The funders had no role in the design of the study; in the collection, analysis, or interpretation of data; in the writing of the manuscript, or in the decision to publish the results.

## References

1. Robinson, A.L.; Boss, P.K.; Solomon, P.S.; Trengove, R.D.; Heymann, H.; Ebeler, S.E. Origins of grape and wine aroma. Part 1. Chemical components and viticultural impacts. *Am. J. Enol. Vitic.* **2014**, *65*, 1–24. [CrossRef]
2. Ebeler, S.E.; Thorngate, J.H. Wine chemistry and flavor: Looking into the crystal glass. *J. Agric. Food Chem.* **2009**, *57*, 8098–8108. [CrossRef] [PubMed]
3. González-Barreiro, C.; Rial-Otero, R.; Cancho-Grande, B.; Simal-Gándara, J. Wine aroma compounds in grapes: A critical review. *Crit. Rev. Food Sci. Nutr.* **2015**, *55*, 202–218. [CrossRef] [PubMed]
4. Swiegers, J.H.; Bartwosky, E.J.; Henschke, P.A.; Pretorius, I.S. Yeast and bacterial modulation of wine aroma and flavour. *Aust. J. Grape Wine Res.* **2005**, *11*, 139–173. [CrossRef]
5. Garde-Cerdán, T.; Ancín-Azpilicueta, C. Review of quality factors on wine ageing in oak barrels. *Trends Food Sci. Technol.* **2006**, *17*, 438–447. [CrossRef]
6. Garde-Cerdán, T.; Lorenzo, C.; Carot, J.M.; Esteve, M.D.; Climent, M.D.; Salinas, M.R. Effects of composition, storage time, geographic origin and oak type on the accumulation of some volatile oak compounds and ethylphenols in wines. *Food Chem.* **2010**, *122*, 1076–1082. [CrossRef]
7. Ling, M.Q.; Xie, H.; Hua, Y.B.; Cai, J.; Li, S.Y.; Lan, Y.B.; Li, R.N.; Duan, C.Q.; Shi, Y. Flavor profile evolution of bottle aged rosé and white wines sealed with different closures. *Molecules* **2019**, *24*, 836. [CrossRef]
8. Maza, M.A.; Martínez, J.M.; Cebrián, G.; Sánchez-Gimeno, A.C.; Camargo, A.; Álvarez, I.; Raso, J. Evolution of polyphenolic compounds and sensory properties of wines obtained from Grenache grapes treated by pulsed electric fields during aging in bottles and oak barrels. *Foods* **2020**, *9*, 542. [CrossRef]
9. Francis, I.L.; Newton, J.L. Determining wine aroma from compositional data. *Aust. J. Grape Wine Res.* **2005**, *11*, 14–126. [CrossRef]
10. Kwiatkowski, M.J.; Skouroumounis, G.K.; Lattey, K.A.; Waters, E.J. The impact of closures, including screw cap with three different headspace volumes, on the composition, colour and sensory properties of a Cabernet Sauvignon wine during two years' storage. *Aust. J. Grape Wine Res.* **2007**, *13*, 81–94. [CrossRef]
11. Liu, D.; Xing, R.-R.; Li, Z.; Yang, D.-M.; Pan, Q.-H. Evolution of volatile compounds, aroma attributes, and sensory perception in bottle-aged red wines and their correlation. *Eur. Food Res. Technol.* **2016**, *242*, 1937–1948. [CrossRef]
12. Smith, P.A.; McRae, J.M.; Bindon, K.A. Impact of winemaking practices on the concentration and composition of tannins in red wine. *Aust. J. Grape Wine Res.* **2015**, *21*, 601–614. [CrossRef]
13. Silva, M.A.; Julien, M.; Jourdes, M.; Teissedre, P.L. Impact of closures on wine post-bottling development: A review. *Eur. Food Res. Technol.* **2011**, *233*, 905–914. [CrossRef]
14. Ugliano, M. Oxygen contribution to wine aroma evolution during bottle aging. *J. Agric. Food Chem.* **2013**, *61*, 6125–6136. [CrossRef] [PubMed]
15. Styger, G.; Prior, B.; Bauer, F.F. Wine flavor and aroma. *J. Ind. Microbiol. Biotechnol.* **2011**, *38*, 1145–1159. [CrossRef] [PubMed]
16. Belda, I.; Ruiz, J.; Esteban-Fernández, A.; Navascués, E.; Marquina, D.; Santos, A.; Moreno-Arribas, M.V. Microbial contribution to wine aroma and its intended use for wine quality improvement. *Molecules* **2017**, *22*, 189. [CrossRef]
17. Falqué, E.; Fernández, E.; Dubourdieu, D. Volatile components of Loureira, Dona Branca, and Treixadura wines. *J. Agric. Food Chem.* **2002**, *50*, 538–543. [CrossRef]

18. Cortés, S.; Blanco, P. Yeast strain effect on the concentration of major volatile compounds and sensory profile of wines from *Vitis vinifera* var. Treixadura. *World J. Microbiol. Biotechnol.* **2011**, *27*, 925–932. [CrossRef]
19. Mirás-Avalos, J.M.; Bouzas-Cid, Y.; Trigo-Córdoba, E.; Orriols, I.; Falqué, E. Amino acid profiles to differentiate white wines from three autochtonous Galician varieties. *Foods* **2020**, *9*, 114. [CrossRef]
20. Vilanova, M.; Escudero, A.; Graña, M.; Cacho, J. Volatile composition and sensory properties of North West Spain white wines. *Food Res. Int.* **2013**, *54*, 562–568. [CrossRef]
21. Versini, G.; Orriols, I.; Dalla Serra, A. Aroma components of Galician Albariño, Loureira and Godello wines. *Vitis* **1994**, *33*, 165–170.
22. Falqué, E.; Fernández, E.; Dubourdieu, D. Differentiation of white wines by their aromatic index. *Talanta* **2001**, *54*, 271–281. [CrossRef]
23. Genisheva, Z.; Oliveira, J.M. Monoterpenic characterization of white cultivars from Vinhos Verdes Appellation of Origin (North Portugal). *J. Inst. Brew.* **2009**, *115*, 308–317. [CrossRef]
24. Moreira, N.; Guedes-de-Pinho, P.; Santos, C.; Vasconcelos, I. Volatile sulphur compounds composition of monovarietal white wines. *Food Chem.* **2010**, *123*, 1198–1203. [CrossRef]
25. Blanco, P.; Mirás-Avalos, J.M.; Suárez, V.; Orriols, I. Inoculation of Treixadura musts with *Saccharomyces cerevisiae* strains: Fermentative performance and influence on the wine characteristics. *Food Sci. Technol. Int.* **2012**, *19*, 177–186. [CrossRef]
26. Bouzas-Cid, Y.; Falqué, E.; Orriols, I.; Mirás-Avalos, J.M. Effects of irrigation over three years on the amino acid composition of Treixadura (*Vitis vinifera* L.) musts and wines, and on the aromatic composition and sensory profiles of its wines. *Food Chem.* **2018**, *240*, 707–716. [CrossRef]
27. Bertrand, A.; Ribéreau-Gayon, P. Determination of volatile components of wine by gas-phase chromatography. *Ann. Falsif. Expert. Chim.* **1970**, *63*, 148–156.
28. Armada, L.; Fernández, E.; Falqué, E. Influence of several enzymatic treatments on aromatic composition of white wines. *LWT Food Sci. Technol.* **2010**, *43*, 1517–1525. [CrossRef]
29. Guth, H. Quantitation and sensory studies of character impact odorants of different white wine varieties. *J. Agric. Food Chem.* **1997**, *45*, 3027–3032. [CrossRef]
30. Ferreira, V. Volatile aroma compounds and wine sensory attributes. In *Managing Wine Quality: Viticulture and Wine Quality*; Reynolds, A.G., Ed.; Woodhead Publishing Limited: Cambridge, UK, 2010; pp. 3–28.
31. Ribéreau-Gayon, P.; Dubourdieu, D.; Doneche, B.; Lonvaud, A. *Handbook of Enology, Volume 1: The Microbiology of Wine and Vinifications*; Dunond: Paris, France, 1998.
32. Simpson, R.F. Some important aroma composition of components of white wine. *Food Technol.* **1979**, *31*, 516–522.
33. Peinado, R.A.; Moreno, J.; Bueno, J.E.; Moreno, J.A.; Mauricio, J.C. Comparative study of aromatic compounds in two young white wines subjected to pre-fermentative cryomaceration. *Food Chem.* **2004**, *84*, 585–590. [CrossRef]
34. Meilgaard, M. Flavor chemistry of beer. Part II: Flavor and threshold of 239 aroma volatiles. *MBBA Tech. Quart.* **1975**, *12*, 151–168.
35. Zoecklein, B.W.; Fugelsang, K.C.; Gump, B.H.; Nury, F.S. *Wine Analysis and Production*; Springer Science: New York, NY, USA, 1994.
36. López, R.; Ferreira, V.; Cacho, J.F. Quantitative determination of the odorants of young red wines from different grape varieties. An assessment of their sensory role. In *Enology 99, Proceedings of the 6th International Symposium on Enology, Bordeaux, France, 10–12 June 1999*; Lonvaud-Funel, A., Ed.; Tec & Doc Éditions: Paris, France, 1999; pp. 148–151.
37. Ferreira, V.; López, R.; Cacho, J.F. Quantitative determination of the odorants of young red wines from different grape varieties. *J. Sci. Food Agric.* **2000**, *80*, 1659–1667. [CrossRef]
38. Étievant, P.X. Wine. In *Volatile Compounds in Food and Beverages*; Maarse, H., Ed.; TNO-CIVO Food Analysis Institute: Zeist, The Netherlands, 1991; pp. 483–546.
39. Salo, P. Determining the odor thresholds for some compounds in alcoholic beverages. *J. Food Sci.* **1970**, *35*, 95–99. [CrossRef]
40. Boidron, J.; Chatonnet, P.; Pons, M. Influence du bois sur certaines substances odorantes des vines. *Connaiss. Vigne Vin* **1988**, *22*, 275–294.
41. R Core Team. *R: A Language for Statistical Computing*; R Foundation for Statistical Computing: Vienna, Austria, 2019.

42. Karbowiak, T.; Gougeon, R.D.; Alinc, J.-B.; Brachais, L.; Debeaufort, F.; Voilley, A.; Chassagne, D. Wine oxidation and the role of cork. *Crit. Rev. Food Sci. Nutr.* **2009**, *50*, 20–52. [CrossRef]
43. Tarko, T.; Duda-Chodak, A.; Sroka, P.; Siuta, M. The impact of oxygen at various stages of vinification on the chemical composition and the antioxidant and sensory properties of white and red wines. *Int. J. Food Sci.* **2020**, *2020*, 7902974. [CrossRef]
44. Lee, D.H.; Kang, B.S.; Park, H.J. Effect of oxygen on volatile and sensory characteristics of Cabernet Sauvignon during secondary shelf life. *J. Agric. Food Chem.* **2011**, *59*, 11657–11666. [CrossRef]
45. Campo, E.; Ferreira, V.; Escudero, A.; Cacho, J. Prediction of the wine sensory properties related to grape variety from dynamic-headspace gas chromatography-olfactometry data. *J. Agric. Food Chem.* **2005**, *53*, 5682–5690. [CrossRef]
46. Iland, P.; Gago, P. *Discovering Australian Wine: A Taster's Guide. An Introduction to Australian Wine, the Vineyards, the Wineries, How to Taste the Wine, Australian Wine Styles, Wine and Food*; Patrick Iland Wine Promotions: Campbelltown, Australia, 2014.
47. Blackman, J.W.; Hopfer, H.; Saliba, A.J.; Schmidtke, L.M.; Barril, C.; Scollary, G.R. Sensory characterization of Hunter Valley Semillon aged in bottle. *Flavour Fragr. J.* **2014**, *29*, 340–349. [CrossRef]
48. Rocha, S.M.; Coutinho, P.; Delgadillo, I.; Dias Cardoso, A.; Coimbra, M.A. Effect of enzymatic aroma release on the volatile compounds of white wines presenting different aroma potentials. *J. Sci. Food Agric.* **2005**, *85*, 199–205. [CrossRef]
49. Oliveira, C.M.; Ferreira, A.C.S.; de Freitas, V.; Silva, A.M.S. Oxidation mechanisms occurring in wines. *Food Res. Int.* **2011**, *44*, 1115–1126. [CrossRef]
50. Robinson, A.L.; Ebeler, S.E.; Heymann, H.; Boss, P.K.; Solomon, P.S.; Trengove, R.D. Interactions between wine volatile compounds and grape and wine matrix components influence aroma compound headspace partitioning. *J. Agric. Food Chem.* **2009**, *57*, 10313–10322. [CrossRef] [PubMed]
51. Sáenz-Navajas, M.P.; Campo, E.; Culleré, L.; Fernández-Zurbano, P.; Valentin, D.; Ferreira, V. Effects of the nonvolatile matriz on the aroma perception of wine. *J. Agric. Food Chem.* **2010**, *58*, 5574–5585. [CrossRef]

MDPI
St. Alban-Anlage 66
4052 Basel
Switzerland
Tel. +41 61 683 77 34
Fax +41 61 302 89 18
www.mdpi.com

*Foods* Editorial Office
E-mail: foods@mdpi.com
www.mdpi.com/journal/foods

www.ingramcontent.com/pod-product-compliance
Lightning Source LLC
Chambersburg PA
CBHW061441210926
51850CB00008B/1578
*9783036516622*